Dust in the Galactic Environment
(Third Edition)

AMERICAN ASTRONOMICAL SOCIETY IOP | ebooks™

AAS Editor in Chief

Ethan Vishniac, Johns Hopkins University, Maryland, USA

About the program:

AAS-IOP Astronomy ebooks is the official book program of the American Astronomical Society (AAS) and aims to share in depth the most fascinating areas of astronomy, astrophysics, solar physics, and planetary science. The program includes publications in the following topics:

Books in the program range in level from short introductory texts on fast-moving areas, graduate and upper-level undergraduate textbooks, research monographs, and practical handbooks.

For a complete list of published and forthcoming titles, please visit iopscience.org/books/aas.

About the American Astronomical Society

The American Astronomical Society (aas.org), established 1899, is the major organization of professional astronomers in North America. The membership (~7,000) also includes physicists, mathematicians, geologists, engineers, and others whose research interests lie within the broad spectrum of subjects now comprising the contemporary astronomical sciences. The mission of the Society is to enhance and share humanity's scientific understanding of the universe.

Editorial Advisory Board

Dust in the Galactic Environment (Third Edition)

D C B Whittet
Rensselaer Polytechnic Institute, Troy, New York, USA

IOP Publishing, Bristol, UK

ISBN 978-0-7503-3275-0 (ebook)
ISBN 978-0-7503-3273-6 (print)
ISBN 978-0-7503-3276-7 (myPrint)
ISBN 978-0-7503-3274-3 (mobi)

DOI 10.1088/2514-3433/ac7204

Version: 20220901

AAS–IOP Astronomy
ISSN 2514-3433 (online)
ISSN 2515-141X (print)

British Library Cataloguing-in-Publication Data: A catalogue record for this book is available from the British Library.

Published by IOP Publishing, wholly owned by The Institute of Physics, London

IOP Publishing, Temple Circus, Temple Way, Bristol, BS1 6HG, UK

US Office: IOP Publishing, Inc., 190 North Independence Mall West, Suite 601, Philadelphia, PA 19106, USA

Cover image: Skies above ESO's Paranal Observatory. Image credit: ESO/P. Horálek.

To my grandson, Edward (b. 2019).

May he and his generation inherit a habitable world.

Contents

Preface xv

Acknowledgments xvi

Author biography xvii

1 Astrophysical Dust: An Overview **1-1**

1.1 Introduction 1-1
1.2 Methods of Investigation 1-3
1.2.1 Interstellar Extinction 1-4
1.2.2 The Reddening of Starlight 1-5
1.2.3 New Light in the Infrared 1-7
1.2.4 Interstellar Polarization 1-8
1.2.5 Spectral Features 1-8
1.2.6 Where the Atoms Went 1-9
1.2.7 Gathering It All Together: Mathematical Models 1-10
1.3 Interstellar Environments 1-11
1.3.1 The Physical State of the Interstellar Medium 1-11
1.3.2 Interstellar Clouds 1-15
1.3.3 H II Regions and Photodissociation Regions 1-17
1.3.4 The Interstellar Environment of the Solar System 1-19
1.4 Significance 1-22
1.4.1 From Cinderella to the Search for Origins 1-22
1.4.2 Dust in Astrochemistry 1-23
1.4.3 Dust as a Proxy for H_2 1-24
1.4.4 Dust as the Universal Thermal Regulator 1-25
1.4.5 Dust and Stellar Evolution 1-25
1.4.6 Starbursts and Luminous Infrared Galaxies 1-26
1.4.7 Back to Basics 1-27
References 1-28

2 The Interaction of Dust and Electromagnetic Radiation **2-1**

2.1 Extinction 2-1
2.1.1 Extinction by Spherical Particles 2-1
2.1.2 Small-Particle Approximations 2-5
2.1.3 Albedo, Scattering Function, and Asymmetry Parameter 2-6
2.1.4 Absorption Features 2-7
2.1.5 Modeling Composite Grains 2-7

2.2 Polarization 2-8
2.2.1 Extinction by Anisotropic Particles 2-8
2.2.2 Polarization Parameters 2-10
2.2.3 Polarization Efficiency 2-12
2.2.4 Circular Polarization 2-12
2.3 Infrared Emission 2-13
2.3.1 Equilibrium Dust Temperatures 2-13
2.3.2 Thermal Transients in Very Small Grains 2-15
2.3.3 Far-Infrared Continuum Emission and Dust Mass 2-18
2.3.4 The Emissivity Function 2-20
2.3.5 Polarized Emission 2-21
2.4 Grain Dynamics 2-22
2.4.1 Radiation Pressure 2-22
2.4.2 The Photoelectric Effect 2-23
2.4.3 Rotational Motion: Suprathermal Spin and Radiative Torques 2-23
References 2-25

3 The Observed Properties of Dust—I. Extinction 3-1

3.1 Observational Methods 3-1
3.2 General Extinction in the Diffuse ISM 3-5
3.2.1 The Average Extinction Curve 3-5
3.2.2 The Ratio of Total to Selective Extinction 3-5
3.2.3 Scattering 3-7
3.2.4 Power-Law Behavior in the Infrared 3-10
3.2.5 Neutral Extinction 3-12
3.2.6 The Dust-to-Gas Ratio 3-13
3.2.7 Mapping the Galactic Distribution 3-15
3.3 Spatial Variations in the Extinction Curve 3-16
3.3.1 Size-dependent Behavior 3-16
3.3.2 Empirical Formulations 3-18
3.3.3 Extinction in Other Galaxies 3-20
3.4 The "Bump" Feature 3-22
3.4.1 Overview of Observed Properties 3-22
3.4.2 A Constraint on Particle Size 3-24
3.4.3 Environmental Effects 3-24
3.4.4 Are the Bump and the FUV Rise Related? 3-25
3.4.5 Implications for the Identity of the Absorber 3-25

3.5	Discrete Structure in the Visible	3-30
	3.5.1 Diffuse Interstellar Bands	3-30
	3.5.2 Broadband Structure	3-31
3.6	Modeling the Extinction Curve	3-33
	3.6.1 The Size Distribution	3-33
	3.6.2 Composition and Structure	3-34
	References	3-36
4	**The Observed Properties of Dust—II. Polarization**	**4-1**
4.1	Polarization and Magnetic Fields	4-1
	4.1.1 Polarization Efficiency and Depolarization	4-2
	4.1.2 Mapping Methodologies	4-3
	4.1.3 Magnetic Fields in Molecular Clouds	4-5
	4.1.4 Magnetic Fields in the Milky Way	4-6
4.2	Spectral Dependence and Environmental Sensitivity	4-8
	4.2.1 Polarization and Extinction Curves Compared	4-8
	4.2.2 The Serkowski Formula	4-10
	4.2.3 Power-Law Behavior in the Infrared	4-11
	4.2.4 Spectral Variations in Position Angle	4-11
	4.2.5 Circular Polarization	4-11
	4.2.6 Effect of Particle Size	4-13
	4.2.7 Polarized Absorption Features	4-14
	4.2.8 Modeling the Polarization Curve	4-16
4.3	Grain Alignment	4-18
	4.3.1 General Principles of Magnetic Alignment	4-19
	4.3.2 Paramagnetic Relaxation: The DG Mechanism	4-20
	4.3.3 Superparamagnetic Alignment	4-22
	4.3.4 Suprathermal Spin: The Purcell Mechanism	4-23
	4.3.5 Radiative Alignment Torques	4-24
	4.3.6 Mechanical Alignment: The Gold Mechanism	4-25
	4.3.7 Alignment in Dense Clouds	4-25
	References	4-27
5	**The Observed Properties of Dust—III. Infrared Absorption Features**	**5-1**
5.1	Basics of Infrared Spectroscopy	5-1
	5.1.1 Vibrational Modes in Solids	5-1

5.1.2 Intrinsic Strengths 5-5
5.1.3 Observational Approach 5-6
5.2 The Diffuse ISM 5-9
5.2.1 Silicates 5-11
5.2.2 Silicon Carbide? 5-14
5.2.3 Hydrocarbons and Organic Refractory Matter 5-15
5.3 The Dense ISM 5-19
5.3.1 An Inventory of Interstellar Ices 5-19
5.3.2 Extinction Thresholds 5-20
5.3.3 Water and Ammonia: The 3 μm Profile 5-22
5.3.4 CO and CO_2: Polar and Apolar Mantles 5-24
5.3.5 The 6.0 and 6.85 μm Features: A Potpourri 5-27
5.3.6 Refractory Dust in the Dense ISM 5-29
References 5-31

6 The Observed Properties of Dust—IV. Continuum and Line Emission 6-1

6.1 Galactic Infrared Continuum Emission 6-2
6.1.1 Morphology 6-2
6.1.2 Spectral Energy Distribution 6-4
6.1.3 Dust and Gas 6-5
6.2 Infrared Spectral Emission Features 6-6
6.2.1 Silicates 6-6
6.2.2 Polycyclic Aromatic Hydrocarbons 6-9
6.2.3 Fullerenes 6-14
6.3 Extended Red Emission 6-15
References 6-17

7 The Observed Properties of Dust—V. Element Depletions 7-1

7.1 The Condensible Elements 7-1
7.1.1 Origins 7-2
7.1.2 The Astration Cycle 7-3
7.1.3 The Solar Abundances 7-4
7.1.4 Abundances in Young Stars 7-7
7.1.5 Abundances from X-Ray Observations 7-8
7.1.6 Converging on a Standard 7-9

7.2 The Observed Depletions 7-10
7.2.1 Methods and Metrics 7-10
7.2.2 Average Depletions in the Milky Way 7-13
7.2.3 Correlation with Condensation Temperature 7-14
7.2.4 Depletions in Other Galaxies 7-17
7.3 Implications for Grain Composition 7-18
7.3.1 Carbon 7-19
7.3.2 Magnesium and Silicon 7-19
7.3.3 The Oxygen Crisis 7-20
7.3.4 Iron 7-22
References 7-24

8 The Life Cycle of Dust—I. Circumstellar Origins 8-1

8.1 Dust Formation in Stellar Outflows 8-1
8.1.1 Theoretical Basis 8-1
8.1.2 The C/O Ratio 8-5
8.1.3 Oxygen-rich Stars 8-6
8.1.4 Carbon Stars 8-7
8.1.5 Planetary Nebulae 8-9
8.1.6 Core-collapse Supernovae 8-9
8.1.7 Exploding White Dwarfs 8-11
8.2 Observations of Stardust 8-11
8.2.1 Infrared Continuum Emission 8-12
8.2.2 Infrared Spectral Features 8-15
8.2.3 Circumstellar Extinction 8-23
8.3 Stardust in Meteorites 8-25
8.3.1 Background 8-25
8.3.2 Composition and Origin: Carbon-rich Grains 8-26
8.3.3 Composition and Origin: Oxygen-rich Grains 8-28
8.3.4 Particle Sizes 8-29
8.4 Stars as Sources of Interstellar Grains 8-30
8.4.1 Mass Loss 8-30
8.4.2 Predicting the Size Distribution 8-33
8.4.3 Dust-to-Gas Ratios 8-33
8.4.4 Production and Destruction Timescales 8-34
References 8-37

9 The Life Cycle of Dust—II. From the ISM to Protostars and Planets **9-1**

9.1 Processes in the Diffuse ISM 9-3
9.1.1 Grain–Grain Collisions 9-3
9.1.2 Sputtering 9-4
9.1.3 Metamorphosis 9-7
9.1.4 Surface Reactions and Molecule Formation 9-9
9.2 Processes in Dark Clouds 9-12
9.2.1 Gas-phase Chemistry 9-12
9.2.2 Mantle Growth and Surface Chemistry 9-14
9.2.3 The Surface Chemistry of CO 9-16
9.2.4 Depletion from the Gas 9-18
9.2.5 The Oxygen Crisis Revisited 9-20
9.3 Processes in Star Formation Regions 9-22
9.3.1 Birth Aggregates: From Quiescent Clouds to Starbursts 9-22
9.3.2 Thermal Processing 9-23
9.3.3 Energetic Processing 9-27
9.3.4 Chemistry in Hot Cores and Corinos 9-28
9.3.5 From Cores to Disks 9-30
9.3.6 Debris Disks 9-35
9.3.7 The Astrochemical Heritage of the Solar System 9-36
References 9-41

10 Conclusion **10-1**

10.1 An Overview of Dust Populations 10-2
10.2 Future Prospects 10-5
10.2.1 The Origin and Structure of Refractory Dust 10-5
10.2.2 Ices and Organics in the Interstellar Medium and Protoplanetary Disks 10-7

Appendix A: Glossary **A-1**

A.1 Units and Constants A-1
A.2 Common Acronyms A-2
A.3 Physical, Chemical and Astrophysical Terms A-3

Preface

Dust is a highly influential ingredient of the cosmos that affects most areas of research in modern astronomy. Small (submicrometer-sized) solid particles pervade interstellar space in the Milky Way and other galaxies, in environments that range from interstellar clouds to protoplanetary disks, from circumstellar outflows to active galactic nuclei. The study of this phenomenon is both an important area of research in its own right and a necessary concomitant for specialists in many other areas of astronomy and related disciplines.

The goal of this text is to provide a thorough overview of the subject, covering general concepts, methods of investigation, important results and their significance, and suggestions for promising avenues of future investigation. The level is appropriate for advanced undergraduate and beginning postgraduate students embarking on related research, and established researchers seeking a primer on this topic. It is assumed that the reader is familiar with basic concepts in stellar and galactic astronomy, such as stellar magnitudes and distance scales and the spectral classification sequence, and has a qualitative familiarity with galactic structure and stellar evolution according to current models.

Advances in our understanding of cosmic dust since publication of the second edition of *Dust in the Galactic Environment* in 2003 are such that this new edition is in large part a different book. These advances have been driven by data from new observational facilities such as the Herschel, Planck, and Spitzer Space Telescopes, which have revolutionized our view of the cosmos, especially at infrared wavelengths where dust is a dominant source of radiation. Parallel developments in laboratory studies of cosmic and terrestrial analog materials and in numerical modeling of all aspects of dust have also advanced the field. This new edition highlights these results and provides a context for future research opportunities.

The author has found cosmic dust to be a challenging and rewarding topic of study. Many decades ago, as a student with ambitions to be an 'astronomer,' I was advised to focus my studies on physics and mathematics, as nothing else in science was deemed to be of much relevance. Since then astronomy has evolved in many ways, from a rather esoteric field to an interdisciplinary one. The 'astro' prefix is now attached to chemistry, geology, and biology, as well as to physics, as we contemplate the nature of exoplanets and their potential to host life. Dust is the bearer of elements and compounds from which planetary systems are made, and as a research topic it reflects the diversity of the field as a whole. The quest to understand it rests on progress not only in the development of new observational facilities but also in advanced sample collection and laboratory analysis techniques, and it spans fields that range from solid state physics to organic chemistry, from meteoritics to small-particle magnetism. Much progress has been made, but there is no shortage of challenging problems. If this book encourages science students and researchers to consider dust particles in the laboratory of space, it will have succeeded in its primary aim.

Acknowledgments

Thanks are due, first and foremost, to my family: to my late parents Doreen and Douglas Whittet, for their love and support throughout my formative years and beyond, and for encouraging my childhood interest in astronomy; to Polly, my wife, soulmate and dearest friend, for her boundless love and support of me and my endeavors; and to my son James, daughter Clair and son-in-law Colin. My grandson, Edward, is the dedicatee of this book. I am so proud of you all.

I am indebted to Leigh Jenkins (Commissioning Editor) and John Navas (Senior Commissioning Manager) at IOP Publishing for encouragement, expertise, and guidance. It has been a pleasure to work with you. But for your enthusiasm and encouragement when I broached the idea of a new edition of this book, it would not have been written. I am also indebted to the production team at IOP.

Many colleagues and friends have contributed over the years to the development of my ideas on cosmic dust. My research benefited immeasurably from interactions with others attracted to this strangely fascinating topic. I wish to record my thanks, especially, to the late Kashi Nandy for stimulating my early interest in the topic; to Walt Duley and Peter Martin for sustaining it during a transformative period of sabbatical leave in Toronto; to Thijs de Graauw for inviting and encouraging my participation in observations with the Infrared Space Observatory; to Rensselaer Polytechnic Institute for a new career opportunity; to my doctoral students and other mentees at Rensselaer for their hard work and dedication to the task of understanding dust in the galactic environment; and to the National Aeronautics and Space Administration (NASA) for financial support of our endeavors. Those who have contributed directly or indirectly to the development of this book in various ways, via helpful comments and discussions or by providing access to data in advance of publication, include B.-G. Andersson, Jean Chiar, Perry Gerakines, Karl Gordon, James Hough, Ant Jones, Larry Nittler, Els Peeters, Julia Roman-Duval, and David Williams. I am also grateful to all who gave permission to reproduce images, acknowledged in the appropriate figure caption.

The text was produced using the TEXstudio LATEX editor on a Lenovo desktop computer running the Windows 10 operating system. Cited literature was sourced from the Astrophysics Data System (ADS), operated by the Smithsonian Astrophysical Observatory with support from NASA, and compiled using the Zotero open-source application developed by Digital Scholar. Many of the graphics were produced by the author using OriginLab software under license to Rensselaer Polytechnic Institute.

Those who know me will not be surprised that I end here on a musical note. The works of Finnish composer Jean Sibelius have been the soundtrack of my life, beginning in the late 1950s when, as a preteen, my dear parents allowed me to stay up 'after bedtime' to watch the BBC TV show *The Sky at Night*, with its Sibelian musical theme. Decades later I learned that Sibelius had a deep interest in astronomy. Perhaps it is infused in the music.

Author biography

Doug Whittet

Doug Whittet is Emeritus Professor of Physics at Rensselaer Polytechnic Institute in Troy, New York. He obtained his bachelor and doctoral degrees in Physics and Astronomy, respectively, from the University of St. Andrews, and was Senior Lecturer and Professor of Astrophysics at the University of Central Lancashire prior to joining the Rensselaer faculty in 1991. He served as Director of the New York Center for Astrobiology from 2008 to 2016. His research interests include the application of spectroscopy and polarimetry to studies of the nature and evolution of cosmic dust, with emphasis on the contributions of interstellar processes to the organic inventories of early solar systems. In the course of his research, he was a principal investigator of observing programs with the Hubble Space Telescope, the Spitzer Space Telescope, and the Infrared Space Observatory.

AAS | IOP Astronomy

Dust in the Galactic Environment (Third Edition)

D C B Whittet

Chapter 1

Astrophysical Dust: An Overview

1.1 Introduction

By terrestrial standards, interstellar space approximates to a perfect vacuum. It has a mean density in the solar neighborhood of our Galaxy, the Milky Way, equivalent to about one atom per cubic centimeter, a factor of 10^{19} less than the density of the Earth's atmosphere at sea level, and a factor of 10^4 less than the lowest densities achieved in terrestrial laboratories. But, as noted in the Hitchhiker's Guide to the Galaxy (Adams 1979), space is big...vastly, hugely, mind-bogglingly big. Dense objects such as stars and planets occupy a tiny fraction of its volume, and the tenuous but voluminous interstellar gas amounts to a significant total mass of material, estimated to be some 15% of that in stars in the Milky Way.

The stellar and interstellar components of galaxies are continually interacting and exchanging material. New stars condense from interstellar clouds and, as they evolve, they bathe the surrounding interstellar medium (ISM) with radiation and streams of ionic and atomic particles (winds). Ultimately, many stars return a substantial fraction of their mass to the ISM, thereby enriching it with heavier chemical elements such as C, O, Mg, Si, and Fe, fused from primordial hydrogen and helium by the nuclear processes occurring in their interiors. A major proportion of these heavier elements are condensed into submicrometer-sized solid particles (dust grains) that account for roughly 1% of the total mass of the ISM and are almost exclusively responsible for its opacity at visible wavelengths. Despite their tiny contribution to the galactic mass (at most 0.2% of that in stars), the ability of such particles to scatter, absorb, and re-radiate starlight ensures that they have a dramatic impact on our view of the universe. For example, the attenuation between us and the center of our Galaxy is such that, at visible wavelengths, only one photon in every 10^{12} reaches our telescopes. The energy absorbed by the grains is re-emitted in the infrared, accounting for some 20% of the total luminosity of the Milky Way.

The influence of interstellar dust may be discerned even with the unaided eye on a dark, moonless night at a time of year when the Milky Way is well-placed for

doi:10.1088/2514-3433/ac7204ch1

observation. The glow of background starlight often appears to be split by dust attenuation in the mid-plane. Southern observers are best placed to view such irregularities (Figure 1.1). The dark patches and rifts were seen by Australian Aborigine observers as a "dark constellation," the Emu, with the dark cloud known as the Coalsack as its head, the dark lane passing through Centaurus, Ara and Norma as its long, slender neck, and the complex system of dark clouds toward the galactic center in Sagittarius as its body and wings. Discoveries in the 20th century enabled us to recognize the Milky Way in Sagittarius as the nuclear bulge of a dusty spiral galaxy, seen from our vantage point within its disk at a distance of a few kiloparsecs from the center. Our view of our home Galaxy is impressively illustrated by wide-field images of the night sky, such as that shown in Figure 1.1. The Milky Way appears to be a fairly typical spiral, with a nucleus and disk surrounded by a spheroidal halo containing globular star clusters. There is a striking resemblance between Figure 1.1 and images of other spiral galaxies of similar morphological type seen edge-on, such as NGC 891, illustrated in Figure 1.2. The visual appearance of such galaxies tends to be dominated by the equatorial dark lane that bisects the nuclear bulge. Obscuration is less obvious (but invariably present) in spirals inclined by more than a few degrees to the line of sight. These observations indicate that dark, absorbing material is a common characteristic of such galaxies, and that this matter is concentrated into disks that are thin in comparison to their radii.

This chapter aims to present a broad overview of the phenomenon of galactic dust and its role in astrophysical processes, as a prologue to the more focused discussions

Figure 1.1. Wide-field image of the southern Milky Way viewed from the site of the Atacama Large Millimeter/submillimeter Array (ALMA) in Chile. The nuclear bulge lies toward the center of the frame. Dark clouds are seen in silhouette (the Coalsack is at lower left). Dusty features associated with the Sco-Oph star-forming region in Gould's Belt (Section 1.3.4) extend below the central bulge. Image credit: ESO/José Francisco (josefrancisco.org).

Figure 1.2. The spiral galaxy NGC 891, similar in size and luminosity to the Milky Way, is seen almost exactly edge-on from our perspective. Starlight from the disk and nuclear bulge is being absorbed and scattered by dust concentrated in the mid-plane, with filamentary structures extending above and below. Patches of blue and red light, emanating from hot young stars and ionized hydrogen gas, respectively, provide evidence of recent star formation in NGC 891. Stars within our own Galaxy appear in the foreground. Image credits: Subaru Telescope (NAOJ), Hubble Legacy Archive, Michael Joner and David Laney (West Mountain Observatory, BYU); processing by Robert Gendler.

in subsequent chapters. The following sections summarize the primary methods of detection, the range of environments to which the grains are exposed, and the significance of dust as a physical and chemical constituent of interstellar matter.

1.2 Methods of Investigation

Of necessity, for the most part, we obtain information on astrophysical dust by remote sensing, using the techniques of observational astronomy. Such investigations may be grouped into two general categories:

- Observations of phenomena that arise directly through the interaction of dust with electromagnetic radiation (Chapter 2), such as absorption, scattering, polarization, and thermal emission (discussed in detail in Chapters 3–6).
- Observations of other phenomena affected by the existence of dust, such as the gas-phase abundances of species that may condense into solids (Chapter 7).

The ideal alternative would be to collect and study actual physical samples of the dust, thus enabling the range of techniques available in terrestrial laboratories to be utilized for detailed analyses. This is, indeed, the primary method for studying dusty material native to our solar system, such as meteorites, cometary dust and lunar regolith. Applications of this approach to studies of interstellar dust encounter problems of sample collection and identification, and concerns that the samples so far available to us might not be pristine or representative of the general population. Progress has been made, nevertheless: the present-day flow of dust into the solar system from the local ISM has been detected (Section 1.3.4), and investigations of presolar grains identified in meteorites—the oldest solids ever studied in a terrestrial laboratory—have greatly enhanced our understanding of grain formation in stellar outflows and supernovae (Chapter 8). Laboratory studies of materials thought to be realistic analogs of interstellar matter are also vital to our interpretation of observational data (Chapters 3–6) and to elucidating the interactions of dust and gas in the ISM (Chapter 9).

The remainder of this section reviews how our understanding of the nature and significance of the dust evolved as the result of observational investigations in the categories itemized above. Sometimes, especially in early work, the discoveries were serendipitous, arising from observations that were not motivated by a desire to investigate the dust itself, or even, in a few cases, by any a priori knowledge of its presence or relevance.

1.2.1 Interstellar Extinction

Figures 1.1 and 1.2 illustrate a primary means by which dust reveals its presence—by its dimming effect on the light of distant stars, a phenomenon termed interstellar extinction.[1] It is interesting to note, however, that a century ago the nature of the dark markings seen toward the Milky Way was controversial. Some astronomers considered them to be mere fluctuations in the spatial distribution of stars rather than the obscuring effect of foreground material, an interpretation in accord with William Herschel's description of one such region as a "hole in the sky." At that time, only the visible part of the spectrum was accessible—the infrared and radio observations that would reveal the presence of thermal emission from dust and line emission from atomic and molecular gas in these apparently empty regions were not yet available. Edward Emerson Barnard, who produced the first comprehensive photographic catalog of dark nebulae (Barnard 1927) had noticed spatial correlations in their distribution relative to bright nebulae illuminated by stars, suggestive of a physical link between them (see Sheehan 1995 for further discussion). However, it was studies of galactic structure that led to general acceptance of the hypothesis that extinction is a pervasive feature of the ISM.

Astronomers working at visible wavelengths attempt to determine the structure of the Milky Way by mapping the spatial distribution of large numbers of stars, and this of course requires accurate measurements of their distances. In the absence of any attenuation along the line of sight, photometric and spectroscopic observations of a star enable its distance to be estimated from the equation

$$V - M_V = 5 \log d' - 5 \qquad (1.1)$$

where V and M_V are the apparent and absolute visual magnitudes, respectively, and d' is the apparent mean distance to the star in parsecs. If dust is present in the line of sight, however, an additional variable is introduced such that the star is dimmed by

[1] Extinction is defined as the combination of two physical processes: absorption (photons are removed from the transmitted beam) and scattering (photons are deflected from the line of sight). See Section 2.1.

extinction, and hence its apparent magnitude is increased.[2] If no correction is made for this effect, Equation (1.1) yields an overestimate of the true distance.

An early demonstration of this anomaly was provided by Trumpler (1930a, 1930b) in a study of open star clusters distributed along the disk of the Milky Way. Estimates of the linear diameters of the clusters revealed an unexpected trend: they appeared to increase systematically in size with distance from the Sun. Realizing that this was most unlikely to be correct, he attributed it to systematic errors in the distances to the clusters, estimated in each case by averaging those deduced for its member stars. If the distance to a cluster of a given angular diameter is overestimated, then so is its linear diameter. To correct for this anomaly, an additional term is needed in the distance equation:

$$V - M_V - A_V = 5 \log d - 5 \tag{1.2}$$

where $A_V \geqslant 0$ is the visual extinction[3] and d is now the true distance. The terms A_V and d are correlated in directions close to the galactic disk because, on average, the more distant the star the greater the accumulation of extinction by dust along the line of sight. The corrections needed to normalize the cluster data amount to a mean accumulation rate of $A_V/d \approx 1$ mag kpc^{-1}.

1.2.2 The Reddening of Starlight

The amount of interstellar extinction toward a star depends not only on the quantity of dust in the line of sight but also on the wavelength of observation. A second problem that puzzled stellar astronomers a century ago was the fact that many stars close to the galactic plane appear *redder* than expected on the basis of their spectral classifications. The spectrum of a star gives an estimate of temperature based on the presence and relative intensities of spectral lines in the stellar photosphere, whereas photometric measures of color, such as the color index $(B - V)$, are indicators of temperature based on the continuum shape and its equivalent blackbody temperature. Many stars that show spectral characteristics indicative of high surface temperature ($T \gtrsim 10{,}000$ K) have color indices appropriate to much cooler stars. This anomaly is easily explained if they are reddened by foreground interstellar dust along the line of sight. By comparing the apparent brightness over a range of wavelengths of intrinsically-similar stars with different degrees of reddening, Trumpler and others showed that interstellar extinction is a roughly linear function of reciprocal wavelength ($A_\lambda \propto \lambda^{-1}$) in the visible region of the spectrum (e.g., Stebbins et al. 1939). This important result implies the presence of submicrometer-sized particles with dimensions roughly comparable with the wavelength of visible light. A particle of this size is expected to contain $\sim 10^9$ atoms if its density is comparable with those of terrestrial solids.[4]

---

[2] Stellar magnitudes are based on an ancient ranking system and are numerically larger for fainter stars.

[3] This terminology apparently arose because the dimming of starlight was described as "absorption" (A) rather than "extinction" in early literature.

[4] The term "smoke" was often used to describe these particles in early literature. Smoke implies the product of combustion, whereas "dust" implies finely powdered matter resulting from the abrasion of larger solids. The former is arguably more appropriate as a description of the particles condensing in stellar atmospheres, an important source of interstellar grains, but "dust" has become firmly established in modern usage.

The process by which interstellar dust reddens starlight is analogous to the reddening of the Sun at sunrise or sunset by particles in the terrestrial atmosphere. A photon encountering a dust grain is either absorbed or scattered (Section 2.1). An absorbed photon is completely removed from the beam and its energy converted into internal energy of the particle, whereas a scattered photon is deflected from the line of sight. Reddening occurs because absorption and scattering are, in general, more efficient at shorter wavelengths in the visible: thus red light is less extinguished than blue light in the transmitted beam, whereas the scattered component is predominantly blue. The appearance of a stellar spectrum over a limited spectral range is not drastically altered by moderate degrees of such reddening, in the sense that the wavelengths and relative strengths of adjacent spectral lines are essentially unchanged: spectral classification therefore gives a reliable measure of the temperature of a star independent of foreground reddening. However, color indices depend on both temperature and reddening, information that can be separated only if the spectral type of the star is known.

The interstellar reddening of a star may be evaluated as

$$E_{B-V} = (B - V) - (B - V)_0 \tag{1.3}$$

in the Johnson photometric system, where $(B - V)$ and $(B - V)_0$ are observed and "intrinsic" values of the color index. E_{B-V} is variously termed the "color excess," "selective extinction," or simply "reddening." As extinction is always greater in the B filter (central wavelength 0.44 μm) than in V (0.55 μm), E_{B-V} is a positive quantity for reddened stars and zero (to within observational error) for unreddened stars. Intrinsic colors are determined as a function of spectral type by studying nearby stars and stars at high galactic latitudes that have little or no reddening. Color excesses may be defined for any chosen pair of photometric passbands by analogy with Equation (1.3). See Bessell (2005) for a review of photometric systems.

The nature of the relationship between color excess and total extinction is important because, in general, only the former may be determined directly from observations (stars close enough to allow accurate distance estimates by the parallax method are exceptions). In the Johnson system, the extinction in the visual passband may be related to E_{B-V} by

$$A_V = R_V E_{B-V} \tag{1.4}$$

where R_V is termed the ratio of total to selective visual extinction. If R_V were a constant, determination of A_V from E_{B-V} would obviously be facile; but in practice it is model-dependent. R_V is expected to vary with the composition, structure and size distribution of the grains, properties that depend on their origin and environment. If an incorrect value of R_V is assumed, the distance to the star inferred from Equation (1.2) will also be in error. Following the discovery of interstellar extinction, much effort was devoted in subsequent years to the empirical evaluation of R_V (e.g., Whitford 1958 and references therein). R_V has been shown to be only modestly variable in diffuse (low-density) regions of the ISM in our Galaxy, and a mean value of $R_V \approx 3.05$ may be assumed as a reasonable approximation for most lines of sight. The origin of this result is discussed in detail in Section 3.2.2.

1.2.3 New Light in the Infrared

The energy gained by a dust grain as it absorbs photons from the ambient radiation field causes internal heating, and if the grain is to maintain a constant (equilibrium) temperature it must lose energy at the same rate. By far the most efficient way for grains to dissipate heat in the near-vacuum of space is by thermal emission of radiation, as first predicted by van de Hulst (1946). A steady state is established if a dust grain emits a power equal to that absorbed at some temperature T_d that depends on its size and composition. Models show that grains in the size range responsible for visible extinction ($\sim$0.1–0.5 μm) reach equilibrium at temperatures $T_d \sim$ 5–20 K under typical interstellar conditions (Section 2.3.1). At these temperatures, the spectrum of the emission peaks at wavelengths in the range 100–300 μm, i.e., in the far-infrared region of the electromagnetic spectrum.

The Earth's atmosphere is opaque to much of the far-infrared spectrum, so confirmation of the predicted emission from cosmic dust had to await the development of airborne and space observatories customized to operate at these wavelengths, beginning in the early 1970s. Subsequent observations led not only to confirmation of the theory but also to new insights into the nature of the dust and its distribution relative to other constituents of the ISM (e.g., Hildebrand 1983). As Neville Woolf (1973) memorably remarked, "the dust that is merely opaque in the visible is self-luminous in the infrared, and so in the midst of this optical darkness there has appeared a great infrared light."

As noted above (Section 1.2.2), dust grains tend to absorb more efficiently at shorter wavelengths, i.e., in the blue, violet, and ultraviolet regions of the electromagnetic spectrum. Because of the quantum nature of light, this is a statistical process, and large differences exist in the energies ($\varepsilon_{ph} = h\nu = hc/\lambda$) of photons contributing most of the energy gain compared with most of the energy loss. Typically, many infrared photons must be emitted to dissipate the energy gained from a single absorption event. Momentary increases in grain temperature (ΔT) thus arise when individual photons are absorbed. In the submicrometer-sized particles responsible for visible extinction, ΔT is expected to be small ($\ll$1 K) provided that the photon energy conducts rapidly throughout the volume of the grain, in which case their instantaneous temperatures do not deviate significantly from the equilibrium temperature determined by the mean rates of energy gain and loss. However, this is no longer true for the smallest (nanometer-sized) interstellar particles, which have much lower heat capacities, such that individual photons can deliver sufficient energy to produce ΔT values as high as $\sim$500 K. These transients result in continuum radiation from dust in the mid-infrared (5–50 μm) with intensity orders of magnitude greater than predicted for grains in thermal equilibrium. Heated grains may dissipate energy by means of spectral line emission as well as continuous emission, and the advent of infrared spectroscopy led to detection of a previously unknown population of ultra-small carbonaceous particles displaying spectral features identified with polycyclic aromatic hydrocarbons (PAHs). See Tielens (2008) for a review.

1.2.4 Interstellar Polarization

The presence of dust renders interstellar space a polarizing as well as an attenuating medium. Starlight scattered by dust in a reflection nebula is partially polarized in a manner analogous to that commonly observed in terrestrial environments, where the efficiency of reflection for an incident beam depends on the orientation of the **E**-vector relative to the plane of the surface. Perhaps more surprising was the discovery that light in the *transmitted* beam from reddened stars is also partially plane polarized, typically at levels of a few percent, as was first demonstrated by Hall (1949) and Hiltner (1949). This observation is most readily explained if interstellar grains have asymmetric optical properties and a preferred orientation in space. The most likely origin of asymmetry is shape: real grains are unlikely to be the perfect spheres often assumed in model calculations! If they are flattened or elongated in some way and become aligned, such that their long axes have a preferred direction, the attenuation of radiation passing through the medium depends on the orientation of the **E**-vector relative to the direction of alignment, resulting in polarization of the transmitted beam.

Grain alignment also causes infrared radiation *emitted* by the particles to be polarized. For each grain, the **E**-vector of the emitted beam has greatest amplitude in the plane containing the longest axis of the grain, so the observed net emission from an ensemble of aligned grains is partially plane polarized. It follows that the position angle of the polarized emission is perpendicular to that of polarization in the transmitted beam, as the amplitude of the **E**-vector for one is greatest when the other is least.

There is a general consensus that alignment is caused by an interaction between the spin of the particles and the ambient magnetic field, as first proposed by Davis and Greenstein (1951). However, the details of the mechanism, and its efficiency with respect to factors such as the composition, size and shape of the grains, the interstellar environment, and the strength of the local magnetic field, have proved difficult to establish. See Section 4.3 for detailed discussion and Andersson et al. (2015) for a review.

1.2.5 Spectral Features

The observational phenomena discussed above do not provide definitive information on the *chemical composition* of the dust grains that produce them. A variety of solids in grains of appropriate size and shape might account for extinction, reddening, and polarization in the visible, and for continuum radiation in the infrared. Spectroscopy provides a potential means to resolve such ambiguities—I say *potential* because unique identifications of spectral signatures in solids are often harder to accomplish than those arising in gaseous atoms and molecules. The observations often indicate the presence of broad classes of materials, such as silicates and hydrocarbons, rather than specific compounds.

Space observatories enable access to the ultraviolet (UV) and X-ray regions of the spectrum. The UV spectrum of dust is dominated by a prominent, broad absorption peak centered at 217.5 nm, often referred to in the literature as the "2175 Å extinction feature" or "bump." Draine (1989) aptly described it as "a dramatic piece of spectroscopic evidence which should have much to tell us about at least a part of the interstellar grain population." The precise nature and origin of this population remains somewhat controversial, but there is general agreement that it includes particles much smaller ($\lesssim$ 20 nm) than those responsible for visible extinction, composed, at least in part, of graphitized carbon (Section 3.4). At X-ray wavelengths, atoms in both dust and gas absorb by the photoelectric effect. Observations of these features are important because they provide a census of the total elemental abundances along a line of sight (Section 7.1.5), and a means of distinguishing the contributions of dust and gas by study of X-ray absorption fine structure (Lee et al. 2009).

The near to mid-infrared region of the spectrum (2–30 μm) is especially rich in features attributed to dust, arising from vibrational modes in the solid particles. Pairs or groups of atoms oscillate at characteristic frequencies, determined by their masses and the strengths of the bonds between them, and these oscillations may result in characteristic absorption or emission features (see Section 5.1). Common examples include linear, spring-like (stretching) vibrations of O–H, C–H, and Si–O bonds in water-ice, hydrocarbons, and silicates, respectively, resulting in spectral features observed near 3.0, 3.4, and 10 μm. Ambiguities arise because these same bonds are present in different materials: almost all organic molecules contain C–H bonds, and all silicates contain Si–O bonds. All is not lost, however, as the vibrations are influenced by the wider molecular environment of the vibrating atoms as well as by the nature of the specific chemical bond; so, for example, the feature produced by C–H bonds in one organic molecule may differ in detail from that produced in another. Laboratory studies of terrestrial analog materials are vital to this area of research, providing data that may enable us to resolve such ambiguities. See Chapter 5 for detailed discussion.

1.2.6 Where the Atoms Went

Spectroscopy of interstellar *gas* equips us with another means of studying dust composition, enabling insight into the chemical elements involved and their relative contributions. As a starting point, a reliable set of standard elemental abundance is required, i.e., values representative of the ISM as a whole, gas and dust combined. This is obtained by a range of techniques and subject to certain assumptions, discussed in Chapter 7. Results from observations of the gas are then compared with the standard for each element. These comparisons show that for many "heavy" elements,[5] especially those most easily condensible into solids at interstellar temperatures, there is a dramatic shortfall in gas-phase abundance, leading us to infer that the "missing" atoms (e.g., Greenberg 1974) are to be found in the solid phase. The amplitude of this "depletion" varies with interstellar environment and is typically greatest in densest regions, as expected if gas and dust exchange material according to local conditions.

[5] In this context, "heavy" implies elements with atomic weight $\geqslant$12, i.e., carbon and above, often referred to collectively as "metals" in the astronomical literature.

Taking abundances, atomic weights and levels of depletion into account, most of the mass of interstellar dust appears to be contributed by a set of just five chemical elements: C, O, Mg, Si, and Fe. Heavier metals such as Cr, Ni, and Ti reside almost entirely in dust but do not contribute significantly to its mass because their abundances are far less. Of the "big five," Mg, Si, and Fe are highly depleted from gas to dust under most conditions, whereas C and O are typically more capricious in their distributions (e.g., Jenkins 2009). Logical candidate grain materials based on these elements include carbonaceous solids (e.g., graphite, amorphous carbon, organic matter), oxygen-rich minerals (silicates, oxides), and ices (solid H_2O, CO, CO_2, etc.).

A further question that should be addressed here is whether hydrogen, the lightest and by far the most abundant of the elements, can contribute significantly to solids under interstellar conditions. For H_2 to freeze, temperatures as low as 3 K would be required—close to the cosmic microwave background temperature, and significantly below those generally associated with even the coldest, densest interstellar clouds. Furthermore, Hoang & Loeb (2020) show that even under such extreme conditions, widespread formation of H_2-rich ice is unlikely because of rapid sublimation induced by collisional heating. It therefore seems safe to conclude that elemental H is included in the solid phase only if it is chemically bonded to a heavier element, e.g., in hydrocarbons or ices, or loosely attached to a solid surface by weak van der Waals forces. Thus, despite its ubiquity, H cannot make a major contribution to dust mass: for example, it accounts for only $\sim 11\%$ of the mass in ices composed of H_2O. All things considered, the total mass of "condensible" elements, i.e., all that *could* be in solids under the most favorable conditions, amounts to about 2% of the total mass of the ISM.

1.2.7 Gathering It All Together: Mathematical Models

The interaction of electromagnetic radiation with small particles may be simulated using mathematical models, which in essence attempt to solve Maxwell's equations with appropriate boundary conditions at grain surfaces (see Chapter 2). Observable phenomena such as the spectral dependence of extinction are predicted for a given set of assumptions on the nature and size of the particles. Results are compared with observational data, and the initial assumptions may then be modified with a view to improving the fit. To model real interstellar grains is, of course, extremely challenging. Assumptions regarding their composition are constrained by the spectroscopic evidence and depletion data discussed above, suggestive of materials such as silicates and various forms of solid carbon. For simplicity, it is often assumed that the grains approximate in shape to homogeneous spheres when modeling extinction, or to spheroids if polarization is to be considered; but in reality grains may assume any number of shapes and structures. Over the years since its discovery, modelers have been developing and refining their codes with a view to converging on a robust, unified theory interstellar dust. This ambitious quest has yet to be fully accomplished, but important advances have been made and it is possible to fit most of the observed properties of the dust with plausible models that do not violate abundance constraints.

A significant area of progress concerns the size distribution of the particles. For spheres of radius a, this is represented numerically by the function $n(a)$, corresponding to the relative number of particles with radii in a small incremental range a to $a + \Delta a$. As noted in Section 1.2.2, the observed extinction varies approximately with λ^{-1} in the visible. For dielectric materials, i.e., those that attenuate primarily by scattering rather than absorption, this implies that the extinction is dominated by particles not much smaller than the wavelength (Section 2.1), placing them in the submicrometer range (~0.1–0.5 μm). Particles much smaller than this produce a much steeper variation, converging to $A_\lambda \propto \lambda^{-4}$ in the "small particle limit" corresponding to Rayleigh scattering. Conversely, much larger particles produce "neutral" extinction with little or no dependence on wavelength. More generally, best fits to the extinction curve over the observed spectral range from the ultraviolet to the infrared are obtained by assuming a power-law form for the size distribution: the relative number of particles varies with respect to size as $n(a) \propto a^{-q}$, where the index q typically has a value of approximately 3.5. It follows that smaller particles greatly outnumber larger ones, although the larger ones produce most of the extinction at the longer wavelengths.

1.3 Interstellar Environments

Hydrogen, the primary constituent of interstellar gas, may exist in any one of three physical states or "phases," molecular (H_2), atomic (H I), or ionized (H II). The properties of the gas are governed by the laws of thermodynamics, and by its interaction with electromagnetic radiation, cosmic rays, shock waves, and gravitational and magnetic fields. The physical processes involved lead to the presence of a vast range of environments, from tenuous, hot plasmas to cold, relatively dense molecular clouds where new stars and planetary systems are born (see Figure 1.3 for examples). This section presents a concise overview of the factors that determine the physical state of the gas. Conditions of density, temperature, and phase that characterize typical environments are identified, including intercloud gas, "diffuse" and "dense" clouds, and H II regions. Finally, the current interstellar environment of our solar system is considered. For more detailed reviews of physical processes in the interstellar medium, the reader is referred to the classic texts by Spitzer (1978), Draine (2011), and Dyson & Williams (3rd edition, 2020).

1.3.1 The Physical State of the Interstellar Medium

The physical properties of interstellar gas are described by temperature (T), number density (n, the number of nuclear particles per unit volume), and phase (molecular, atomic, or ionized). For an ideal gas, the thermodynamic pressure is

$$p = nkT \tag{1.5}$$

where k is Boltzmann's constant. Thus, for pressure-equilibrium between regions with different temperatures, the product nT should be constant (with a value typically $\sim 3 \times 10^9\ \mathrm{m^{-3}}$ K; Mathis 2000). However, disturbances such as energetic stellar winds and supernova shock waves cause turbulence in the ISM

Figure 1.3. Contrasting interstellar environments. Above: a section of the Veil Nebula in Cygnus. Wisps of hot, tenuous gas are seen as the shock wave from a supernova explosion that occurred some 10,000 years ago expands into the surrounding ISM. The bright star 52 Cygni (above center) lies in the foreground. Below: the Lupus 3 star formation region. The blue nebula at center is caused by scattering of starlight by dust; opaque regions to left and right contain cold dust and molecular gas yet to be processed into stars. Image credits: Jschulman555 / CC-BY-SA-3.0 (Wikimedia Commons) and ESO/R. Colombari, respectively.

(e.g., Struck & Smith 1999), such that pressure equilibrium is not well established. Also, the magnetic field contributes to the pressure of an ionized or partially-ionized gas, and this may sometimes exceed the thermodynamic pressure. Nevertheless, a general trend roughly in accord with Equation (1.5) is observed for typical environments, illustrated in Figure 1.4.

The temperature in a given environment depends on the balance of heating and cooling mechanisms. The dust is heated primarily by absorption of energetic photons from the interstellar radiation field, and cooled by thermal emission of photons at longer wavelengths. The gas temperature is regulated not only by photon absorption but also by collisions with photoelectrons ejected from small grains, by binding energy released by molecule formation, and by turbulent motions driven by shocks, and it is cooled principally by spectral line emission (Bakes & Tielens 1994; Draine & McKee 1993; Draine 2011). Cosmic rays also contribute to the heating of both dust and gas, and are significant especially in molecular clouds where the radiation field is strongly attenuated. The transitions responsible for cooling the gas are generally driven by collisional excitation, such that the rate varies as n^2. A small increase in density can therefore lead to a substantial increase in heat loss; this lowers the temperature and pressure, causing contraction of the region and further increase in density. The ISM thus tends naturally to separate into cool, relatively dense regions ("clouds") and hot, rarefied regions (the "intercloud medium").[6] Gas temperatures greatly exceed dust temperatures at low densities, but they tend to converge in the densest clouds.

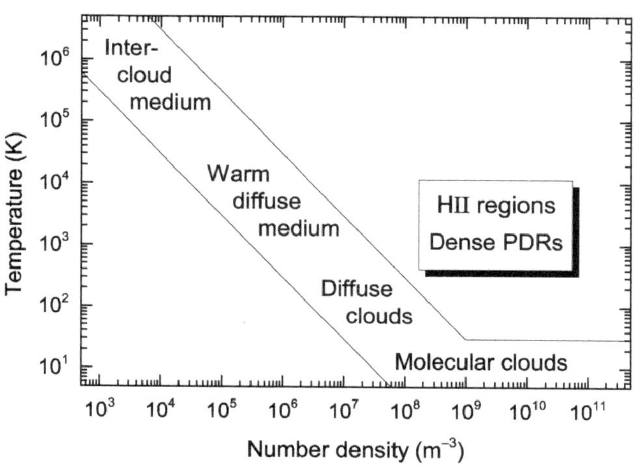

Figure 1.4. A schematic plot of number density versus kinetic temperature for interstellar hydrogen gas, illustrating the loci of typical environments. Regions between the diagonal lines roughly follow the trend predicted by Equation (1.5) for $p/k \sim 3 \times 10^9\ m^{-3}$ K.

[6] It is convenient to describe relatively dense regions of the ISM as "clouds" but it should be remembered that they are not clouds in the terrestrial sense, arising as local accumulations of matter driven by the various factors noted above rather than as condensation of gas into particles. See also Section 1.3.2.

The physical state of the gas in a given region of the ISM is strongly influenced by the local intensity of the interstellar radiation field. Photons of the highest energies ($h\nu \geqslant 11.2$ eV) drive photoionization and photodissociation of hydrogen:

$$\mathrm{H} + h\nu \rightarrow \mathrm{p} + \mathrm{e} \quad (h\nu \geqslant 13.6 \text{ eV}) \tag{1.6}$$

and

$$\mathrm{H}_2 + h\nu \rightarrow \mathrm{H} + \mathrm{H} \quad (11.2 \leqslant h\nu \leqslant 13.6 \text{ eV}). \tag{1.7}$$

Analogous reactions can be driven by other energy sources such as cosmic ray impacts and shocks. Note, however, that dissociation of H_2 occurs at discrete energies in the range quoted. Interstellar H_2 normally exists in the ground state, and there are no allowed transitions to higher states that result in dissociation. The reaction summarized by Equation (1.7) is actually a two-step process, in which excitation by photon absorption to a higher level is followed by spontaneous transition to an excited vibrational level of the ground state from which dissociation can occur (see Spitzer 1978, p. 124). Under equilibrium conditions, ionization and dissociation reactions are balanced by recombination. Protons and electrons may recombine to form H I, with release of kinetic and binding energy typically in the form of several lower-energy photons. Recombination of atomic H to form H_2 requires a third body to absorb binding energy and occurs almost entirely on grain surfaces (Section 9.1.4). The equilibrium state of a given region of the ISM is thus strongly influenced by its density: high densities promote recombination and provide shielding against ionizing and dissociating radiation; low densities inhibit recombination and increase exposure to the interstellar radiation field.

The phase structure of the ISM may be characterized conveniently in terms of four discrete components: cold (molecular), cool (atomic), warm (atomic or partially ionized), and hot (ionized) gas. Their average properties are listed in Table 1.1, and their approximate distribution in the T versus n plane is shown in Figure 1.4. This representation is loosely based on the so-called three-phase model of McKee and Ostriker (1977), with subdivision of cold gas into atomic and molecular phases. In this model, the structure is regulated by supernova explosions. On average, a given point in the ISM is swept out by an expanding remnant once every few million years, leading to disruption of existing cloud structure and the establishment of low-density bubbles of hot gas; thus, supernova explosions are primarily responsible for maintaining the intercloud medium in its hot, ionized state. The intercloud H II and warm diffuse components of the ISM should be distinguished from H II regions

Table 1.1. A Four-component Model for the Interstellar Medium.

Description	Phase	T (K)	n (m^{-3})	ϕ
Cold molecular	H_2	~15	$>1 \times 10^8$	<0.01
Cool atomic	H I	~80	$\sim 3 \times 10^7$	~0.03
Warm diffuse	H I, H II	~8000	$\sim 3 \times 10^5$	~0.1–0.5
Hot ionized	H II	$\sim 10^6$	$\sim 5 \times 10^3$	~0.5–0.8

and photodissociation regions (Figure 1.4) resulting from irradiation of much denser gas by individual OB stars or associations (Section 1.3.3).

In addition to phase, temperature and number density, Table 1.1 lists the volume filling factor ϕ of each component, defined as the average fractional volume of space that it occupies in the galactic disk. Along a line of sight that intercepts many different regions, each typified by one of the four "standard" environments, the average number density is

$$\langle n \rangle = \sum \phi_i n_i \tag{1.8}$$

where ϕ_i and n_i, are the volume filling factor and number density, respectively, of the ith component. The filling factors sum to unity ($\sum \phi_i = 1$) if all components of the ISM are accounted for. Values of ϕ are uncertain and subject to spatial variability: as examples, molecular clouds are more common at galactocentric distances $R_G \sim 3$ to 7 kpc in the galactic disk than in the solar neighborhood ($R_G \sim 8.5$ kpc), and hot ionized gas is more pervasive in the galactic halo and in interarm regions of the disk than in spiral arms. However, the contrast in density between the hotter and cooler components is sufficiently large that an important general conclusion may be drawn on the basis of the rough estimates for ϕ listed in Table 1.1: although the volume of the ISM is mostly filled by tenuous, relatively hot and mostly ionized gas, the number density, and hence the mass, is dominated by the cooler, denser atomic and molecular phases, i.e., by *clouds*.

1.3.2 Interstellar Clouds

Historically, the term "cloud" (Latin: nebula) was adopted to describe the most obvious visual manifestations of interstellar matter observed in the Milky Way. Later, it was also applied to clumps of interstellar gas identified spectroscopically, each individual cloud along a line of sight being distinguished by its discrete Doppler motion. However, it would be simplistic to describe the ISM as an assemblage of clouds of various shapes and sizes enclosed in a uniform intercloud medium. The ISM is not merely inhomogeneous but also hierarchical, displaying fractal behavior ("clumps within clumps") over a wide range of size scales. Clouds of similar size and mass may have very different morphologies, and clouds of different size and mass may give rise to similar environments. In the modern view, clouds are considered to be peaks in the density distribution on size scales that correspond to observed concentrations of interstellar gas and dust (Elmegreen 2002).

Given this caveat, it might seem unrealistic to attempt to characterize "representative" interstellar environments. However, the phase structure of the ISM does allow a useful distinction to be made: it is convenient to adopt the labels "diffuse" and "dense" to describe clouds in which the gas at peak density is predominantly atomic and molecular, respectively. These are distinguished numerically by the parameter $f(H_2)$, the fractional abundance of hydrogen in molecular form, defined as

$$f(H_2) = 2n(H_2)/n_H \tag{1.9}$$

where

$$n_H = n(H) + 2n(H_2) \tag{1.10}$$

is the total number density of H nucleons in atoms and molecules for the common isotopic form. In a diffuse cloud, $f(H_2)$ typically ranges from zero to ~0.3; in the core of a dense cloud it approaches unity. An idealized, schematic representation of a cloud of each type is shown in Figure 1.5. Each lacks any internal source of luminosity and is externally-heated by the interstellar radiation field (ISRF), which permeates the intercloud medium virtually unattenuated as it contains little dust and the plasma is optically thin to ionizing radiation. The general properties of diffuse and dense clouds may be summarized as follows.

A *diffuse cloud* is defined as one of moderate density ($n_H \sim 10^7$–10^8 m^{-3}) and extinction ($A_V \sim 0.1$–1.0) in which the dominant phase is H I. A typical example might have dimensions ~5 pc and mass ~30 $M_\odot$, but with wide variations from cloud to cloud. The core of a diffuse cloud (Figure 1.5) is opaque to radiation beyond the Lyman limit ($h\nu \geqslant 13.6$ eV), but somewhat transparent to radiation of energy in the range 11.2–13.6 eV that can dissociate H_2 (Equation (1.7)). Cool H I gas is encased in a halo of warm gas, the outer layers of which are partially ionized (Equation (1.6)) by hard ultraviolet photons in the ISRF, which it strongly absorbs. The warm gas is predominantly neutral, however, and heated by soft X-ray photons ($h\nu \sim 40$–120 eV) emitted by the hot intercloud gas. The least massive, most tenuous diffuse clouds may lack a cool phase entirely, but these tend to dissolve into the surrounding hot medium and are thus short-lived. Denser diffuse clouds are by no means devoid of gas-phase molecules: species such as CO, OH, CH, CH^+, CN, H_3^+, and HCO^+, as well as H_2, may be present with small but detectable abundances; see Snow & McCall (2006) for a review.

A *dense cloud* contains regions sufficiently dense ($n_H > 10^8$ m^{-3}) that virtually all of the H I is predicted to be converted to H_2 ($f(H_2) \rightarrow 1$) by grain-surface catalysis on timescales of a few million years, short compared with their expected lifetimes (the terms "dense cloud" and "molecular cloud" are often used interchangeably in

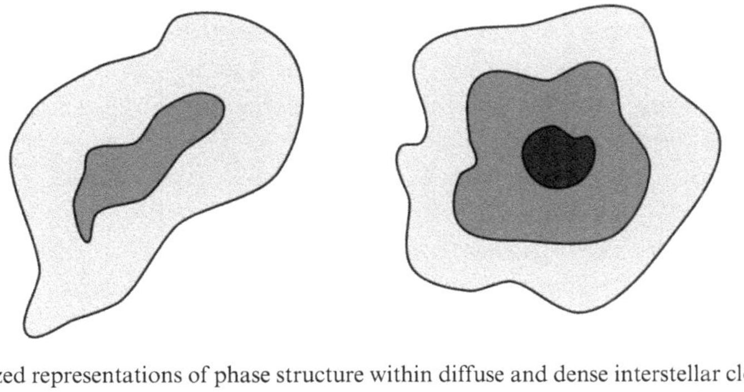

Figure 1.5. Idealized representations of phase structure within diffuse and dense interstellar clouds. The diffuse cloud (left) contains cool atomic gas (dark gray) encased in an outer layer of warm neutral or partly ionized gas (light gray). The dense cloud (right) contains an additional molecular zone (black). Image credit: Reproduced courtesy of Perry Gerakines.

the literature). Such conditions are found in structures ranging from compact (<1 pc), low mass (<50 $M_\odot$) clumps to giant molecular clouds ranging up to $\sim$50 pc and $\sim 10^6$ $M_\odot$ in size and mass. Dense molecular gas is opaque to both ionizing and dissociating radiation (Equations (1.6) and (1.7)). The molecular region is effectively self-shielded from the external ISRF by the outer layers of the cloud itself, which remain predominantly atomic (Figure 1.5), and in which dissociating photons are attenuated by both gas and dust. But although the interior is shielded from the external ISRF, relativistic particles (cosmic rays) may permeate this environment and thereby induce appreciable levels of ionization. Cosmic-ray ionization is important as it influences both chemical and physical processes within the cloud. Ions are more reactive than neutral species, and therefore participate more readily in chemical processes that form new molecules (Section 9.2.1). The magnetic field within a cloud controls the flow of ions, and also indirectly that of neutral species through collisions between ions and neutrals (Section 4.1.3). The propensity of a gravitationally-bound cloud to collapse and form stars is strongly influenced by this effect.

1.3.3 H II Regions and Photodissociation Regions

Interstellar clouds are in a continuous state of evolution. Clumps in the densest regions may undergo gravitation collapse to give birth to new stars, thereby creating internal sources of energy that disrupt and ultimately dissipate the parent cloud. The level of disruption imposed by each newly-formed star depends on its mass. The most massive stars are the most luminous and the quickest to evolve: for example, a 15 $M_\odot$ star (spectral type B0) has a luminosity of 13,000 $L_\odot$ and a main-sequence life-expectancy of only 10 Myr, less than typical cloud lifetimes of 10–100 Myr. Such stars emit most of their radiant energy in the ultraviolet, resulting in photoionization and photodissociation of the surrounding medium to form nebulae (e.g., Figure 1.6) of ionized gas (H II regions) and photodissociation regions (in which H is mostly atomic) that are typically very much denser than diffuse phases of the ISM with similar temperatures and ionization states (Figure 1.4 and Table 1.1).

The physics of H II regions is discussed in detail elsewhere (e.g., Chapter 15 in Draine 2011), and only a few important results are summarized here. Ions are formed by reaction (1.6), and any excess photon energy (above the ionization potential) contributes to the kinetic energy of the products, thus maintaining the plasma at high temperature, typically $\sim$10,000 K. Other species may be ionized as well as hydrogen, of course, but H II is the dominant phase. Strong Coulomb attraction between protons and electrons leads to recombination, i.e., the inverse of reaction (1.6), with energy released in the form of photons at discrete wavelengths, observed as spectral emission lines from the nebula. Equilibrium is established when rates of photoionization and recombination are balanced; the volume of gas a star can maintain in an ionized state is then limited by the rate (S_*) at which it emits ionizing photons, and by the density of the gas. In an idealized, continuous, dust-free medium of uniform number density n, this volume is a sphere (known as the Strömgren sphere) with radius

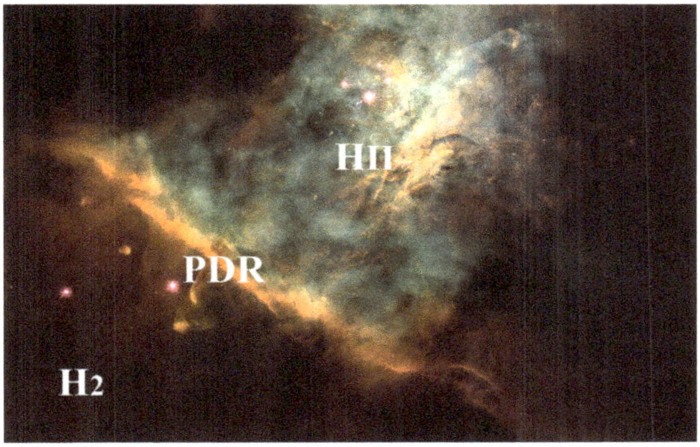

Figure 1.6. Mosaic of images taken with the Hubble Space Telescope, illustrating H II and photodissociation (PDR) regions in the Orion Nebula. The Trapezium group of luminous stars (above center) is the primary source of ionizing radiation. The bright, diagonal ridge of nebulosity, known as the Orion Bar, occurs as energetic photons meet the outer layers of a dense molecular cloud (lower left), forming a narrow transition zone in which molecules are subject to photodissociation. The kinetic temperature of the gas ranges from ~ 10,000 K in the H II region to ~ 10 K in the molecular cloud. Image credit: C. R. O'Dell (Rice University), and NASA/ESA; labeling by the author.

$$R_S = \left(\frac{3S_*}{4\pi n^2 \beta}\right)^{\frac{1}{3}} \tag{1.11}$$

where β is the rate coefficient for recombination. This coefficient is temperature-dependent, but plasma temperatures in H II regions do not vary greatly, and a constant value of $\beta \approx 2 \times 10^{-19}$ m^3 s^{-1} may be assumed as a reasonable approximation. For a star of spectral type O5, we have $S_* \approx 5 \times 10^{49}$ s^{-1}, so a Strömgren radius $R_S \approx 3.1 \times 10^{16}$ m $\approx$ 1 pc is predicted in a dense medium with $n = 10^9$ m^{-3}. For a B0 star, S_* is a factor of 100 less and hence R_S is a factor of $(100)^{\frac{1}{3}} \approx 5$ less. For stars later than B0 in spectral type, S_* declines rapidly and hence so does R_S. Note that $R_S \propto n^{-\frac{2}{3}}$, thus the Strömgren sphere becomes larger in a more rarefied medium.

These estimates are based on the assumption that H I gas is the only significant absorber of ionizing radiation. However, H II regions will generally also contain significant quantities of refractory dust particles, such as graphitic carbon and silicates, which are hardy enough to survive in this environment. Carbonaceous particles, in particular, are efficient absorbers at the relevant wavelengths, and their presence reduces the effective size of the H II region by depleting the supply of ionizing photons. The presence of dust also influences the density structure of the region, through the effect of radiation pressure (Section 2.4.1). Dust grains are accelerated outward from the radiation source, and their motion also affects the gas distribution through collisions. The gas density thus tends to be reduced around the star and enhanced toward the outer boundaries of the nebula.

Newly born OB stars that remain embedded in the parent molecular cloud generally form compact H II regions, the sizes of which are limited by the availability of ionizing photons, as discussed above. Photons penetrating beyond the ionization front have energies <13.6 eV, sufficient to dissociate H_2 (reaction 1.7) and other molecules, but not to ionize H I. The thickness of this photodissociation region (PDR) is limited by the density of the gas and its attenuation of photons with energies in the range 11.2–13.6 eV: the transition from hot, fully ionized gas at its inner edge to cold, undisturbed molecular gas beyond its outer edge can be abrupt, as in the case of the Orion Bar, illustrated in Figure 1.6. See Hollenbach & Tielens (1997) for a review of the physics and chemistry of dense PDRs.

When an OB star emerges into lower density phases of the ISM, its sphere of influence increases dramatically. Ionization of ambient interstellar gas creates diffuse H II regions that can be orders of magnitude larger and more rarefied than compact H II regions seen in molecular clouds. Diffuse H II regions may be density-limited rather than photon-limited, i.e., their boundaries may be determined by the physical limits of the cloud rather than the supply of ionizing photons. At intercloud densities, the nominal Strömgren radii for the most luminous O-type stars approach galaxian dimensions; such stars contribute to the flux of high-energy photons in the general interstellar radiation field.

1.3.4 The Interstellar Environment of the Solar System

As a middle-aged star of moderate mass, the Sun has easily outlived the molecular cloud that gave birth to it some 4.6 Gyr ago, and has subsequently migrated into the general stellar disk population of the Milky Way. We cannot identify where it formed, but its current location is of obvious interest. Our vantage point within our home Galaxy governs our view of the universe, and the local interstellar medium is the screen through which we observe it.

The "big picture" view is illustrated schematically in Figure 1.7. The Milky Way is a large and somewhat typical barred-spiral galaxy (Bland-Hawthorn & Gerhard 2016) with a stellar disk of radius ~ 15 kpc. The Sun lies in the disk, close to the mid-plane (within ~ 25 pc) and ~ 8 kpc from the center. It is situated near the inner edge of a minor feature known as the Orion Arm (or Orion Spur), which appears to be a bridging structure between the Perseus and Carina-Sagittarius spiral arms. On scales of a few kiloparsecs, the region around the Sun encompasses the entire range of physical environments (Table 1.1), including both "arm" and "interarm" regions of the Milky Way, and numerous regions of active star formation. On scales below 500 pc, however, the distribution of matter is dominated less by spiral structure than by a local feature known as Gould's Belt, a disk-like system of young stars and interstellar matter tilted at about 20° to the plane of the galactic disk. Gould's Belt contains many of the closest and best-studied regions of active star formation, including those in Perseus, Taurus, Orion, Lupus, Ophiuchus, Scorpius, and Serpens (Ward-Thompson et al. 2007; Dunham et al. 2015). Associated dust is responsible for an asymmetry in the extinction toward the galactic bulge, noticeable in Figure 1.1.

Closer to the Sun, within about 100 pc, is a region of hot, tenuous, ionized intercloud gas ($T \sim 10^6$ K, $n \sim 5 \times 10^3$ m^{-3}) known as the Local Bubble (see Frisch

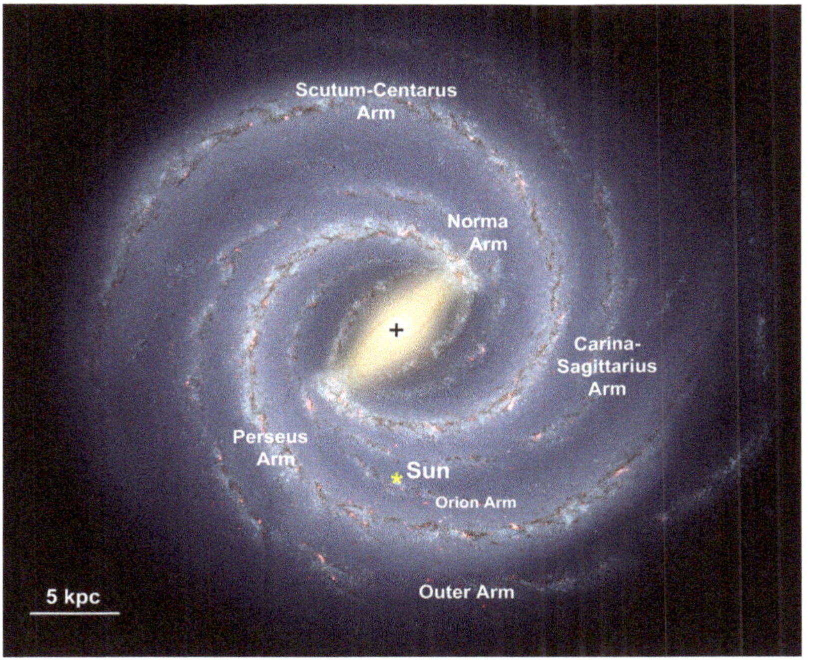

Figure 1.7. Schematic impression of the Milky Way Galaxy as it might appear from a vantage point toward the North Galactic Pole, constructed from infrared and radio observations. Spiral features are labeled, and the locations of the Galactic Center (black cross) and our Sun (yellow asterisk) are indicated. Red areas within the spiral arms represent H II regions generated by recent star formation. Image credit: NASA/JPL-Caltech; labeling by the author.

et al. 2011 for a review). The properties, distribution, and kinematics of the local gas are determined from observations of interstellar absorption lines in the spectra of nearby stars. Results are consistent with observations of dust-related phenomena, which show, for example, that stars closer than 100 pc typically exhibit little or no interstellar reddening ($E_{B-V} < 0.01$; Section 1.2.2): evidently, the solar system does not currently reside in the vicinity of an interstellar cloud of appreciable opacity. This result is not surprising: estimates of filling factors (Table 1.1) imply that the probability of finding the solar system in an interstellar H I or H_2 cloud at any given time is no more than a few percent. The Local Bubble may be a product of stellar activity, cleared out by supernova shocks and stellar winds originating in the star-forming clouds of Gould's Belt. Embedded within the Local Bubble are pockets of cooler, denser, partially-ionized gas with properties characteristic of the warm-diffuse phase of the ISM ($T \sim 7000$ K, $n \sim 10^5$ m^{-3}). The Sun lies close to one such region, illustrated in Figure 1.8.

The Local Cloud is flowing past us at a heliocentric speed of about 25 km s^{-1}, in a direction consistent with outward pressure emanating from the nearest regions of recent star formation in Gould's Belt (those in constellations Scorpius, Ophiuchus, and Lupus, which lie off-frame toward the top in Figure 1.8). The Sun interacts with

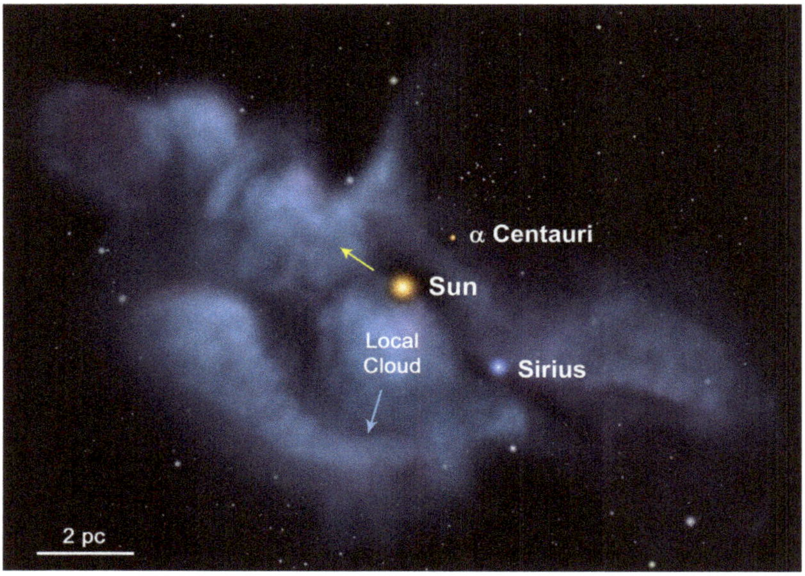

Figure 1.8. The distribution of interstellar matter close to the Sun, shown in projection onto the galactic plane with the direction of the galactic center toward the top. The yellow arrow indicates the Sun's direction of motion relative to the local rest frame, and the blue arrow indicates the approximate average motion of the Local Cloud. Image credit: NASA/Goddard/Adler/U. Chicago/Wesleyan; labeling by the author.

this flow of interstellar gas to form the heliosphere, which is in fact a comet-shaped (rather than spherical) structure, sculpted by the solar motion and the ram pressure of the solar wind. The heliospheric boundary lies at a distance of about 100 AU from the Sun in the upstream direction (well beyond the orbit of Neptune) and extends much further in the downstream direction. A unique aspect of this environment is the opportunity it provides to investigate interstellar matter by means of data collected *in situ* as well as by astronomical observations.

Space missions such as Ulysses, Galileo, and Stardust enabled gas and dust entering the solar system from the local ISM to be studied. The flow of neutral atomic elements such as helium is a useful tracer of the gas as it penetrates into the inner heliosphere, its direction and speed broadly consistent with the motion of the local cloud as determined by stellar spectroscopy. Dust detectors carried by these missions recorded impacts with interplanetary grains, those of interstellar provenance distinguished from those native to the solar system on the basis of kinematics: interstellar grains have much higher average speeds and a much narrower range of trajectories (see Mann 2010 for a review). The collection process is destructive, because of the high impact-speeds, but it enables the masses and sizes of individual particles to be estimated. The size distribution of the detected particles peaks at a radius of $a \sim 0.3\ \mu$m and has an extreme range of 0.02–10 μm.

These results suggest a bias toward larger sizes compared with those inferred from astronomical observations and modeling of more widely-distributed dust in the ISM

(Section 1.2.7). The apparent dearth of very small grains in the local sample may be explained by a selection effect: the forces exerted on particles entering the heliosphere by radiation pressure and the interplanetary magnetic field selectively prevent smaller ones from reaching the inner solar system. The detection of grains with $a > 1\ \mu$m is nevertheless surprising as it is not predicted by observations of interstellar extinction, and could be inconsistent with known abundance constraints if applicable to the general population. The local dust may be atypical, however, perhaps as the result of recent processing by shocks that created the Local Bubble: see Frisch et al. (1999) and Frisch & Slavin (2013) for further discussion.

A final question to consider is whether the detected flow of interstellar dust into the solar system can be reconciled with observational evidence, noted above, that the local ISM is essentially translucent, producing minimal extinction and reddening of starlight. The inward flux implies that the contribution of dust to the local mean density of the ISM amounts to $\sim 5 \times 10^{-24}$ kg m^{-3} (Mann 2010), which corresponds to about 40%–50% of the average density determined from observations of interstellar reddening toward stars within a few kiloparsecs of the Sun (see Section 3.2.6). This is consistent with the "warm diffuse" status inferred from other observations of the Local Cloud. The rate of extinction per unit distance scales with dust density, suggesting a mean value of 0.5 mag kpc^{-1} for the local ISM. Using this result and a scale length of 10 pc for the Local Cloud and associated material close to the Sun (Figure 1.8), if follows that the visual extinction suffered by radiation passing through it amounts to only $A_V \sim 0.005$ magnitudes, which is negligible. Thus, there is no conflict between these diverse results.

1.4 Significance

1.4.1 From Cinderella to the Search for Origins

The earliest studies of interstellar dust, specifically those of extinction and reddening (Section 1.2), were motivated merely by a need to negate its effect—i.e., to correct photometric data for its presence—rather than by an intrinsic interest in the dust itself or an appreciation of its broader significance. According to Gaustad (1971) it was once the case, as far as a typical (male) astronomer of the era was concerned, that "...if you simply tell him the reddening law, particularly the ratio of total to selective extinction, he can unredden his clusters, correct the distance moduli, find the turnoff points, determine the age of the Galaxy, and be happy!" Times have changed. Dust is now recognized as a vital ingredient of the cosmos, a revelation that came about largely through exploration of new regions of the electromagnetic spectrum during the latter part of the 20th century. Some of these developments are summarized here and discussed further in subsequent chapters. Perhaps of greatest significance, however, is the simple realization that dust is the primary repository in the interstellar medium (Section 1.2.6) for the chemical elements required to make terrestrial planets and life. The metals and mineral-forming elements that make up the bulk of the Earth, together with the carbon needed to make organic molecules and the oxygen needed to make water, were created in stars and delivered to the cloud that gave birth to our planetary system largely in the form of solid particles.

"We are stardust," wrote Joni Mitchell,[7] and this is literally true at the elemental level. The quest to understand the origin and evolution of interstellar dust is thus part of the quest for our origins.

1.4.2 Dust in Astrochemistry

For many years, the only molecules known to exist in interstellar space were the radicals CH, CH^+, and CN, identified by their characteristic gas-phase absorption lines in stellar spectra at blue-visible wavelengths (McKellar 1940; Douglas & Hertzberg 1960). The list has grown dramatically since the late-1960s, thanks to the availability of facilities for spectroscopy at wavelengths beyond the visible, and the number of confirmed interstellar and circumstellar molecules now exceeds 200.[8] The most abundant molecular constituents of interstellar gas are diatomics such as H_2 and CO; but the list includes many polyatomics, the majority of which are organic, and some of which may have prebiotic significance. Examples of the latter include HCN, H_2CO, HCOOH, and NH_2CH_3, which are chemical precursors of nucleobases, sugars, and amino acids. Molecule-specific detections in the solid phase (Section 1.2.5) include ices of H_2O, CO, CO_2, CH_3OH and several other species, discussed in detail in Chapter 5. These detections demonstrate the complexity of interstellar chemistry and present a challenge to our understanding of chemical processes occurring in extreme environments: see Herbst (2014) for a review. Astrochemical models have been developed that attempt to explain the relative abundances of the observed molecules and to identify the principal formation and destruction mechanisms. Dust is a key player in the models, as both a physical mediator and an active chemical ingredient.

The primary roles of dust in astrochemistry may be summarized as follows:

- Surface reactions on dust provide essential routes to molecule formation, most notably for H_2 (the most abundant gas-phase molecule) and H_2O (the primary constituent of interstellar ices).
- As a major repository for heavy elements, dust regulates their availability for chemical reactions occurring in the gas phase; the inventory evolves as dust and gas exchange material according to environment.
- In dense clouds, grains provide substrates for accumulation of molecular ices, which may undergo further chemical evolution in the solid phase.
- Extinction by dust helps to protect molecules from photodissociation within interstellar clouds by attenuating the ambient radiation field.

Grain-surface catalysis is, indeed, the *only* viable means of efficient H_2 production at interstellar densities (Hollenbach & Salpeter 1971). A simple two-body encounter of H atoms in the gas phase cannot produce a stable molecule because there is no means of releasing binding energy in the absence of a third particle. However, H atoms that become physically attached to grain surfaces have no such problem:

[7] From "Woodstock" (Ladies of the Canyon), Reprise Records, 1970.
[8] See https://en.wikipedia.org/wiki/List_of_interstellar_and_circumstellar_molecules.

when two migrating surface atoms combine, excess binding energy is absorbed into the grain lattice and the resulting molecule is ejected from the surface. Astrochemical models based on networks of gas-phase reactions between H_2 molecules formed in this way and elements already present in the gas are quite successful in accounting for a number of commonly observed interstellar molecules, but H_2 production on dust is the vital first step. See Sections 9.1.4 and 9.2.1 for further discussion.

When O and OH attach themselves to grain surfaces, they are liable to be hydrogenated to form H_2O. In contrast to H_2, a molecule such as H_2O that includes at least one "heavy" element is unlikely to be ejected from a grain at low temperature when binding energy is released. Similarly, other common polyatomic molecules that may form by hydrogenation or oxidation reactions on grain surfaces, such as CH_4, CH_3OH, CO_2, and NH_3, tend to remain attached. The formation rate increases rapidly with the density of the cloud. Thus, in dense clouds, the grains become nucleation centers for the growth of *icy mantles*, and these mantles may grow to become major reservoirs of molecular material (Section 9.2.2). As well as sites of accumulation, the grains may become crucibles for chemical evolution in situations where an energy source is available to drive chemical reactions, and sublimation of the products contributes to the chemical complexity of the gas (Section 9.3.4). At all levels of density, the exchange of material between gas and dust is essential to our understanding of the chemical composition and evolution of the ISM.

The form of the observed grain size distribution (Section 1.2.7) indicates that small grains vastly outnumber large ones, and because of this it is the smaller grains that provide most of the available surface area for catalytic chemistry. Small grains are also important because they attenuate starlight most efficiently in the mid to far-ultraviolet, and thereby afford greatest protection from photodissociative radiation. It follows that physical processes affecting the size distribution of the dust also affect both the radiative environment and the chemistry. Increased collision rates within clouds may result not only in mantle formation but also in rapid grain–grain coagulation (Section 9.1.1), leading to a reduction in the number of small grains responsible for ultraviolet opacity. These changes may, in turn, influence the chemical evolution of dense clouds.

1.4.3 Dust as a Proxy for H_2

Unlike the majority of known interstellar molecules, H_2 lacks spectral signatures at microwave or radio wavelengths, and observations of its infrared features are limited to shocked or heated gas. The principal means of detection in cold gas is by means of ultraviolet absorption lines in stellar spectra (Savage et al. 1977). This is problematic because these observations are possible only in lines of sight with modest levels of ultraviolet extinction (roughly equivalent to $A_V < 2$), and the sample is therefore limited to diffuse clouds and the outer layers of dense clouds. Thus, the principal molecular ingredient in the cores of dense clouds cannot be observed directly. The assumption that $f(H_2) \rightarrow 1$ in these environments (Section 1.3.2) is based on extrapolation from data obtained at lower densities, together with calculations of

H_2 formation rates, rather than on direct observation. This assumption is reasonable and generally accepted, but we must rely on observations of other constituents to evaluate the mass and distribution molecular gas.

Thermal emission from dust is one of two commonly-used proxies for molecular gas in regions where H_2 cannot be detected directly. The intensity of infrared and submillimeter continuum emission from a cloud is proportional to the mass of emitting dust (Hildebrand 1983). Hence, observations of this emission provide information on the distribution of mass in a cloud, subject to an assumed dust-to-gas ratio (Section 2.3.3). The other proxy is gas-phase CO, the next most abundant molecule after H_2, observable in dense clouds by means of spectral line emission at millimeter wavelengths. In this case, results depend on an assumed ratio of integrated CO emission intensity to H_2 column density (see Bolatto et al. 2013 for a review). This ratio is sensitive to environmental factors in the coldest clouds, as a significant fraction of the CO may be frozen onto grains where, in solid form, it is undetected by the millimeter-wave observations (see Section 9.2.4).

1.4.4 Dust as the Universal Thermal Regulator

Because of its capacity to absorb and re-emit radiation, dust has a profound influence as a thermal regulator in a wide range of astrophysical environments. It is a vital ingredient in models for many sources of infrared radiation, ranging from interplanetary dust in our solar system to interstellar clouds, protostars and protoplanetary disks, H II regions, photodissociation regions, late-type stellar winds, planetary nebulae, supernova remnants, starburst regions, galactic nuclei, and entire galaxies (e.g., Neugebauer et al. 1984). Models for these diverse phenomena have in common the presence of dust heated by the absorption of ambient radiation. In many cases, emission from the dust dominates the emergent spectral energy distribution in the infrared and constitutes an important cooling mechanism; some specific examples are noted in the following sections. Modeling such sources is often challenging and may depend critically on the optical properties assumed for the grains and on their size distribution. Near and mid-infrared emission in excess of that expected from grains in thermal equilibrium must be considered in situations where the smallest particles are subject to transient heating. See Sections 2.3.2, 6.1, and 6.2 for detailed discussions.

1.4.5 Dust and Stellar Evolution

The life cycles of stars and dust are intricately linked. Stars may be associated with dust at any phase of their evolution, from protostellar youth to cataclysmic old age. Stars are born in dense clouds when clumps of gas and dust become gravitational bound, and the grains mediate their subsequent collapse by acting as radiators of excess heat energy that might otherwise inhibit or even halt the contraction. Magnetic fields also play vital roles in moderating the star formation process, and polarimetric observations of dust grains provide a means of mapping them (Section 4.1). As a protostar contracts, rotation flattens its circumstellar envelope into a disk that contains raw materials for planet-formation (Section 9.3). Planetary systems are now known to be common products of accretion in these disks; but in a number of cases, of which Vega is the best-known example, circumstellar debris disks persist (or recur) during the main-sequence phase of the star's evolution. In the case of our Sun, an apparently normal main-sequence star, a modest circumstellar dust population is maintained by cometary activity and asteroidal abrasion.

After leaving the main sequence, stars may evolve to become significant sources of new grain material. The atmospheres of luminous red giant and supergiant stars contain environments especially conducive to the rapid nucleation and growth of refractory dust grains. Once formed, the grains are driven outward by stellar radiation pressure, and their motion also influences the dynamics of the gas in the stellar wind through collisions. Typically, these stars are shedding mass at rates that exceed those associated with the winds of main-sequence stars by factors of 10^6 or more: in some cases, the outflow may be sufficient to influence the ultimate fate of a star during the late phases of its evolution. Cataclysmic events such as nova or supernova outbursts may lead to further episodes of grain formation. The outflows and expanding remnants merge into the surrounding ISM and contribute to the interstellar inventory of heavy elements. See Chapter 8 for further discussion.

1.4.6 Starbursts and Luminous Infrared Galaxies

The star formation process naturally tends to create luminous stars deep within the regions of galaxies that contain the highest concentrations of interstellar gas and dust. Hence, the first radiation to emerge from these regions is inevitably dominated by infrared emission resulting from absorption and reradiation of starlight by the dust grains. The contributions such regions make to the total luminosity of a galaxy therefore increase with the rate of star formation. Currently-quiescent spiral galaxies such as the Milky Way have modest star formation rates (a few solar masses per year) and emit no more than about 30% of their total energy output in the infrared. However, many galaxies have estimated star formation rates that are factors of 30–1000 times higher. This group includes optically-detected "starburst" galaxies and luminous infrared galaxies (LIGs), the latter discovered largely as the result of infrared all-sky surveys (see Sanders & Mirabel 1996 for a review). These surveys have shown that the majority of the most luminous galaxies in the observable universe emit most of their energy in the mid to far-infrared. Some of this luminosity may emanate from synchrotron radiation emitted by relativistic electrons, and from emission by dust heated by other source of energy, such as active galactic nuclei, as well as from stars. However, there is general agreement that star formation is the primary cause in a large majority of cases: the observed emission is dominated by regions of rapid star formation, in which dust is heated by local concentrations of massive young stars.

Starbursts must be transient phases in galactic evolution, as the star formation rates are such that all available material would otherwise be consumed on timescales of a few hundred million years, much shorter than galactic lifespans. They are thought to be triggered by tidal interactions and mergers of gas-rich spiral galaxies, causing disruption and compression of their interstellar media that typically result in rapid star formation. Many of the most luminous LIGs have active nuclei in which heated circumnuclear dust also contributes to the infrared flux, suggesting an evolutionary link between starburst and active galaxies. Thermal emission from dust has been detected in galaxies out to redshifts of $z \approx 5$ to 10 (Burgarella et al. 2020). The radiation we detect from these galaxies was emitted $\gtrsim$13 Gyr ago, when the universe was still very young. The observations thus imply that the chemical elements needed to make dust became available soon after the Big Bang, generated by core-collapse supernova explosions in the first generations of stars (Section 8.1.6).

1.4.7 Back to Basics

Finally, the original motivation to study interstellar dust remains important: there is an ever-present need to compensate for its effect on astronomical observations and thereby to recover the intrinsic properties of affected sources. In broad terms, the observed properties of the dust are now fairly well understood. Empirical methods and mathematical models have been developed to describe them, as discussed in the following chapters. Although these models are incomplete and by no means definitive, it is nevertheless possible to make reasonable predictions as to how dust modifies the observations of a given source in a given situation, at least within our own Galaxy and its nearest neighbors. One does not necessarily need a complete understanding of dust properties to predict its behavior and to evaluate its effect.

This optimistic statement must be qualified by an important caveat: because the ISM is a complex system of contrasting environments (Section 1.3), widespread application of general results can be misleading. The line of sight to a source obscured by dust inevitably passes through a range of environments. The amplitudes of metrics such as extinction and polarization vary stochastically with direction and distance, and their spectral dependencies are determined by the size, structure, and composition of the grains. These factors depend in turn on both the history and the current setting of the particles. The observed extinction may arise mostly in a single cloud, or may be the summation of contributions arising in any number of clouds of differing morphologies, densities, and opacities. Ideally, to correct for its presence, the average extinction curve specific to the line of sight should be determined over all wavelengths of interest, but this may not be practical or even possible in most cases.

The more realistic approach is to assume an average extinction curve that seems most applicable to the line of sight, in terms of the range of environments included, based on other measurements. An important step toward standardization of extinction corrections in the local Galaxy was taken by Cardelli et al. (1989). It was shown that the most significant environmental effects that lead to changes in the extinction curve, specifically those affecting the grain size distribution, are systematic and may be characterized with reasonable precision by a single parameter, the ratio of total to selective extinction (R_V), defined in Section 1.2.2. Observations have shown this quantity to be only modestly variable in lines of sight that sample diffuse regions of the ISM, with an average value of about 3.05 (Section 3.2.2). This is an important and, indeed, satisfactory result. Any major variations in R_V with direction along the disk of the Milky Way would impact investigations of galactic structure and the extragalactic distance scale (e.g., Johnson 1968), but no such behavior has been substantiated. Significant deviations above the mean are limited almost entirely to dense clouds and H II regions, where R_V values in the range 4–6 are not unusual and attributed to grain growth in these environments (Section 3.3.1). Deviations to values significantly below the diffuse-ISM average are less common but not unknown.

On the wider stage, the extent to which cosmologically-distant galaxies are obscured by dust is vital to our understanding of the structure and evolution of the universe (Ostriker & Heisler 1984). Dust reveals its presence in distant galaxies most readily by thermal infrared emission, as previously discussed, but its attenuating effect at shorter wavelengths is more difficult to quantify. Extinction by foreground dust in our own Galaxy is generally small and easily evaluated in directions away from the galactic disk (Section 3.2.7); but accounting for the extinction arising beyond the Milky Way—within or between intervening galaxies along the line of sight, as well as in the target galaxies themselves—is a more challenging problem. If it is large, extinction may bias our surveys of objects at high redshift; it may hide evidence of star formation in the most distant galaxies and, indeed, of galaxy formation itself. See Mirabel et al. (1998) and Wolfe et al. (2005) for further discussion.

References

Adams, D. 1979, The Hitchhiker's Guide to the Galaxy (London: Pan Books)

Andersson, B.-G., Lazarian, A., & Vaillancourt, J. E. 2015, ARA&A, 53, 501

Bakes, E. L. O., & Tielens, A. G. G. M. 1994, ApJ, 427, 822

Barnard, E. E. 1927, in Atlas of Selected Regions of the Milky Way, ed. E. B. Frost, & M. R. Calvert (Washington, DC: Carnegie Institute)

Bessell, M. S. 2005, ARA&A, 43, 259

Bland-Hawthorn, J., & Gerhard, O. 2016, ARA&A, 54, 529

Bolatto, A. D., Wolfire, M., & Leroy, A. K. 2013, ARA&A, 51, 207

Burgarella, D., Nanni, A., Hirashita, H., et al. 2020, A&A, 637, A32

Cardelli, J. A., Clayton, G. C., & Mathis, J. S. 1989, ApJ, 345, 245

Davis, L., & Greenstein, J. L. 1951, ApJ, 114, 206

Douglas, A. E., & Morton, J. R. 1960, ApJ, 131, 1

Draine, B. T. 1989, in IAU Symp. 135, Interstellar Dust ed. L. J. Allamandola, & A. G. G. M. Tielens (Dordrecht: Kluwer), 313

Draine, B. T. 2011, Physics of the Interstellar and Intergalactic Medium (Princeton, NJ: Princeton Univ. Press)

Draine, B. T., & McKee, C. F. 1993, ARA&A, 31, 373

Dunham, M. H., Allen, L. E., Evans, N. J. II, et al. 2015, ApJS, 220, 11

Dyson, J. E., & Williams, D. A. 2020, The Physics of the Interstellar Medium (3rd ed.; London and New York: CRC Press)

Elmegreen, B. G. 2002, ApJ, 564, 773

Frisch, P. C., & Slavin, J. D. 2013, EP&S, 65, 175

Frisch, P. C., Dorschner, J. M., Geiss, J., et al. 1999, ApJ, 525, 492

Frisch, P. C., Redfield, S., & Slavin, J. D. 2011, ARA&A, 49, 237

Gaustad, J. E. 1971, in Dark Nebulae, Globules and Protostars, ed. B. T. Lynds (Tucson, AZ: University of Arizona Press), 91
Greenberg, J. M. 1974, ApJ, 189, L81
Hall, J. S. 1949, Sci, 109, 166
Herbst, E. 2014, PCCP, 16, 3344
Hildebrand, R. H. 1983, QJRAS, 24, 267
Hiltner, W. A. 1949, Sci, 109, 165
Hoang, T., & Loeb, A. 2020, ApJ, 899, L23
Hollenbach, D. J., & Salpeter, E. E. 1971, ApJ, 163, 155
Hollenbach, D. J., & Tielens, A. G. G. M. 1997, ARA&A, 35, 179
Jenkins, E. B. 2009, ApJ, 700, 1299
Jenkins, E. B. 2009, ApJ, 700, 1299
Johnson, H. L. 1968, in Nebulae and Interstellar Matter, ed. B. M. Middlehurst, & L. H. Aller (Chicago: Chicago Univ. Press), 167
Lee, J. C., Xiang, J., Ravel, B., Kortright, J., & Flanagan, K. 2009, ApJ, 702, 970
Mann, I. 2010, ARA&A, 48, 173
Mathis, J. S. 2000, in Allen's Astrophysical Quantities, ed. A. N. Cox (4th ed.; New York: Springer), 523
McKee, C. F., & Ostriker, J. P. 1977, ApJ, 218, 148
McKellar, A. 1940, PASP, 52, 187
Mirabel, I. F., Vigroux, L., Charmandaris, V., et al. 1998, A&A, 333, L1
Neugebauer, G., Beichman, C. A., Young, E. T., et al. 1984, Sci, 224, 14
Ostriker, J. P., & Heisler, J. 1984, ApJ, 278, 1
Sanders, D. B. 1996, ARA&A, 34, 749
Savage, B. D., Bohlin, R. C., Drake, J., & Budich, W. 1977, ApJ, 216, 291
Sheehan, W. 1995, The Immortal Fire Within: The Life and Work of Edward Emerson Barnard (Cambridge: Cambridge Univ. Press)
Snow, T. P., & McCall, B. J. 2006, ARA&A, 44, 367
Spitzer, L. 1978, Physical Processes in the Interstellar Medium (New York: Wiley)
Stebbins, J., Huffer, C. M., & Whitford, A. E. 1939, ApJ, 90, 209
Struck, C., & Smith, D. C. 1999, ApJ, 527, 673
Tielens, A. G. G. M. 2008, ARA&A, 46, 289
Trumpler, R. J. 1930a, LicOB, 14, 154
Trumpler, R. J. 1930b, PASP, 42, 214
van de Hulst, H. C. 1946, Recherches Astronomiques de l'Observatoire d'Utrecht, 11, 1
Ward-Thompson, D., Di Francesco, J., Hatchell, J., et al. 2007, PASP, 119, 855
Whitford, A. E. 1958, AJ, 63, 201
Wolfe, A. M., Gawiser, E., & Prochaska, J. X. 2005, ARA&A, 43, 861
Woolf, N. J. 1973, in IAU Symp. 52, Interstellar Dust and Related Topics, ed. J. M. Greenberg, & H. C. van de Hulst (Dordrecht: Reidel), 485

D C B Whittet

Chapter 2

The Interaction of Dust and Electromagnetic Radiation

This chapter discusses the physical processes upon which several of the primary methods used to detect and study cosmic dust are based. The observable outcomes of the interaction between dust and electromagnetic radiation include dimming and reddening of light from distant stars and galaxies, reflection nebulae and diffuse background radiation caused by scattering, thermal infrared emission from heated dust, and polarization of both transmitted starlight and thermal dust emission, arising when the particles are optically asymmetric and systematically aligned. For an introductory overview of these topics, see Chapter 1 (Sections 1.2.1–1.2.4). Effects of irradiation on grain dynamics are discussed in the final section.

2.1 Extinction

Extinction of the transmitted beam occurs whenever electromagnetic radiation propagates through a medium containing small particles. In this context, "small" means ranging from macromolecular dimensions up to sizes comparable with the longest wavelengths being considered (from a few nanometers to a few micrometers). The transmitted beam is reduced in intensity by the combination of two physical processes, absorption and scattering: the energy of an absorbed photon is converted into internal energy of the particle; a scattered photon is deflected from the line of sight. Both absorption and scattering generally modify the spectrum of the transmitted beam as well as causing a reduction in its total intensity. Depending on the composition of the particle, absorption may be a major component of extinction across the spectrum or mostly limited to spectral features at discrete wavelengths.

2.1.1 Extinction by Spherical Particles

We begin by considering the optical properties of small, homogeneous spheres. As a representation of real interstellar grains, this is obviously highly idealized: the

doi:10.1088/2514-3433/ac7204ch2

polarization of starlight provides direct evidence that at least some of the dust has anisotropic optical properties, a topic discussed in detail in Section 2.2. However, homogeneous spheres are a reasonable and mathematically convenient starting point for situations where polarization is not being considered. Calculations indicate that the optical properties of spheroids are closely similar to those of spheres of equal volume when averaged over all orientation angles. Methods for calculating the extinction of more complex forms are summarized in Section 2.1.5.

Consider spherical dust grains of radius a, distributed with number density n_d per unit volume, in a cylindrical column of length L and unit cross-sectional area, extending along the line of sight from a distant point source (e.g., a star) to the observer. The incremental change in intensity of the starlight at a given wavelength resulting from extinction produced in a discrete element of the cylindrical column of length dL is given by

$$\frac{\mathrm{d}I}{I} = -n_\mathrm{d} C_\mathrm{ext} \mathrm{d}L \tag{2.1}$$

where C_ext is the extinction cross-section. Integrating Equation (2.1) over the entire pathlength gives

$$I = I_0 \, e^{-\tau} \tag{2.2}$$

where I_0 is the initial value of I (at $L = 0$), and

$$\begin{aligned} \tau &= \int n_\mathrm{d} \mathrm{d}L \cdot C_\mathrm{ext} \\ &= N_\mathrm{d} C_\mathrm{ext} \end{aligned} \tag{2.3}$$

is the optical depth of extinction caused by the dust. The quantity N_d in Equation (2.3) is the column density of the dust, i.e., the total number of dust grains in the unit column ($N_\mathrm{d} = n_\mathrm{d} L$). For comparison with observations, the reduction in intensity may be expressed as extinction in magnitudes (Section 1.2.1), defined at some wavelength λ by

$$A_\lambda = -2.5 \log\left(\frac{I}{I_0}\right). \tag{2.4}$$

Combining Equations (2.2) and (2.4) yields a simple proportionality between extinction and optical depth,

$$A_\lambda = 1.086 \tau_\lambda \tag{2.5}$$

where the constant is 2.5 log(e). It is convenient to express the extinction in terms of an efficiency factor Q_ext, defined as the ratio of extinction cross-section to geometric cross-section:

$$Q_\mathrm{ext} = \frac{C_\mathrm{ext}}{\pi a^2}. \tag{2.6}$$

Hence,

$$A_\lambda = 1.086 N_d \pi a^2 Q_{ext}. \tag{2.7}$$

If, instead of grains of constant radius a, we consider a size distribution such that $n(a)\,da$ is the number of grains per unit volume in the line of sight with radii in the range a to $a + da$, then Equation (2.7) is replaced by

$$A_\lambda = 1.086 \pi L \int a^2 Q_{ext} n(a)\,\mathrm{d}a. \tag{2.8}$$

Thus, for a given grain model with an assumed composition and size distribution, the predicted spectral dependence of extinction A_λ is determined by evaluation of Q_{ext}.

The extinction efficiency is the sum of corresponding efficiency factors for absorption and scattering,

$$Q_{ext} = Q_{abs} + Q_{sca}, \tag{2.9}$$

and these efficiencies are dependent on two quantities, a dimensionless size parameter,

$$X = \frac{2\pi a}{\lambda} \tag{2.10}$$

and a composition parameter, the complex dielectric function of the grain material,

$$\epsilon = \epsilon_1 - \mathrm{i}\epsilon_2. \tag{2.11}$$

The dielectric function is directly related to the complex refractive index (m) of the material:

$$\epsilon = m^2 \tag{2.12}$$

where

$$m = n - \mathrm{i}k. \tag{2.13}$$

In this equation, the real part (n), is related to the velocity of the radiation in the medium of the solid, and the imaginary part (k) is an extinction coefficient, related to the decay in amplitude of the incident electric field.

To compute Q_{abs} and Q_{sca} for an assumed grain material, the optical functions n and k must be specified over all relevant wavelengths. Candidate materials may be divided loosely into two broad categories according to their optical properties: dielectric and non-dielectric. Hypothetically, a pure dielectric is non-absorbing ($k = 0$) with a real refractive index ($m = n$). Ices and silicates are examples of solids that behave as dielectrics to a reasonable approximation ($k \ll 0.1$) over extended regions of the electromagnetic spectrum remote from their absorption bands. The refractive index of a good dielectric may be represented empirically by the Cauchy equation

$$n \simeq c_1 + c_2 \lambda^{-2} \tag{2.14}$$

where c_1 and c_2 are constants, and in such cases n is only weakly dependent on λ ($c_1 \gg c_2$). For strongly absorbing, non-dielectric materials such as metals and graphitic carbon, k may be of the same order as n, and both n and k may vary rapidly with wavelength.[1]

Calculated and experimentally-measured values of n and k are available online and in the published literature for a wide range of astrophysically relevant solids (e.g., Henning et al. 1999; Jäger et al. 2003; Gerakines & Hudson 2020). Armed with such data for an assumed grain composition, the problem of determining $Q_{\rm ext}$ at a given value of X (Equation (2.10)) is that of solving Maxwell's equations, with appropriate boundary conditions at the grain surface, to describe the response of the target material to the oscillating electric and magnetic fields. A solution was first formulated by Mie (1908) and independently by Debye (1909), resulting in what is now known as the Mie Theory. Excellent accounts of Mie theory and its applications appear in van de Hulst (1957) and Bohren & Huffman (1983), to which the reader is referred for detailed discussion.

Figures 2.1 and 2.2 illustrate the results of Mie theory calculations for weakly-absorbing spherical grains of constant refractive index $m = 1.5 - 0.05$i. Values of efficiency factor $Q_{\rm ext}$ and its absorption and scattering components (Equation (2.9)) are plotted against $X = 2\pi a/\lambda$. Figure 2.2 is an enlargement of Figure 2.1 near the origin. It is helpful to regard these plots in terms of the variation in extinction with λ^{-1} for constant grain radius (or with grain radius for constant wavelength). A few general features are evident. $Q_{\rm ext}$ increases monotonically with X for $0 < X < 4$.

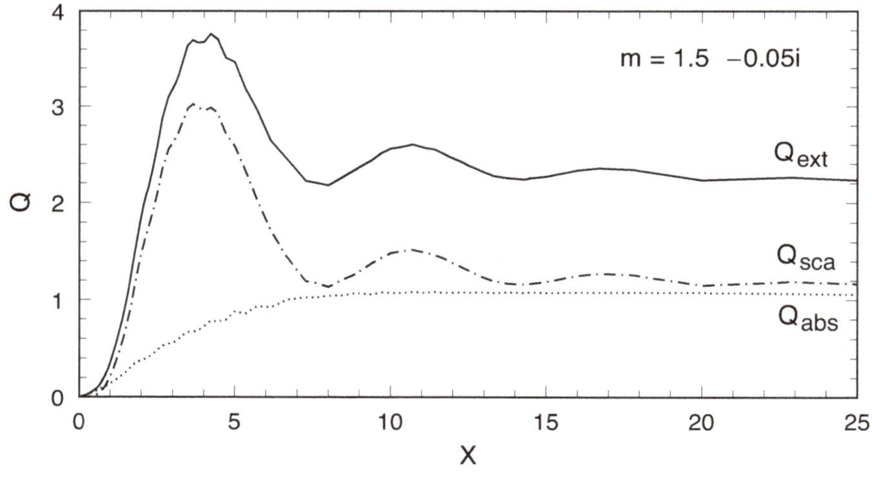

Figure 2.1. Results of Mie theory calculations for spherical grains of refractive index $m = 1.5 - 0.05$i. Efficiency factors $Q_{\rm ext}$, $Q_{\rm sca}$, and $Q_{\rm abs}$ (solid, dotted–dashed, and dotted curves, respectively) are plotted against the dimensionless size parameter $X = 2\pi a/\lambda$.

[1] The parameters n and k are widely referred to in the literature as "optical constants," but this is misleading given their often strong dependence on wavelength.

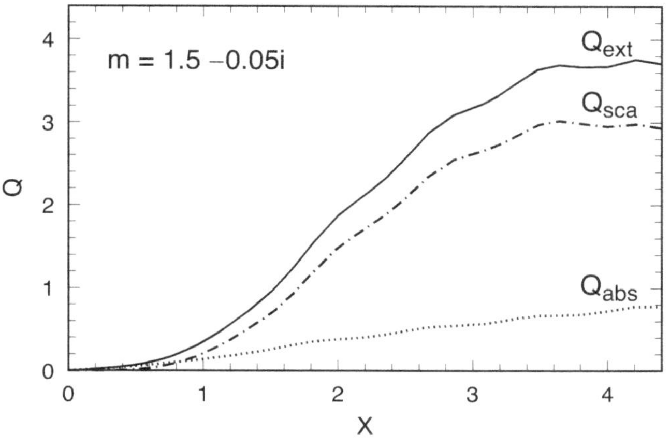

Figure 2.2. An enlargement of Figure 2.1 showing the initial rise in extinction efficiency with X near the origin. If grains of constant radius $a = 0.2$ μm are considered, this corresponds to the predicted extinction curve (Q_{ext} versus λ^{-1}) from infrared through visible wavelengths.

In this domain, extinction is dominated by scattering for the chosen refractive index, and its magnitude is sensitive to the precise value of X. For $1 < X < 3$, Q_{ext} increases almost linearly with X. At higher values of X, peaks arise in the scattering component of the extinction, caused by resonances between wavelength and grain size, and these will disappear when the contributions of grains with many different radii in a size distribution are summed (Equation (2.8)). Q_{ext} becomes almost constant as X becomes large, indicating that the extinction is neutral (wavelength-independent) for grains much larger than the wavelength.

2.1.2 Small-Particle Approximations

When the particles are very small compared with the wavelength ($X \ll 1$), useful approximations may be used to give simple expressions for the efficiency factors (see Chapter 5 in Bohren & Huffman 1983):

$$Q_{sca} \simeq \frac{8}{3}\left[\frac{2\pi a}{\lambda}\right]^4 \left|\frac{m^2-1}{m^2+2}\right|^2 \tag{2.15}$$

and

$$Q_{abs} \simeq \frac{8\pi a}{\lambda}\,\mathrm{Im}\left\{\frac{m^2-1}{m^2+2}\right\}. \tag{2.16}$$

A situation in which these approximations become valid is termed the Rayleigh limit. For a pure dielectric, m is real and almost constant with respect to wavelength, as discussed above. In this case, we have $Q_{sca} \propto \lambda^{-4}$ (Rayleigh scattering) and $Q_{abs} = 0$. More generally, the quantity $(m^2 - 1)/(m^2 + 2)$ is often only weakly dependent on wavelength for materials that are not strongly absorbing, in which case $Q_{sca} \propto \lambda^{-4}$ and $Q_{abs} \propto \lambda^{-1}$ to good approximations. The wavelength-dependence of extinction by small particles may thus be quite different depending on whether absorption or scattering is dominant.

2.1.3 Albedo, Scattering Function, and Asymmetry Parameter

Quantities describing the scattering properties of the grains may be calculated from Mie theory. The albedo is defined

$$\alpha = \frac{Q_{\rm sca}}{Q_{\rm ext}} \tag{2.17}$$

and is bounded by the limits $0 \leqslant \alpha \leqslant 1$ (since $0 \leqslant Q_{\rm sca} \leqslant Q_{\rm ext}$), the extreme values representing perfect absorbers and pure dielectrics, respectively. More generally, a dust model based on a mix of absorbing and dielectric particles will predict some form for the dependence of α on wavelength. We may use Equations (2.17) and (2.9) to express the efficiency factors $Q_{\rm sca}$ and $Q_{\rm abs}$ in terms of α:

$$\begin{aligned} Q_{\rm sca} &= \alpha Q_{\rm ext} \\ Q_{\rm abs} &= (1 - \alpha) Q_{\rm ext}. \end{aligned} \tag{2.18}$$

It follows that if the extinction and albedo can be determined observationally over the same spectral range, the contributions of absorption and scattering may be easily separated.

The angular redistribution of light resulting from scattering by dust may be described in terms of a scattering function, $S(\theta)$, where θ is the scattering angle relative to the direction of incidence. Assuming axial symmetry, $S(\theta)$ is defined such that the intensity of light scattered into unit solid angle in direction θ is $I_0 S(\theta)$, where I_0 is the incident intensity. The scattering cross section, defined as $C_{\rm sca} = \pi a^2 Q_{\rm sca}$ by analogy with Equation (2.6), is related to $S(\theta)$ by

$$C_{\rm sca} = 2\pi \int_0^{\pi} S(\theta) \sin\theta \, {\rm d}\theta. \tag{2.19}$$

The asymmetry of scattering is specified by the parameter $g(\theta)$, defined as the mean value of $\cos\theta$ weighted with respect to $S(\theta)$:

$$\begin{aligned} g(\theta) &= \langle \cos\theta \rangle \\ &= \frac{\int_0^{\pi} S(\theta) \sin\theta \cos\theta \, {\rm d}\theta}{\int_0^{\pi} S(\theta) \sin\theta \, {\rm d}\theta} \\ &= \frac{2\pi}{C_{\rm sca}} \int_0^{\pi} S(\theta) \sin\theta \cos\theta \, {\rm d}\theta. \end{aligned} \tag{2.20}$$

The asymmetry parameter is bounded by the limits $-1 \leqslant g(\theta) \leqslant 1$, positive and negative values indicating forward and back-scattering, respectively. Calculations for dielectric spheres show that $g(\theta) \simeq 0$ in the small-particle limit, which

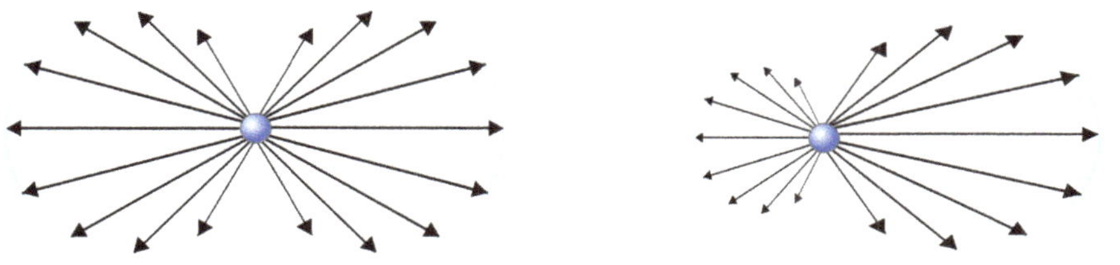

Figure 2.3. Schematic examples of scattering functions for spherical particles. The direction of incidence is from the left. Particles much smaller than the wavelength ($X \ll 1$) produce Rayleigh scattering, which is symmetrical in forward and backward directions (left). Particles comparable in size with the wavelength ($X \sim 1$) produce Mie scattering, which is predominantly in the forward direction (right). Image credit: Sharayanan/CC-BY-SA-3.0, Creative Commons.

corresponds to isotopic scattering (see Figure 2.3). As the ratio of grain diameter to wavelength increases from 0.3 to 1.0 (corresponding to $1 \lesssim X \lesssim 3$), a range of particular relevance for studies of interstellar dust, the value of $g(\theta)$ increases rapidly from 0.15 to 0.75 (Witt 1989); hence, $g(\theta)$ is a potentially sensitive measure of grain size.

2.1.4 Absorption Features

In general, the transfer of energy accompanying absorption of a photon by a solid results in excitation of an electron to a higher energy state or an incremental increase in the vibrational motion of atoms in the lattice. The absorptive properties of solids are closely linked to their electrical conductivities. Electrons in conductors such as metals and graphite may occupy a near-continuum of energy states and may therefore absorb continuously over a wide spectral range. Conductors may also exhibit distinct absorption features arising from quantized energy states of oscillating surface electrons (plasmons; e.g., Bohren & Huffman 1983). The occurrence of such absorptions in graphite provides a possible explanation of the ultraviolet absorption "bump" in the extinction curve (Section 3.4). In pure semiconductors or insulators, however, electronic transitions occur only for photon energies $\varepsilon_{\rm ph} \geqslant \varepsilon_{\rm g}$, where $\varepsilon_{\rm g}$ is the energy gap between valence and conduction bands; otherwise they absorb strongly only at lower energies corresponding to vibrational frequencies characteristic of specific molecular bonds (Section 5.1). Silicates are pertinent examples, exhibiting absorption features corresponding to vibrations of the Si–O bond in the mid-infrared and continuous absorption due to electron excitation in the far-ultraviolet, but behaving as good dielectrics over the spectral range in between. Infrared spectroscopy of vibrational modes in solids is one of the primary techniques for studying grain composition and evolution. Rotational spectroscopy is also important in astrochemistry as a vital technique for studying gas-phase molecules, but has no direct relevance to the dust as rotational modes are suppressed in solids.

2.1.5 Modeling Composite Grains

The assumption that the grains are homogeneous and spherical (Section 2.1.1) is a convenient generalization. It facilitates calculations and may be physically reasonable in some situations. However, the growth and destruction processes to which real dust grains are subjected in the interstellar medium (Chapter 9) are likely to produce irregular particles that are inhomogeneous in structure, composition, and shape. Growth by grain–grain coagulation may result in porous aggregates that contain voids as well as a mixture of materials, and grains of all shapes and sizes may be resurfaced by accumulations of atoms and molecules adsorbed from the gas. How can the extinction produced by such particles be calculated? Solutions based on classical Mie theory are possible only for certain special cases, such as spheroids or cylinders commonly used to model observations of interstellar polarization (Section 2.2), and concentric spheres or spheroids that might be representative of ice-coated (core/mantle) grains inside molecular clouds. An extension of classical theory to particles of inhomogeneous composition is possible by means of effective medium theory (EMT), in which a grain is represented by a system of cells composed of different materials. An average dielectric function is calculated to represent the bulk particle, and Mie calculations are then applied to enable the extinction produced by an assemblage of such particles to be determined. See Mathis & Whiffen (1989), Ossenkopf (1991), and Wolff et al. (1994, 1998) for further discussion and examples.

An important alternative method is the discrete dipole approximation (DDA), first proposed by Purcell & Pennypacker (1973). In this approach, a composite grain of arbitrary shape is represented by an array of dipole elements: each dipole has an oscillating polarization in response to both incident radiation and the electric fields of the other dipoles in the array, and the superposition of dipole polarizations leads to extinction and scattering cross sections. The calculations are computationally-intensive but well within the capabilities of modern desktop computers. Results are generally consistent with Mie calculations in special cases where the two can be directly compared. The reliability increases with the number of dipoles used to represent the assumed grain structure, with a high density required especially at surfaces or edges. For examples of the application of this method to studies of interstellar dust, see Draine (1988), Fogel & Leung (1998), Shen et al. (2008), and Ysard et al. (2018).

2.2 Polarization

2.2.1 Extinction by Anisotropic Particles

In this section, we extend the discussion of extinction by small particles (Section 2.1) to include the production of linear polarization by *dichroic* extinction, i.e., extinction that varies with the orientation of the **E**-vector in the transmitted beam. A beam of initially-unpolarized light propagating through a dusty medium will become partially plane polarized if (i) individual dust particles are optically anisotropic, and (ii) there is net alignment of their axes of anisotropy. The processes responsible for alignment are reviewed in Chapter 4 (Section 4.3). The most likely source of optical anisotropy is asymmetry in the shape of the particles. Another possibility is anisotropy intrinsic to the grain material itself: graphite, for example, displays intrinsic anisotropy arising from its layered crystalline structure; however, such particles tend naturally to be asymmetric in shape as well. An endless variety of shapes and structures may be represented in the interstellar grain population, and generalization is inevitable when attempting to model their polarizing properties. It is convenient to assume simple, axially-symmetric forms such as cylinders or spheroids, as their extinction cross sections may be calculated by extension of the Mie theory for spheres (Section 2.1.1). A more sophisticated approach is to use methods such as the discrete dipole approximation (Section 2.1.5) to simulate the optical properties of grains of any desired shape.

To illustrate how dust polarizes starlight, consider an idealized ensemble of elongated grains (Figure 2.4), aligned with respect to the ambient magnetic field direction such that their long axes are parallel to each other and orthogonal to the line of sight. We define $Q_{\|}$ and $Q_{\perp}$ as the values of the extinction efficiency $Q_{\rm ext}$ (Section 2.1.1) when the **E**-vector is parallel and perpendicular to the long axis of a grain, respectively. The anisotropy in physical shape introduces a corresponding anisotropy in extinction: because the **E**-vector in the incident beam "sees" an apparently larger grain in the parallel direction, we have $Q_{\|} \geqslant Q_{\perp}$. As an example, calculated values of $Q_{\|}$ and $Q_{\perp}$ are plotted in Figure 2.5 against the dimensionless size parameter $X = 2\pi a/\lambda$ for long dielectric cylinders of radius a. The quantity $\Delta Q = Q_{\|} - Q_{\perp}$ is a measure of the resulting polarization, which is small relative to the extinction (note the scaling in Figure 2.5). ΔQ is reduced if, as is generally the case, the direction of alignment is not orthogonal to the line of sight, and becomes zero for propagation along the axes of symmetry of the grains.

It is informative to compare the behavior of polarization and extinction in Figure 2.5. Both become very small at wavelengths long compared with particle size

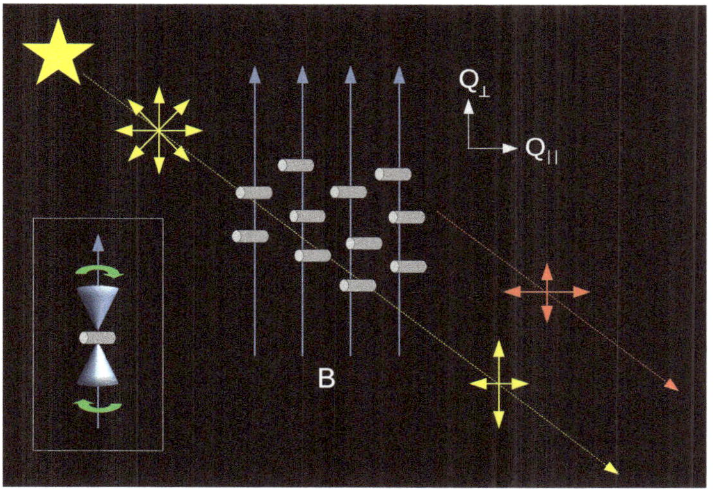

Figure 2.4. Schematic illustration of polarization produced by aligned, asymmetric grains (represented here by gray cylinders). Blue arrows indicate the direction of the magnetic field (**B**), and the grains align on average with their long axes perpendicular to **B**. The difference between extinction efficiencies $Q_{\|}$ and $Q_{\perp}$ causes initially-unpolarized starlight (yellow) to become partially plane polarized. Thermal emission from the grains (red) is also partially plane polarized, but in the orthogonal direction (see Section 2.3.5). The inset illustrates precession of a spinning grain's angular momentum around the magnetic field.

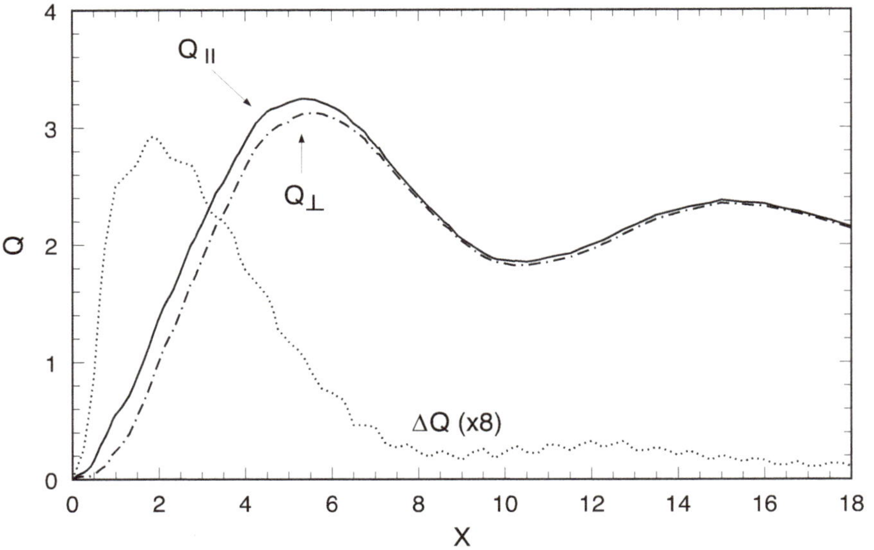

Figure 2.5. Extinction efficiency factors $Q_{\|}$ (continuous curve) and $Q_{\perp}$ (dashed curve) plotted against $X = 2\pi a/\lambda$ for cylinders of radius a and refractive index $m = 1.33 - 0.05\text{i}$. The cylinders are assumed to be very long ("infinite") in comparison to their radii. Note that $Q_{\|} \geqslant Q_{\perp}$ for all values of X. The dotted curve is the difference, $\Delta Q = Q_{\|} - Q_{\perp}$, scaled up by a factor of 8 for display. The equivalent calculation of Q_{ext} for spheres (Section 2.1.1) would show a dependence on X qualitatively similar to the mean of $Q_{\|}$ and $Q_{\perp}$.

($X \to 0$); however, in contrast to extinction, the polarization also becomes small at wavelengths short compared with particle size ($X \gtrsim 8$ in this example). A peak in ΔQ appears at an intermediate value of X. Comparing calculations for dielectric cylinders of differing refractive index n, the peak occurs when $X(n-1) \sim 1$, and this approximation may be expressed as a wavelength of maximum polarization for a characteristic grain size:

$$\lambda_{\text{max}} \sim 2\pi a(n-1). \tag{2.21}$$

2.2.2 Polarization Parameters

Linear polarization observed at some wavelength λ may be specified by a position angle and an amplitude. The position angle measures the plane of vibration of the **E**-vector in the transmitted beam projected on to the celestial sphere, relative to some reference direction. For studies of our Galaxy, it is convenient to choose the orientation of the Milky Way to specify the reference direction: the galactic position angle (θ_G) is defined as the angle between the plane of vibration of the **E**-vector and the direction of the North Galactic Pole, increasing counterclockwise on the sky, and bounded by the limits $0 \leqslant \theta_G \leqslant 180°$. Note that the position angle measures the *orientation* of the magnetic field responsible for alignment, but not its *polarity*, hence the 180° range. Thus, strictly speaking, polarization is not a true vector quantity although it can usually be treated as one.

The polarization amplitude is related to ΔQ, the amplitude of the variation in the extinction efficiency (Section 2.2.1). Considering parallel and perpendicular cases in turn, the extinction produced by a column density N_d of aligned dust grains is

$$\begin{aligned} A_{\parallel} &= 1.086 N_d \sigma_d Q_{\parallel} \\ A_{\perp} &= 1.086 N_d \sigma_d Q_{\perp} \end{aligned} \tag{2.22}$$

by analogy with Equation (2.7), where σ_d is the cross-sectional area of a dust particle in the plane of the wave front. The amplitude of the resulting linear polarization, expressed as a differential extinction in magnitudes, is

$$\begin{aligned} m_p &= A_{\parallel} - A_{\perp} \\ &= 1.086 N_d \sigma_d \Delta Q. \end{aligned} \tag{2.23}$$

Calculated values of $Q_{\parallel}$ and $Q_{\perp}$ for a given grain model (size, shape, composition) thus enable m_p to be evaluated as a function of wavelength for comparison with observations. However, the amplitude is usually expressed as a fractional or percentage polarization in the astronomical literature. The fractional polarization is defined

$$p = \frac{I_{max} - I_{min}}{I_{max} + I_{min}} \tag{2.24}$$

where I_{max} and I_{min} are observed maximum and minimum intensities with respect to orientation of the analyzer. The corresponding definition of the polarization in magnitudes is

$$m_p = 2.5 \log \frac{I_{max}}{I_{min}}. \tag{2.25}$$

If the polarization is sufficiently small, as is usually the case, these measures are effectively proportional: using a series expansion for the logarithm in Equation (2.25),

$$\frac{I_{max}}{I_{min}} \simeq 1 + \left\{ \frac{\ln 10}{2.5} \right\} m_p, \tag{2.26}$$

neglecting m_p^2 and higher powers, and with $I_{max} \simeq I_{min}$ in Equation (2.24), it follows that

$$p \simeq 0.5 \left\{ \frac{I_{max}}{I_{min}} - 1 \right\} \simeq 0.4605\, m_p. \tag{2.27}$$

A more general description of linear polarization is provided by the Stokes parameters Q and U, defined by the relations

$$\begin{aligned} Q/I &= p \cos 2(\theta_G - 90) \\ U/I &= p \sin 2(\theta_G - 90) \end{aligned} \tag{2.28}$$

where $I = I_{max} + I_{min}$ (e.g., Spitzer 1978, p. 178). This formulation is helpful for large-scale studies of galactic polarization and magnetic fields as it enables data for different stars to be averaged. Observed values of Q/I and U/I for lines of sight that pass through several clouds correspond to the summation of individual components in each case, provided that p is small. If the radiation includes a circularly polarization component (Section 2.2.4), this is specified by an additional parameter V, defined by analogy with Equation (2.24), replacing I_{max} and I_{min} with intensities of right-handed and left-handed polarizations.

2.2.3 Polarization Efficiency

The ratio of polarization to extinction is a measure of the polarization efficiency of the medium along a line of sight. The total extinction is given by

$$A = 1.086 N_d \sigma_d \left\{ \frac{Q_{\parallel} + Q_{\perp}}{2} \right\} \tag{2.29}$$

and combining Equations (2.23) and (2.29) yields

$$\frac{m_p}{A} = 2 \left\{ \frac{\Delta Q}{Q_{\parallel} + Q_{\perp}} \right\}. \tag{2.30}$$

In practice, this ratio depends not only on the nature of the grains, and the efficiency with which they are aligned, but also on purely geometrical factors, i.e., the direction of net alignment relative to the plane of the wave front, and whether all grains along the line of sight share the same direction of alignment (see Section 2.2.4 below). A useful upper limit for m_p/A may be calculated by considering the most efficient polarizing medium conceivable, taken to be an ensemble of long (needle-like) cylinders with diameters comparable to the wavelength, perfectly aligned such that their long axes are parallel to each other and perpendicular to the line of sight. Considering dielectric particles in the visual passband, this limit is estimated to be $[m_p/A]_V \lesssim 0.3$, corresponding to

$$\frac{p_V}{A_V} \lesssim 14\% \text{ mag}^{-1}. \tag{2.31}$$

As expected, observed values of p_V/A_V are always well below this theoretical limit (see Section 4.1.1).

2.2.4 Circular Polarization

So far we have considered only the linear component of interstellar polarization, arising from the dependence of extinction on the vibrational plane of the **E**-vector for uniform grain alignment. However, if the direction of alignment varies along the line of sight, a non-linear component is also produced. To illustrate this, consider again an idealized situation where initially-unpolarized radiation from a distant star is partially plane polarized by transmission through a single cloud with uniform grain alignment. If the beam then encounters a second cloud of similar opacity and uniform alignment in a *different* direction, the emergent light is in general elliptically polarized, the ISM behaving as an inefficient wave plate. In effect, this introduces a weak component of circular polarization and simultaneously causes a reduction ("depolarization") in the amplitude of the linear component, yielding an anomalously low polarization efficiency. In the extreme case where the alignment axes of the two clouds are orthogonal, high extinction can be produced with little no net polarization. Magnetic fields are implicated in the alignment process (Figure 2.4) and real interstellar clouds frequently exhibit complex magnetic field structure; thus, changes in alignment geometry along a line of sight can arise because of twisted magnetic field lines within a single cloud, as well as systematic differences from one cloud to another.

Observations of linear and circular polarization and their dependencies on wavelength in the same lines of sight place useful constraints on grain models. These and other observations of interstellar polarization are discussed in detail in Chapter 4.

2.3 Infrared Emission

2.3.1 Equilibrium Dust Temperatures

Interstellar grains exchange energy with their environment as a result of absorption and emission of radiation, collisions, cosmic ray exposure, and exothermic surface reactions such as hydrogen recombination. In general, the equilibrium (steady state) dust temperature is determined primarily by radiative processes in diffuse regions of the ISM, collisions becoming more important with increasing density in molecular clouds. As photons are absorbed, energized electrons may occasionally be ejected from the grain (the photoelectric effect; Section 2.4.2), but most of the incident energy is dissipated into vibrational modes within the solid, i.e., into heat. This is balanced by cooling as the result of thermal emission from the grain.

Consider a spherical particle in the diffuse ISM, remote from any individual source of luminosity and heated by absorption from the ambient interstellar radiation field (ISRF). If the particle behaves as a perfect blackbody, its equilibrium temperature, T_b, is determined by the Stefan–Boltzmann law

$$U = \int_0^\infty u_\lambda \mathrm{d}\lambda = \frac{4\sigma}{c} T_b{}^4 \tag{2.32}$$

where U is the total energy density of the ISRF and u_λ is its spectral distribution with respect to wavelength. Assuming a mean value of $U \approx 8 \times 10^{-14}$ J m^{-3} for diffuse clouds (e.g., Mathis 2000), Equation (2.32) yields $T_b \approx 3.2$ K (close to the cosmic microwave background temperature), a result first deduced by Eddington (1926). If instead the particle behaves as an imperfect absorber, the power absorbed from the ISRF is

$$W_{\mathrm{abs}} = c(\pi a^2) \int_0^\infty Q_{abs}(\lambda)\, u_\lambda \mathrm{d}\lambda \tag{2.33}$$

where a is the radius and Q_{abs} is the efficiency factor for absorption (Section 2.1.1). Note that no energy would be absorbed if the grain were composed of perfectly dielectric material ($Q_{abs} = 0$), but all real solids absorb to some extent: even good dielectrics may have intrinsic absorption features at specific wavelengths (Section 2.1.4) and additional absorptions may be induced by the presence of impurities. The power radiated by the heated grain is

$$W_{rad} = 4\pi(\pi a^2)\int_0^\infty Q_{em}(\lambda)B_\lambda(T_d)d\lambda \tag{2.34}$$

where T_d is the dust temperature, $Q_{em}(\lambda)$ is the emissivity, i.e., the efficiency factor for emission from the grain, and

$$B_\lambda(T) = \frac{2hc^2}{\lambda^5}\frac{1}{\exp(hc/\lambda kT) - 1} \tag{2.35}$$

is the Planck function. It follows from Kirchhoff's law that $Q_{abs}(\lambda)$ and $Q_{em}(\lambda)$ are, in fact, identical functions of wavelength at a given temperature, and we may replace them in Equations (2.33) and (2.34) by a single function, Q_λ. For equilibrium between the rates of gain and loss of internal energy, we have $W_{abs} = W_{rad}$ and thus

$$\int_0^\infty Q_\lambda u_\lambda d\lambda = \frac{4\pi}{c}\int_0^\infty Q_\lambda B_\lambda(T_d)d\lambda. \tag{2.36}$$

For a perfect blackbody ($Q_\lambda = 1$ at all wavelengths), Equation (2.36) yields the Eddington result (see Equation (2.32)). More generally, this formulation allows T_d to be estimated for an assumed grain model, using calculations based on Mie theory (Section 2.1.1) for grains of appropriate composition and size to specify $Q_\lambda \equiv Q_{abs}$, together with a suitable mathematical representation of u_λ (Mathis et al. 1983; Hocuk et al. 2017).

Predicted temperatures for dust in diffuse (translucent) regions of the ISM in the solar neighborhood of the Galaxy are typically in the range 10–20 K for realistic grain materials, i.e., substantially higher than the blackbody temperature ($T_b \approx 3.2$ K). This difference arises from the fact that real grains have divergent values of Q_λ in the pertinent spectral regions. Absorption, represented by the left-hand side of Equation (2.36), occurs mostly in the ultraviolet (UV), where stellar radiation dominates the ISRF, whereas emission, represented by the right-hand side, occurs mostly in the far-infrared (FIR). For any plausible grain model, we find $Q_{UV} \gg Q_{FIR}$ at all relevant wavelengths.

Strongly absorbing solids such as graphitic carbon are predicted to reach equilibrium at somewhat warmer temperatures (~20 K) than weak absorbers such as silicates (~15 K). These estimates are insensitive to size if the grains are homogeneous spheres and sufficiently large that transient heating (Section 2.3.2) can be ignored. Similarly, shape appears to have little or no effect on predicted temperatures for realistic grain models. Greenberg & Shah (1971) found that non-spherical particles may reach equilibrium at somewhat lower temperatures than spheres of equal volume, but the difference is generally very small and always less than 10% for dielectric spheroids and cylinders; only highly flattened or elongated metallic particles show more significant deviations.

Spatial and environmental changes in the spectral energy distribution of the ISRF, represented by u_λ in Equation (2.36), naturally result in corresponding changes in T_d. Systematic behavior with respect to galactocentric radius arises from variations in the stellar population density, such that higher dust temperatures are predicted in regions closer to the galactic center, and inversely (Mathis et al. 1983). Within a starless dense cloud heated only by the external ISRF, the ambient UV flux is strongly attenuated, leading to lower predicted grain temperatures. As collision rates increase with increasing density, the temperatures of the gas and dust tend to converge, and the optical properties of the grains may be modified as they become coated with ice mantles. Models for the transport of radiation within such regions allow dust temperatures to be calculated as functions of optical depth (e.g., Leung 1975; Mathis et al. 1983; Hocuk et al. 2017), yielding T_d values typically in the range 5–10 K for ice-mantled silicate or graphite grains at extinctions $A_V \gtrsim 10$. Note that grains at such low temperatures radiate predominantly in the extreme FIR to submillimeter region of the spectrum (200–600 μm) and are essentially undetectable in emission at shorter wavelengths.

The results discussed above apply to regions remote from individual stars. In the vicinity of a luminous star, particularly one with a spectrum that peaks in the UV (spectral type O or B), enhancement of the ambient energy density naturally leads to higher predicted dust temperatures. A useful rule of thumb that allows rough estimates of T_d for a weak absorber is given by the equation

$$T_d \sim T_* \, w^{\frac{1}{5}} \tag{2.37}$$

where the star is treated as a "dilute" blackbody of temperature T_* with a dilution factor w (van de Hulst 1946). This formula predicts a realistic value of $T_d \sim 15$ K for dust in the diffuse ISM if the ISRF is approximated by a stellar temperature $T_* = 10{,}000$ K with a dilution factor $w = 10^{-14}$. Now consider a grain situated in an H II region (Section 1.3.3) at a distance of, say, $r = 0.5$ pc from a star of radius $R_* = 1.1 \times 10^{12}$ m and surface temperature $T_* = 40{,}000$ K (corresponding to spectral type O6). The dilution factor is $w = (R_*/2r)^2 = 1.27 \times 10^{-9}$, and inserting these values into Equation (2.37) gives $T_d \sim 665$ K, corresponding to peak emission at $\lambda \sim 5$ μm. Ultimately, the temperature a grain can reach is limited by its sublimation temperature.

2.3.2 Thermal Transients in Very Small Grains

So far, we have assumed that dust temperatures are determined by the balance between time-averaged rates of absorption and emission of energy (Equations (2.33) and (2.34)) and that individual quantum events are unimportant. Two conditions must hold for this assumption to be valid: (i) the total thermal energy content of a grain must be large compared with that received when a single energetic photon is absorbed; and (ii) the energy of the absorbed photon must be distributed throughout the full volume of the grain (i.e., the grain must reach internal thermal equilibrium) on a timescale less than that for photon emission. But, as already noted, the interstellar dust population includes a wide range of sizes, from nanoparticles implicated by observations of extinction in the far-ultraviolet to micrometer-sized grains detected in the flow of interstellar dust entering the solar system. Let us assume initially that condition (ii) holds for all sizes, and examine the inevitable breakdown of condition (i) as progressively smaller particles are considered.

The thermal properties of solids are described by the theory of Debye (1912). For a material with Debye temperature Θ, the heat capacity at constant volume varies as $C_V \propto T^3$ in the low-temperature approximation ($T \ll \Theta$). Values of $\Theta \sim 500$ K are expected for typical candidate grain materials. Purcell (1976) showed that the internal energy content of a grain of radius a (μm) and temperature T_d (K), arising from vibrational modes (i.e., heat) is then given in electron volts (eV) by

$$H = \int_0^{T_d} C_V \, dT \approx 17a^3T_d^4 \tag{2.38}$$

assuming that the bulk heat capacity is applicable to small particles and surface effects can be ignored. If the grain absorbs a single photon of energy ε_{ph} (eV) and this energy is internalized within the solid ($H \rightarrow H + \varepsilon_{ph}$) we have

$$T_2 \approx \left\{T_1^4 + \frac{\varepsilon_{ph}}{17a^3}\right\}^{\frac{1}{4}} \tag{2.39}$$

where T_1 and T_2 are the dust temperatures before and after absorption.

For illustration, consider the amplitude ($\Delta T = T_2 - T_1$) of the increase in temperature induced by absorption of individual 10 eV photons in grains of different size, assuming an initial temperature $T_1 = 15$ K. The increase is minimal ($\Delta T < 0.5$ K) for relatively large grains ($a > 0.05$ μm), but becomes progressively larger for smaller grains: e.g., $\Delta T \approx 13$ K for $a = 0.01$ μm and $\Delta T \approx 32$ K for $a = 0.005$ μm. The heated grain subsequently cools by emission of lower-energy photons. Small grains thus undergo a sequence of stochastic temperature fluctuations or "spikes," each associated with the arrival of an energetic photon, as illustrated schematically in Figure 2.6 for $a = 0.005$ μm. Each spike is followed by a relatively smooth decline, as the rate at which infrared photons are emitted may be typically 10^3 times greater than the rate at which energetic photons are absorbed. Note that the time-averaged temperature of the grain is greater than the lowest temperature reached in between major absorption events.

At higher temperatures, Equations (2.38) and (2.39) are no longer appropriate. The heat capacity becomes insensitive to temperature for $T > \Theta$, and is given by $C_V = 3Nk$, where $3N$ is the number of degrees of freedom in the grain. The amplitude of the temperature increase produced by a photon of energy ε_{ph} (eV) is then

$$\Delta T \approx 4 \times 10^3 \, \varepsilon_{ph}/N. \tag{2.40}$$

For a grain of radius 0.001 μm, $N \sim 40$ and a transient of amplitude $\Delta T \sim 1000$ K is predicted for a 10 eV photon (Sellgren 1984).

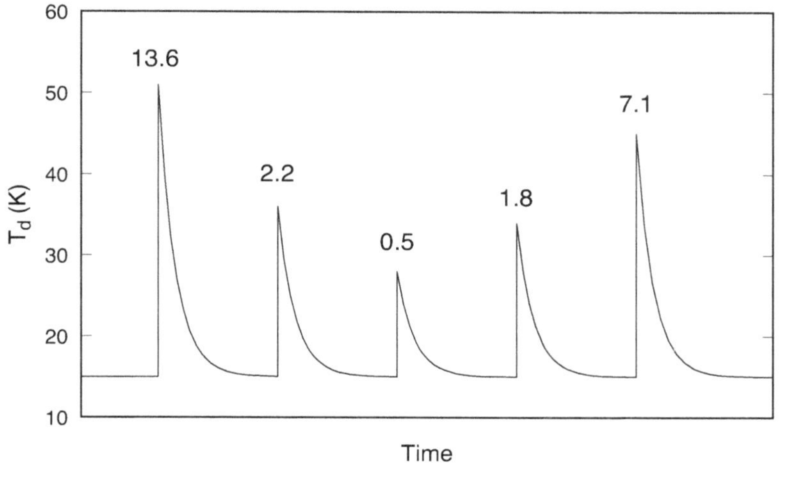

Figure 2.6. A schematic representation of predicted temperature fluctuations in a small grain (radius $a = 0.005$ μm) induced by sequential absorption of individual photons. Each fluctuation is labeled by the photon energy in eV. (Adapted from Aannestad & Kenyon 1979).

We now return to the second condition noted at the beginning of this section. Calculations based on Equation (2.39) predict temperature spikes to be negligible in a relatively large grain ($a > 0.05$ μm), but this is valid only if the energy of an absorbed photon is distributed throughout the full volume of the grain on a timescale less than that for radiative cooling. Whether this is so depends on the structure and thermal properties of the particle as well as on its size. Real interstellar grains may not behave as homogeneous solids in which vibrational energy disperses rapidly throughout the lattice when a photon is absorbed. As previously noted, grains assembled by accretion tend to be porous, containing voids and subparticles that may be in poor thermal contact with one another. Furthermore, amorphous carbon, a likely constituent of the dust, contains intrinsic structure that naturally inhibits rapid internal dispersion of heat. Calculations suggest that loosely-connected graphitic units within larger amorphous carbon grains may behave as quasi-independent particles, undergoing thermal transients of amplitude $\sim$100 K on absorption of individual UV photons (Duley & Williams 1988).

All of these factors affect the spectral energy distribution that emerges from an assemblage of particles reflecting the diversity in size, composition, and structure expected to occur in the ISM. Components subject to stochastic heating, whether independent small grains or isolated units within a larger one, produce net continuum emission that is "warmer" in terms of its equivalent blackbody temperature, peaking at shorter wavelengths, compared with that from large grains in thermal equilibrium with the same environment. Stochastic heating may also induce excitation of energy levels, most notably vibrational modes in polycyclic aromatic hydrocarbons, that result in spectral-line emission. These considerations provide a basis for interpreting the observed infrared emission spectrum of dust, reviewed in Chapter 6.

2.3.3 Far-Infrared Continuum Emission and Dust Mass

The emission from dust carries information not only on the nature of the particles but also on the distribution of mass (Section 1.4.3), both in the Milky Way and in distant galaxies (e.g., Hildebrand 1983; Eales et al. 2012; Berta et al. 2016). The key region of the spectrum to consider is the far-infrared through submillimeter ($\lambda > 100$ μm). At these wavelengths, several useful conditions are met: (i) the dust is optically thin ($\tau_\lambda \ll 1$), thus observations sample the emission from at all depths in a cloud with equal efficiency; (ii) small-particle approximations (Section 2.1.2) are appropriate for even the largest grains in the size distribution; and (iii) we need consider only grains in thermal equilibrium with the ambient radiation field. The effects of transient heating (Section 2.3.2) on the observed emission spectrum occur primarily at shorter wavelengths, and the smallest particles that experience the largest thermal transients contribute little to the total dust mass (Bianchi 2013).

With these conditions in mind, consider a cloud containing $\mathcal{N}$ spherical dust grains of uniform size and composition, each in a state of thermal equilibrium at temperature T_d. The flux density at wavelength λ detected by an external observer is

$$F_\lambda = \mathcal{N}\left\{\frac{\pi a^2}{d^2}\right\} Q_\lambda B_\lambda(T_\mathrm{d}) \tag{2.41}$$

where Q_λ and a are the emissivity and radius of the grains, d is the distance to the cloud, and $(\pi a^2/d^2)$ is the solid angle subtended by each individual grain. Schematic illustrations of the functions F_λ, Q_λ, and B_λ for $T_\mathrm{d} = 15$ K are shown in Figure 2.7. A real interstellar cloud will, of course, contain grains of many sizes. However, in the small-particle approximation, the quantity Q_λ/a is independent of a and depends only on the refractive index m of the grain material at wavelength λ (see Equation

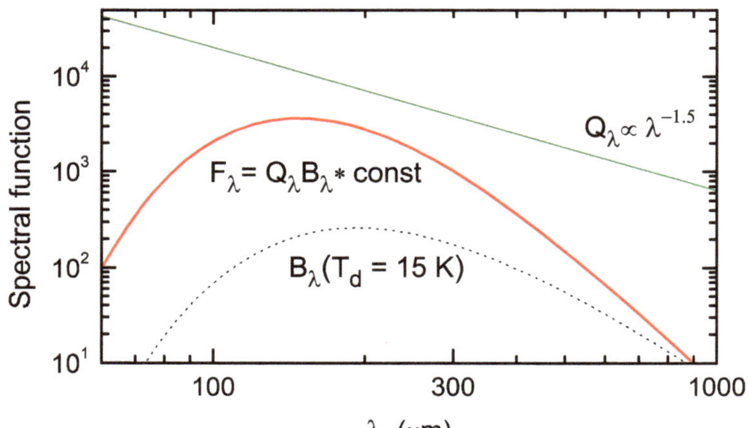

Figure 2.7. Schematic illustration of the emission spectrum (red curve) for an assemblage of dust grains at constant temperature $T_\mathrm{d} = 15$ K, represented by the product of the unmodified Planck function (dotted curve) and a power-law emissivity function of constant index $\beta = 1.5$ (green diagonal line; see Section 2.3.4). Vertical placement is arbitrary.

(2.16), recalling that the efficiency functions for absorption and emission are the same). The contribution of each grain to the observed flux spectrum in Equation (2.41) is thus proportional to $a^2Q_\lambda = a^3\,(Q_\lambda/a)$, i.e., to its *volume*.

The total volume of dust in the cloud is $V = \mathcal{N}v$, where $v = \frac{4}{3}\pi a^3$ is the volume of an individual grain. Substituting to eliminate $\mathcal{N}$ in Equation (2.41), the total dust volume is

$$V = \left\{\frac{F_\lambda d^2}{\pi a^2 Q_\lambda B_\lambda(T_\mathrm{d})}\right\} v, \tag{2.42}$$

and if the grains are composed of solids of mean density ρ_s, Equation (2.42) yields the total dust mass:

$$M_\mathrm{d} = V\rho_\mathrm{s} = \frac{4\rho_\mathrm{s} F_\lambda d^2}{3B_\lambda(T_\mathrm{d})}\left\{\frac{a}{Q_\lambda}\right\}. \tag{2.43}$$

Equation (2.43) may be expressed conveniently in terms of a mass absorption coefficient, κ_λ:

$$M_\mathrm{d} = \frac{F_\lambda d^2}{\kappa_\lambda B_\lambda(T_\mathrm{d})} \tag{2.44}$$

where

$$\kappa_\lambda = \frac{\pi a^2 Q_\lambda}{v\rho_\mathrm{s}} = \frac{3}{4\rho_\mathrm{s}}\left\{\frac{Q_\lambda}{a}\right\}. \tag{2.45}$$

M_d is thus independent of grain size in the small-particle limit and may be deduced from the flux spectrum, subject to adoption of a suitably weighted average of κ_λ at each observed wavelength for an assumed grain model. Uncertainties in κ_λ place important limits on the accuracy of the results (e.g., Mennella et al. 1998; Demyk et al. 2017a, 2017b; Ysard et al. 2018).

Another potential source of error concerns the temperature distribution of the emitting grains. In a quiescent dense cloud containing no internal sources of luminosity, attenuation of the external radiation field results in predicted internal dust temperatures of 5–10 K at extinctions $A_V \gtrsim 10$ (Section 2.3.1). At such low temperatures, grains emit almost entirely at $\lambda \gtrsim 200\ \mu$m and are therefore missed by observations at shorter wavelengths. However, dust in the outer layers of the cloud are warmed by the external ISRF to temperatures approaching those in the diffuse ISM (15–20 K), so their emission peaks at significantly shorter wavelengths ($100 \lesssim \lambda \lesssim 150\ \mu$m). Thus, clouds of different mass may have similar surface brightness at, for example, 100 μm, the longest wavelength detected by the Infrared Space Observatory. More recent space telescopes such as Herschel and Planck have provided essential coverage of the longer FIR-submillimeter wavelengths needed to account for dust at the coldest temperatures.

Clouds that contain embedded sources of luminosity are naturally subject to much greater variations in temperature. Individual stars produce local hot-spots, but if the cloud is dense the most energetic radiation is quickly attenuated. Absorption and re-emission events in dust surrounding the source convert ultraviolet radiation into infrared, to which the cloud is relatively transparent, and this heats the dust elsewhere in the cloud to more moderate temperatures that depend on the total internal luminosity. The observed spectrum of a cloud, whether internally or externally heated, may be modeled (Equation (2.41)) by superposition of isothermal increments over a suitable range of temperature.

2.3.4 The Emissivity Function

The discussion above illustrates the importance of the emissivity as a property of the dust that influences the emergent spectrum. In general, it displays power-law behavior with respect to wavelength in the FIR, represented by

$$Q_\lambda = Q_{\lambda_0}\left(\frac{\lambda_0}{\lambda}\right)^\beta \tag{2.46}$$

where β is the spectral index, and Q_{λ_0} is the value of Q_λ determined at some reference wavelength λ_0. The spectral index is sensitive to the nature of the grain material. A value of $\beta = 1$ is predicted in the idealized case of grains with constant refractive index in the small-particle limit (Equation (2.16)). Expected values for likely constituents of the dust (see below) lie typically in the range $1 \lesssim \beta \lesssim 2$.

If β remains constant across the spectrum, Q_λ modifies the dust emission in a straight-forward, predictable way, as shown in Figure 2.7, shifting the peak to shorter wavelength relative to that of the Planck function at the same temperature. By analogy with the Wien displacement law, the peak in the spectrum emitted by grains at temperature T_d occurs at

$$\lambda_\mathrm{peak} \approx \left(\frac{5}{\beta + 5}\right)\frac{b}{T_\mathrm{d}} \tag{2.47}$$

where $b = 2989$ μm K is a constant. Setting $\beta = 0$ recovers the normal form of Wien's law for a blackbody. Equation (2.47) predicts $\lambda_\mathrm{peak} \approx 150$ μm for the example illustrated in Figure 2.7 ($\beta = 1.5$, $T_\mathrm{d} = 15$ K), compared with 200 μm for the blackbody component.

In general, measured or theoretically predicted values of the spectral index depend on the conductivity and crystallinity of the material. As examples, available data suggest $\beta \approx 1$ for amorphous carbon, $\beta \approx 2$ for highly conductive materials such as metals and graphite, and values in the range 1.6–2.1 for amorphous silicates (e.g., Day 1981; Blanco et al. 1991; Coupeaud et al. 2011). Thus, in principle, the slope of the far-infrared continuum emission from dust provides a means of constraining grain models. If observations are available at sufficiently long wavelengths, the blackbody component converges to the Rayleigh–Jeans approximation

$$B_\lambda(T) \approx \frac{2ckT}{\lambda^4}. \tag{2.48}$$

Then with $Q_\lambda \propto \lambda^{-\beta}$ (Equation (2.46)) and assuming that β remains constant, the observed spectrum behaves as a simple power law, $F_\lambda \propto Q_\lambda B_\lambda(T) \propto \lambda^{-(\beta+4)}$, from which β is readily estimated. However, a problem arises if, as expected, the dust being sampled contains a mixture of materials. Different grain populations may, depending on their optical properties, reach equilibrium at different temperatures, and so their relative contributions to the observed spectrum may vary with wavelength. If each population is characterized by a different value of β, the assumption of a single power-law emissivity is no longer valid. Indeed, both observational and experimental evidence indicates systematic changes in the effective value of β at the longest wavelengths (e.g., Demyk et al. 2017a). To make, progress it is necessary to develop integrated models that simultaneously satisfy all available constraints.

2.3.5 Polarized Emission

The observed polarization of starlight demonstrates the presence of aligned dust grains with asymmetric optical properties in the ISM, and, as expected, the thermal emission from the dust is also polarized. The **E**-vector of the emitted beam has greatest amplitude in the plane containing the longest axis of the grain. Thus, the observed net emission from an ensemble of aligned grains is partially plane polarized with a position angle orthogonal to that of polarization in the transmitted beam from a background star observed through the exact same ensemble, as illustrated schematically in Figure 2.4.

The polarized emission may be specified in terms of parameters analogous to those used to describe the polarization of starlight (Section 2.2.2). The position angle is again determined by the plane of vibration of the **E**-vector relative to galactic north, and the amplitude $P_{\rm em}$ is evaluated from the maximum and minimum flux densities measured at some appropriate wavelength with respect to the orientation of an analyzing element in the detector:

$$P_{\rm em} = \frac{F_{\rm max} - F_{\rm min}}{F_{\rm max} + F_{\rm min}}. \tag{2.49}$$

At a given grain temperature, the flux densities are proportional to the emissivities (Equation (2.41)) and hence, by Kirchhoff's law, to the absorption efficiencies; therefore Equation (2.49) may be written

$$P_{\rm em} = -\frac{\Delta Q}{Q_{\parallel} + Q_{\perp}} \tag{2.50}$$

using the same notation as in Section 2.2.1, where the minus sign arises from the fact that the position angles for polarized emission and absorption are orthogonal. Comparing Equation (2.50) with Equation (2.30), we infer that $P_{\rm em}$ is equivalent to a polarization efficiency for absorption; if the polarization is small, this may be expressed as

$$P_{em}(\lambda) = -\frac{p_\lambda}{\tau_\lambda} \quad (2.51)$$

at some wavelength λ, with reference to Equations (2.5) and (2.27).

Comparisons between observations of polarized dust emission and polarized starlight in the same lines of sight place constraints on grain properties (e.g., Hildebrand 1988; Hildebrand & Dragovan 1995; Planck Collaboration 2015). Starlight polarization is usually measured in the visible or near-infrared: both p_λ and τ_λ become very small at the longer infrared wavelengths required to measure $P_{em}(\lambda)$. The polarized emission at wavelength λ_1 and polarization by differential extinction at wavelength $\lambda_2 \ll \lambda_1$ may then be related by

$$P_{em}(\lambda_1) = \xi \frac{p(\lambda_2)}{\tau(\lambda_2)} \quad (2.52)$$

where the factor ξ depends on the dielectric function and shape of the particles for an assumed grain model (Hildebrand & Dragovan 1995). See Chapter 4 for further discussion.

2.4 Grain Dynamics

2.4.1 Radiation Pressure

Incident radiation exerts a force on any matter with which it interacts, arising from the exchange of momentum between the target and the oscillating electromagnetic field. This force is usually negligible, but may become significant when grains are subject to intense and directionally inhomogeneous radiation fields (e.g., emanating from a specific nearby source or group of sources). If radiation pressure is dominant, its net effect is to drive grains away from the source of radiation. Notable situations in which this can occur include H II regions around luminous OB stars (Section 1.3.3) and the dusty circumstellar envelopes of red giants and supergiants (Section 8.4.1). In our own planetary system, solar radiation pressure moderates the influx of dust grains arising from its motion relative to the local interstellar medium (Section 1.3.4).

The rate at which a plane wave of intensity I carries linear momentum across a unit area normal to the direction of propagation is I/c. A grain that intercepts a portion of the wave experiences a net force of magnitude $(I/c)\, C_{pr}$, where C_{pr} is the cross section for radiation pressure, defined as the effective area over which the pressure is exerted. We may also define an efficiency factor, $Q_{pr} = C_{pr}/\pi a^2$, the ratio of radiation pressure cross section to geometric cross section for grains of radius a. Q_{pr} depends on both the absorption and scattering characteristics of the grains, and is related to the corresponding efficiency factors (Section 2.1) by

$$Q_{pr} = Q_{abs} + \{1 - g(\theta)\} Q_{sca} \quad (2.53)$$

where $g(\theta)$ is the scattering asymmetry parameter, defined in Equation (2.20). Mie theory calculations for a chosen grain model thus yield values of Q_{pr}, and the results may be used to predict the forces acting on grains according to their size and composition in a given radiative environment. In a gaseous medium the drift speed of the grains is limited by frictional drag exerted by the gas and, in a circumstellar envelope, by the star's gravitational pull. Radiation pressure is an important factor in models for mass loss from dust-producing stars, further discussed in Section 8.4.1.

2.4.2 The Photoelectric Effect

Absorption of energetic photons ($\varepsilon_{\rm ph} \gtrsim 10$ eV) by dust may excite electrons sufficiently to eject them from the grain material, which then becomes positively charged. As noted in Section 1.3.1, this process affects the thermodynamics of the gas because the energy carried by photoelectrons is a significant source of gas-phase heating in diffuse regions of the ISM. As for the dust, in the presence of a magnetic field **B**, the component of its velocity **v** in the direction transverse to **B** is subject to the Lorentz force

$$\mathbf{F} = q(\mathbf{v} \times \mathbf{B}) \tag{2.54}$$

where q is the net charge of the grain. Magnetic fields thereby exert a significant influence on the dynamics of charged particles, causing them to spiral along magnetic field lines rather than cross them.

The photoelectric yield, defined as the probability that absorption of a photon of a given energy will release an electron from a grain, depends on the size and composition of the particle, and on the magnitude of any pre-existing electrostatic charge (e.g., Weingartner & Draine 2001). The degree to which a grain can thereby accumulate charge is effectively self-regulating: there is an increase in net positive charge with each successive photoelectric emission event, thus increasing the Coulomb barrier that further electrons must overcome to escape from the grain, while simultaneously increasing the probability of recombination arising from random encounters with free electrons. Mathematical models enable rates of photoelectric emission and electron capture to be predicted as functions of particle size, composition, and the spectral energy density of the radiation field. Best estimates for likely dust constituents in typical diffuse regions of the ISM indicate that larger ($a \gtrsim 0.01$ μm) grains tend to maintain positive potentials, whereas smaller ones have a higher probability of being neutral or negatively charged (Draine 2011).

2.4.3 Rotational Motion: Suprathermal Spin and Radiative Torques

Several processes may contribute to the rotational motion of a dust grain in the ISM. First consider a grain immersed in atomic gas with a Maxwell–Boltzmann distribution of velocities at temperature $T_{\rm gas}$. Random collisions with gas atoms impart impulsive torques that give the grain rotational as well as translational energy, and its angular speed (ω) is limited by rotational friction with the gas. If the collisions are elastic, the rotational kinetic energy of the grain about a principal axis with moment of inertia I is

$$E_{\rm rot} = \tfrac{1}{2} I \langle \omega^2 \rangle = \tfrac{3}{2} k T_{\rm rot} \tag{2.55}$$

where $T_{\rm rot}$ is the rotation temperature of the grain. A typical value of the mean angular speed may be estimated by taking an average moment of inertia $I = \frac{2}{5} m_{\rm d} a^2$ (appropriate to a sphere of radius a and mass $m_{\rm d}$); Equation (2.55) then gives

$$\omega_{\rm rms} \approx \left(\frac{1.8 k T_{\rm rot}}{a^5 \rho_{\rm s}} \right)^{\frac{1}{2}} \tag{2.56}$$

where $\rho_{\rm s}$ is the density of the solid grain. Under conditions of local thermodynamic equilibrium ($T_{\rm rot} \approx T_{\rm gas}$), the grains undergo "Brownian" rotation, and Equation (2.56) shows that this is most rapid for the smallest grains. Considering grains of density $\rho_{\rm s} = 2000$ kg m^{-3} immersed in cool atomic gas ($T_{\rm gas} \sim 80$ K; Table 1.1), Equation (2.56) yields values of $\omega_{\rm rms} \sim 3 \times 10^5$ rad s^{-1} for $a = 0.1$ μm and $\omega_{\rm rms} \sim 1 \times 10^8$ rad s^{-1} for $a = 0.01$ μm.

Additional factors ensure that real interstellar grains are often subject to "spin-up," i.e., they rotate at suprathermal speeds with $T_{\rm rot} \gg T_{\rm gas}$ (Purcell 1979). Processes that may act on grains with irregular surfaces to enhance their angular speeds include emission of photoelectrons (Section 2.4.2), impulsive ejection of H_2 molecules formed at active surface sites (Section 4.3.4), and radiative torques. Harwit (1970) first drew attention to the possibility that absorption of starlight by a grain may result in a significant transfer of angular momentum, provided that the radiation field is anisotropic. A situation where this may occur in the ISM is near the surface of an interstellar dark cloud, where grains are shielded from the ISRF in the direction toward the cloud, and exposed to it in the opposite direction. Harwit considered the intrinsic angular momentum of the photons themselves, but subsequent calculations suggested that its contribution is overwhelmed by the randomizing effect of collisions with gas atoms and by isotropic emission of low-energy photons from the grains (Purcell & Spitzer 1971; Martin 1972).

Subsequent work led to the proposal of a more realistic mechanism by which photons may impart angular momentum (Dolginov & Mytrophanov 1976). Consider a grain that is helical in shape: such a particle absorbs and scatters left-handed and right-handed circular components of polarization in a transmitted beam differently (i.e., their extinction cross-sections are different), and so directionally anisotropic irradiation by unpolarized light (having equal LH and RH components) may nevertheless impose a radiative torque that causes spin. A helix is not, of course, a plausible shape for a real interstellar grain; but an irregular grain of arbitrary shape will, in general, have some non-zero average "helicity," resulting in optical properties analogous to those of a chiral molecule.

Radiative torques are now widely accepted as important and potentially dominant sources of suprathermal spin. Calculations show that the efficiency of angular momentum transfer varies with grain size and with the spectrum as well as the intensity of the illuminating radiation (Draine & Weingartner 1996). The effect is greatest for grains with sizes $a > \lambda/2$. The spectrum of the mean interstellar radiation field drops precipitously in the extreme ultraviolet owing to absorption by atomic hydrogen beyond the Lyman limit (91.2 nm), hence radiative torques become less effective for grains smaller than $a \sim 0.05$ μm. For larger grains, spin-up is sufficient to overcome the randomizing effects of collisions and isotropic photon emission. Radiative torques thus appear to be more effective than other potential sources of spin enhancement, such as photoelectron emission and H_2 formation, and for this reason they are considered vital to a quantitative understanding of the mechanism responsible for grain alignment, a topic we return to in Chapter 4 (Section 4.3).

References

Aannestad, P. A., & Kenyon, S. J. 1979, Ap&SS, 65, 155

Berta, S., Lutz, D., Genzel, R., Förster-Schreiber, N. M., & Tacconi, L. J. 2016, A&A, 587, A73

Bianchi, S. 2013, A&A, 552, A89

Blanco, A., Fonti, S., & Rizzo, F. 1991, InfPh, 31, 167

Bohren, C. F., & Huffman, D. R. 1983, Absorption and Scattering of Light by Small Particles (New York: Wiley)

Coupeaud, A., Demyk, K., Meny, C., et al. 2011, A&A, 535, A124

Day, K. L. 1981, ApJ, 246, 110

Debye, P. 1909, AnPh, 30, 59

Debye, P. 1912, AnPh, 39, 789

Demyk, K., Meny, C., Lu, X.-H., et al. 2017a, A&A, 600, A123

Demyk, K., Meny, C., Leroux, H., et al. 2017b, A&A, 606, A50

Dolginov, A. Z., & Mytrophanov, I. G. 1976, Ap&SS, 43, 291

Draine, B. T. 1988, ApJ, 333, 848

Draine, B. T. 2011, Physics of the Interstellar and Intergalactic Medium (Princeton, NJ: Princeton Univ. Press), 296

Draine, B. T., & Weingartner, J. C. 1996, ApJ, 470, 551

Duley, W. W., & Williams, D. A. 1988, MNRAS, 231, 969

Eales, S., Smith, M. W. L., Auld, R., et al. 2012, ApJ, 761, 168

Eddington, A. S. 1926, RSPSA, 111, 423

Fogel, M. E., & Leung, C. M. 1998, ApJ, 501, 175

Gerakines, P. A., & Hudson, R. L. 2020, ApJ, 901, 52

Greenberg, J. M., & Shah, G. A. 1971, A&A, 12, 250

Harwit, M. 1970, Natur, 226, 61

Henning, Th. 1999, A&AS, 136, 405

Hildebrand, R. H. 1983, QJRAS, 24, 267

Hildebrand, R. H. 1988, QJRAS, 29, 327

Hildebrand, R. H., & Dragovan, M. 1995, ApJ, 450, 663

Hocuk, S., Szűcs, L., Caselli, P., et al. 2017, A&A, 604, A58

Jäger, C., Il'in, V. B., Henning, T., et al. 2003, JQSRT, 79–80, 765

Leung, C. M. 1975, ApJ, 199, 340

Martin, P. G. 1972, MNRAS, 158, 63

Mathis, J. S. 2000, in Allen's Astrophysical Quantities, ed. A. N. Cox (4th ed.; New York: Springer), 523

Mathis, J. S., Mezger, P. G., & Panagia, N. 1983, A&A, 128, 212

Mathis, J. S., & Whiffen, G. 1989, ApJ, 341, 808
Mennella, V., Brucato, J. R., Colangeli, L., et al. 1998, ApJ, 496, 1058
Mie, G. 1908, AnPh, 25, 377
Ossenkopf, V. 1991, A&A, 251, 210
Planck Collaboration, 2015, A&A, 576, A106
Purcell, E. M. 1976, ApJ, 206, 685
Purcell, E. M. 1979, ApJ, 231, 404
Purcell, E. M., & Pennypacker, C. R. 1973, ApJ, 186, 705
Purcell, E. M., & Spitzer, L. 1971, ApJ, 167, 31
Sellgren, K. 1984, ApJ, 277, 623
Shen, Y., Draine, B. T., & Johnson, E. T. 2008, ApJ, 689, 260
Spitzer, L. 1978, Physical Processes in the Interstellar Medium (New York: Wiley)
van de Hulst, H. C. 1946, RAOU, 11, 1
van de Hulst, H. C. 1957, Light Scattering by Small Particles (New York: Wiley)
Weingartner, J. C., & Draine, B. T. 2001, ApJS, 134, 263
Witt, A. N. 1989, in IAU Symp. 135, Interstellar Dust, ed. L. J. Allamandola, & A. G. G. M. Tielens (Dordrecht: Kluwer), 87
Wolff, M. J., Clayton, G. C., Martin, P. G., & Schulte-Ladbeck, R. E. 1994, ApJ, 423, 412
Wolff, M. J., Clayton, G. C., & Gibson, S. J. 1998, ApJ, 503, 815
Ysard, N., Jones, A. P., Demyk, K., Boutéraon, T., & Koehler, M. 2018, A&A, 617, A124

AAS | IOP Astronomy

Dust in the Galactic Environment (Third Edition)

D C B Whittet

Chapter 3

The Observed Properties of Dust—I. Extinction

The physical basis for interstellar extinction—the attenuation of starlight resulting from absorption and scattering by dust grains along the line of sight—is summarized in the previous chapter (Section 2.1). The focus of this chapter is on observational results and their interpretation. The following sections include a description of the methods used to obtain information, a review of the principal results, and an assessment of their implications for our understanding of grain properties.

3.1 Observational Methods

In general, the spectrum of a source observed through a dusty medium is a combination of two components: the intrinsic spectrum of the source itself, modified by the extinction spectrum imposed by dust along the line of sight. The goal is to separate these components to yield the extinction spectrum (termed the extinction curve), independent of the source. To accomplish this apparently simple task, the form of the intrinsic spectrum must be inferred. If the source is a normal star with a reliably-determined spectral type, this is straightforward, at least in principle. For observations limited to broadband photometry in standard passbands, databases of intrinsic color indices for unreddened stars are available (e.g., Drilling & Landolt 2000; Ducati et al. 2001) to enable calculations of color excesses such as E_{B-V} (Section 1.2.2). More detailed investigations typically make use of spectrophotometry capable of resolving structure of width $\Delta\lambda \lesssim 10$ nm. The available wavelength range extends from the mid-infrared to the far-ultraviolet, respective limits being set by the inherent decline in stellar flux with increasing wavelength in the infrared and the opacity of interstellar atomic hydrogen gas beyond the Lyman limit at 91.2 nm in the ultraviolet.

Alternative methods are available to extract extinction curves from the data:

1. Compare the observed spectrum of the target star with corresponding observations of a standard star of the same spectral type that has little or no extinction (the "pair" method; e.g., Fitzpatrick & Massa 1990).

doi:10.1088/2514-3433/ac7204ch3

2. Use theoretical models of stellar spectra to represent the intrinsic component of the observed spectrum (“extinction without standards”; e.g., Fitzpatrick & Massa 2005).

An example of the pair method is shown in Figure 3.1. Hot main-sequence stars (spectral classes O–B9 and luminosity class V) are generally preferred for such investigations. Their spectra are relatively simple and predictable, with less star-to-star dispersion within a given class, compared with post-main-sequence stars and those of later spectral types. Also, their intrinsic luminosity and frequent spatial association with dusty regions render OB stars most suitable for probing interstellar extinction, especially at visible and ultraviolet (UV) wavelengths. However, care must be taken to avoid stars with circumstellar shells as these generally have non-typical spectra that might include emission lines and/or infrared continuum emission in excess of a normal photosphere. Given this caveat, and given that relatively few OB stars are close enough to the Sun or at high enough galactic latitude to have negligible reddening, investigations of extinction in our home Galaxy using the pair method tend to rely on a rather small set of well-studied standards.

The observed apparent magnitude of the target star at wavelength λ is given by

$$m(\lambda) = M(\lambda) + 5 \log d - 5 + A(\lambda) \tag{3.1}$$

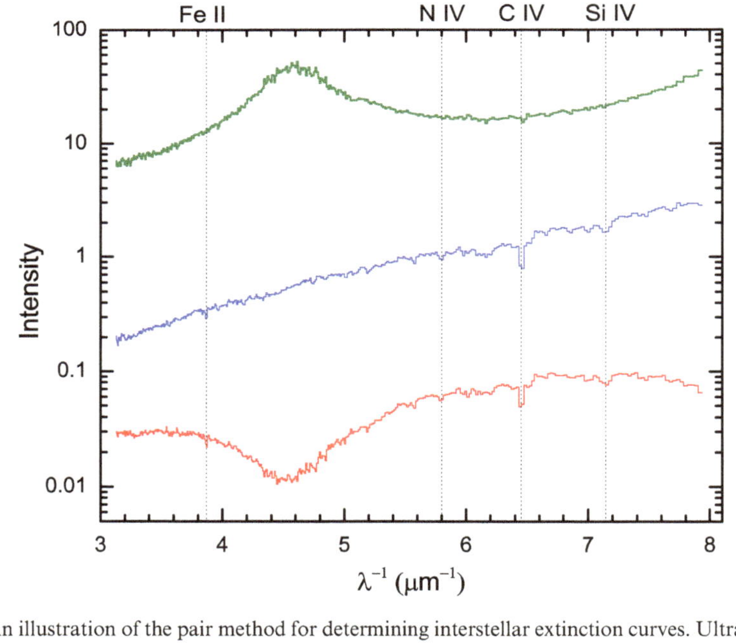

Figure 3.1. An illustration of the pair method for determining interstellar extinction curves. Ultraviolet spectra of a reddened star (HD 34078, spectral type O9.5 V, $E_{B-V} = 0.54$) and an almost unreddened star of the same spectral type (HD 38666, $E_{B-V} = 0.03$) are shown as red and blue curves. Vertical dashed lines indicate the positions of representative stellar spectral features. The intensity scale is logarithmic in arbitrary units (equivalent to magnitude). The extinction curve (green) is the intensity ratio (equivalent to magnitude difference) of the two spectra. Based on data from the IUE Atlas of Low Dispersion Spectra (Heck et al. 1984).

where $M(\lambda)$ is the intrinsic spectrum of the star (in the form of an absolute magnitude), d is its distance, and $A(\lambda)$ is the spectral dependence of extinction that we seek to determine (see Section 1.2.1). The function $M(\lambda)$ is represented by observations of a standard star (pair method) or by a theoretical model. If a standard star is used, the difference between the apparent magnitudes (Equation (3.1)) for target and standard, denoted by subscripts 1 and 2, respectively, is

$$\begin{aligned}\Delta m(\lambda) &= m_1(\lambda) - m_2(\lambda)\\ &= A(\lambda) + 5\log\left(\frac{d_1}{d_2}\right)\end{aligned} \tag{3.2}$$

subject to the assumptions that $M_1(\lambda) = M_2(\lambda)$ and $A(\lambda) = A_1(\lambda) \gg A_2(\lambda)$, i.e., the intrinsic spectra of the two stars are identical and the standard star has negligible extinction.

The final term in Equation (3.2) is independent of wavelength and constant for a given pair of stars, but uncertainties in their distances and in the absolute flux calibration of the observations generally preclude an exact transformation from $\Delta m(\lambda)$ to $A(\lambda)$. An analogous situation arises when a model spectrum is used (see Fitzpatrick & Massa 2005 for detailed discussion and a comparison of the two methods). Nevertheless, if the spectral matching is accurate, the quantity $\Delta m(\lambda)$ provides a reliable representation of the *spectral dependence* of $A(\lambda)$, independent of stellar parameters and limited only by the quality of the observational data. Figure 3.1 illustrates the cancelation of *stellar* spectral lines to produce a relatively smooth extinction curve in which residual fluctuations are determined primarily by the signal-to-noise ratios of the paired spectra. In contrast, a broad absorption feature centered near 4.6 μm^{-1}, conspicuous in the reddened star and weak or absent in the standard star, results in a prominent extinction peak attributed to a component of the dust.

It is helpful to eliminate the wavelength-independent component of $\Delta m(\lambda)$ in Equation (3.2) by means of a simple "normalization" with respect to two standard wavelengths, λ_1 and λ_2. The normalized extinction is defined:

$$\begin{aligned}E_{\rm norm} &= \frac{\Delta m(\lambda) - \Delta m(\lambda_2)}{\Delta m(\lambda_1) - \Delta m(\lambda_2)}\\ &= \frac{A(\lambda) - A(\lambda_2)}{A(\lambda_1) - A(\lambda_2)}\\ &= \frac{E(\lambda - \lambda_2)}{E(\lambda_1 - \lambda_2)}\end{aligned} \tag{3.3}$$

where $E(\lambda_1 - \lambda_2)$, the difference in extinction between the selected wavelengths, is equivalent to a color excess (Section 1.2.2). These wavelengths are usually chosen to match those of standard passbands such as B and V, so $E_{\rm norm} = E_{\lambda-V}/E_{B-V}$ (see Figure 3.2 for an example). Normalization allows extinction curves for stars with different color excesses to be superposed and compared: effectively, the reddening in the line of sight is then standardized to unit color excess ($E_{B-V} = 1$) and the slope of

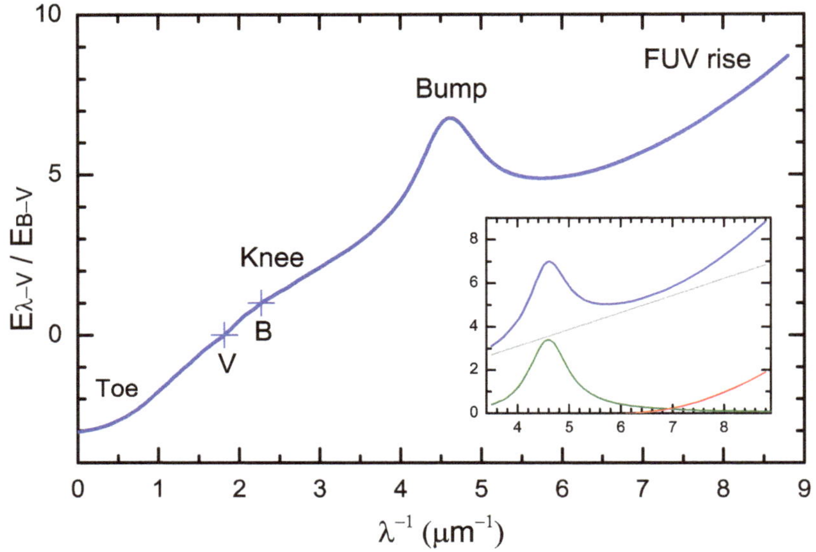

Figure 3.2. The average interstellar extinction curve (blue) for diffuse regions of the galactic disk within ~3 kpc of the Sun (data from Fitzpatrick et al. 2019). The V and B normalization points and various features discussed in the text are labeled. Decomposition into components of UV extinction based on the FM method (Section 3.3.2) is shown in the inset: bump profile (green), FUV rise (red), and linear continuum (gray).

the curve between the standard wavelengths is fixed irrespective of the number of dust grains in the line of sight. Calculated extinction curves (Section 2.1; Equation (2.8)) deduced for a given grain model may be normalized in the same way to allow direct comparison between observations and theory.

A caveat should be noted here: because the Johnson B and V passbands traditionally used for normalization are quite broad (widths ~100 nm), the *effective* wavelengths of observation may differ appreciably from the nominal values (440 and 550 nm), depending on the spectrum of the source. Thus, an extinction curve normalized using broadband B and V photometry may not be identical to one normalized to the precise B and V wavelengths using higher-resolution spectrophotometry for the same source. See Fitzpatrick et al. (2019) for detailed discussion and analysis.

The methods described in this section work well as probes of extinction enabled by observations of OB stars over wavelengths ranging from the far-ultraviolet to the near-infrared ($0.1 \lesssim \lambda \lesssim 3\ \mu$m). Extension to longer wavelengths is hindered by the fact that OB stars are relatively faint in the mid-infrared, requiring observations of extreme sensitivity to yield reliable data (Gordon et al. 2021). Another approach is to target much larger and more diverse samples of highly reddened and intrinsically red sources that typically lack precise spectral classifications. Although often too faint to be studied at ultraviolet or even blue-visible wavelengths, such sources are easy targets in the red and infrared. At wavelengths corresponding to the Rayleigh–Jeans part of their spectra, stars with normal photospheres have effectively the same intrinsic colors to a first approximation (Majewski et al. 2011). Using data from (e.g.) the Two Micron All-Sky Survey and the Spitzer Space Telescope, analyses of color–magnitude and color–color diagrams enable intrinsic properties to be constrained and extinction data to be extracted. See Indebetouw et al. (2005), Zasowski et al. (2009), Schlafly et al. (2016), and Wang & Chen (2019) for examples.

3.2 General Extinction in the Diffuse ISM

The dust responsible for extinction in spiral galaxies tends to be confined to their disks (e.g., Figure 1.2), typically within about 100 pc of the midplane. Our vantage point within the Milky Way is from a position close to the midplane, about 8 kpc from the nucleus (Section 1.3.4). The ISM is cloudy, and most of the dust is located in clouds. Thus, the accumulation of extinction along a typical line of sight toward a distant star is stochastic, in the sense that it depends on the number of clouds that it happens to intersect, and their individual opacities. In directions away from the galactic plane (galactic latitudes $|b| \gtrsim 30°$), the extinction is generally independent of pathlength and often negligible, but may be significant if the line of sight intersects a high-latitude cloud. In directions close to the disk of the Milky Way ($|b| \lesssim 15°$), a typical line of sight may pass though many clouds.

3.2.1 The Average Extinction Curve

Studies of large numbers of stars have shown that the extinction curve displays a similar general form in many lines of sight. Regional variations are evident, nevertheless, and these are important for investigations of grain properties and their response to environment, as discussed in detail in Sections 3.3 and 3.4. Here we focus on the average extinction curve for diffuse regions of the ISM (those in which hydrogen is predominantly atomic rather than molecular; Section 1.3) in the solar neighborhood of the galactic disk, as determined from observations of OB stars within about 3 kpc of the Sun. This provides a valuable benchmark for comparison with data for individual lines of sight, for other regions and environments in the Milky Way, for other galaxies, and for the results of model calculations.

The average extinction curve, shown in Figure 3.2 in the normalized form $E_{\lambda-V}/E_{B-V}$ versus λ^{-1}, displays several distinctive features. It is approximately linear in the visible region from 1 μm^{-1} to 2.2 μm^{-1}, with changes in slope occurring in the near-infrared and blue regions (the "toe" and the "knee," respectively). This behavior is broadly consistent with predictions based on Mie theory for dielectric grains with dimensions of the same order as the wavelength (Section 2.1). The most prominent characteristic of the curve is a broad peak centered near 4.6 μm^{-1} in the mid-ultraviolet, known as the 217 nm feature or "bump," discussed in detail in Section 3.4. A trough beyond the bump is followed by a steep, non-linear rise into the far-ultraviolet (FUV, $\lambda^{-1} > 6$ μm^{-1}).

3.2.2 The Ratio of Total to Selective Extinction

The behavior of extinction in the infrared (the "toe" region in Figure 3.2) is significant for several reasons. It affects our ability not only to make precise reddening corrections at these wavelengths but also to calibrate the entire (relative) extinction curve in terms of total (absolute) extinction. It also provides additional tools with which to explore grain properties (size and composition), discussed in Section 3.2.4.

A parameter commonly used to calibrate the conversion from relative to absolute extinction is the ratio $R_V = A_V/E_{B-V}$, introduced in Section 1.2.2. The normalized extinction curve is related to R_V by

$$\frac{E_{\lambda-V}}{E_{B-V}} = \frac{A_\lambda - A_V}{E_{B-V}} = R_V\left\{\frac{A_\lambda}{A_V} - 1\right\}. \tag{3.4}$$

If, as expected, the extinction becomes negligible as the wavelength becomes very long (see Section 3.2.5), R_V may be expressed as a limiting value

$$R_V = -\left[\frac{E_{\lambda-V}}{E_{B-V}}\right]_{\lambda\to\infty} \tag{3.5}$$

and may therefore be evaluated by extrapolating the observed extinction curve ($\lambda^{-1} \to 0$). An offset power law provides a reasonable mathematical representation of the extinction curve in the toe region ($\lambda^{-1} \lesssim 1\ \mu\text{m}^{-1}$). Fits of the form

$$\frac{E_{\lambda-V}}{E_{B-V}} = k_{\text{IR}}\lambda^{-\beta} - R_V \tag{3.6}$$

may therefore be used to guide extrapolation, where k_{IR} is a scaling constant, on the assumption that the spectral index β does not vary with λ. Typical values of these parameters in the diffuse ISM are $k_{\text{IR}} \approx 1.2$ and $\beta \approx 1.8$ (Martin & Whittet 1990; see also Fitzpatrick & Massa 2009 for further discussion and caveats). The resulting average value of R_V for diffuse–ISM dust within ~3 kpc of the Sun is deduced to be[1]

$$\langle R_V \rangle = 3.05 \pm 0.10. \tag{3.7}$$

R_V is not constant in the diffuse ISM, but this average is generally representative. For comparison, lines of sight that intersect dense clouds often have substantially higher values, typically in the range $3.5 \lesssim R_V \lesssim 5.5$ (Section 3.3.1).

Estimates of R_V for individual lines of sight can be made without a need to construct detailed extinction curves. The approximation $R_V \simeq 1.1\ E_{V-K}/E_{B-V}$ is commonly used to evaluate R_V toward a reddened star of known spectral type from photometry in just three passbands. In effect, this approximation corresponds to Equation (3.5) with λ set to the K passband (2.2 μm), using a scaling constant consistent with theoretical or empirical models to represent extrapolation to infinite wavelength. One could easily devise alternative approximations for different passbands and with different scaling constants, but K is the preferred infrared

[1] Note that R_V is somewhat sensitive to the type of data used. The extinction curve in Figure 3.2, which uses spectrophotometric data normalized to the precise V and B wavelengths, yields $R_V = 3.02$, compared with 3.10 for the same stars using broadband photometry (Fitzpatrick et al. 2019).

passband for this purpose as it has the longest wavelength in the "nonthermal" infrared: at longer wavelengths, thermal background radiation increases rapidly, limiting the accuracy of photometry with ground-based telescopes. Moreover, a vast database of stellar photometry in the K band is available from the Two Micron All-Sky Survey. It should be noted, however, that systematic errors can arise for stars that emit excess infrared emission in the K band, as typically occurs if circumstellar shells of hot, ionized gas are present. The color excess ratio E_{V-K}/E_{B-V} is then anomalously large (because the K magnitude is numerically less), leading to an overestimate of R_V. Such stars can usually be identified and eliminated by investigation of their optical spectra, which typically show emission lines, or from their distribution in infrared color–color diagrams such as $J - H$ versus $H - K$.

Objects subject to extreme amplitudes of extinction, such as stars within or behind dense cloud cores or at great distances in the galactic disk, are often too faint to be observable at visible wavelengths. In such cases, information on extinction is limited to the infrared, and R_V cannot be determined directly. Nevertheless, it is often desirable to quantify the *visual extinction* toward an infrared source, and this can be estimated from its observed infrared colors. The reddening may be expressed in terms of a color excess such as E_{J-K} if a spectral type is indicated by the infrared data (the fractional error caused by any mismatch becomes progressively less significant with increasing extinction). We may then express A_V in terms of E_{J-K}:

$$A_V = rE_{J-K} \tag{3.8}$$

where r is the ratio of total visual to selective infrared extinction, analogous to R_V. This parameter is evaluated from observations of stars with full visual-infrared extinction coverage (He et al. 1995; Whittet et al. 2001), yielding a mean value of $r = 5.9 \pm 0.2$ in the diffuse ISM. More generally, an empirical law relating r to R_V across different environments is given by

$$r = \frac{a}{b - R_V^{-1}} \tag{3.9}$$

where a and b are constants: values of $a \simeq 2.38$ and $b \simeq 0.73$ are consistent with available data for both diffuse and dense clouds. A quantitative estimate of r is thus possible for a given target if the probable value of R_V for the region can be deduced from other observations.

3.2.3 Scattering

The observed extinction (Figure 3.2) is the sum of contributions arising from absorption and scattering (Section 2.1). In principle, it is possible to separate these contributions, and thereby gain additional insight into the nature of the dust, by studying observational phenomena that are unique products of the scattering component. These include the diffuse galactic light, reflection nebulae illuminated by individual stars or star clusters, diffuse scattered radiation from dark clouds illuminated by the general interstellar radiation field, and scattering halos around X-ray sources (e.g., Gordon 2004; Mattila et al. 2018; Valencic & Smith 2015).

The diffuse galactic light (DGL) is a widespread background glow arising from scattering of starlight by dust in the disk of our Galaxy. Being intrinsically faint and subject to contamination by other sources of diffuse emission, it is one of the most challenging of all dust-related phenomena to study. Potential sources of contamination include galactic and extragalactic stellar background radiation, zodiacal light and, in the case of ground-based or airborne observations, telluric airglow. In common with other scattering phenomena, it tends to be brightest at shorter wavelengths, and almost all of the available observations were carried out in the ultraviolet using space-borne telescopes and detectors. Analysis of the data depends on an idealized plane-parallel model for radiative transfer in the galactic disk, and requires detailed knowledge of the spectral dependence of the illuminating radiation. Moreover, stars contributing most intensely to this radiation field at different wavelengths may have different spatial distributions, and the resultant spectrum may thus be dependent on geometry. See the reviews by Witt (1988, 1989) and Gordon (2004) for further discussion.

Reflection nebulae (RNe) have the advantage of being much brighter than the DGL. The observations are also generally much easier to analyze, especially in cases where the source of illumination is a single star, as both the geometry and the spectrum of the incident radiation are tightly constrained. Being relatively bright in the visible as well as the ultraviolet, observations of RNe both overlap and extend the spectral coverage afforded by observations of the DGL. As RNe commonly occur in relatively dense regions, however, the dust they contain might not be typical of the diffuse ISM. Further extension of the spectral coverage into the infrared is enabled by observations of scattered radiation from dark clouds, again with the caveat that the grains may differ from those in the diffuse ISM. Observations of scattering from clumps within dark clouds ("core-shine"), in particular, are indicative of grain growth in these dense environments (Pagani et al. 2010; Saajasto et al. 2018).

The relevant quantities determined from these observations are the albedo, α, and the asymmetry parameter, $g(\theta)$, defined in Section 2.1.3. The spectral dependence of the albedo is shown in Figure 3.3, combining data from observations of DGL, RNe, and dark clouds. Results from these diverse phenomena are mostly in agreement to within the estimated uncertainties. Data in the spectral range from the near-infrared to the near-ultraviolet (0.4–3 μm^{-1}) are generally consistent with a mean value of $\alpha = 0.62 \pm 0.1$, with no obvious evidence for a systematic trend. In the ultraviolet, a minimum corresponding to the 4.6 μm^{-1} extinction peak is clearly detected. Uncertainties are greatest in the far-ultraviolet: the data suggest a precipitous drop in albedo between 7 and 8 μm^{-1}, implying a rapid onset of absorption. However, inconsistencies in the data and the lack of any known theoretical basis for so dramatic a change preclude firm conclusions (Gordon 2004).

The observations enable the asymmetry factor to be investigated over a similar spectral range. Results suggest a mean value of $g(\theta) = 0.75 \pm 0.15$, with no clear evidence of structure corresponding to the 4.6 μm^{-1} feature or, indeed, any substantive, systematic variation with wavelength. This result indicates strongly forward-directed scattering by grains with diameters $2a \sim \lambda$ (Witt 1988). Scattering

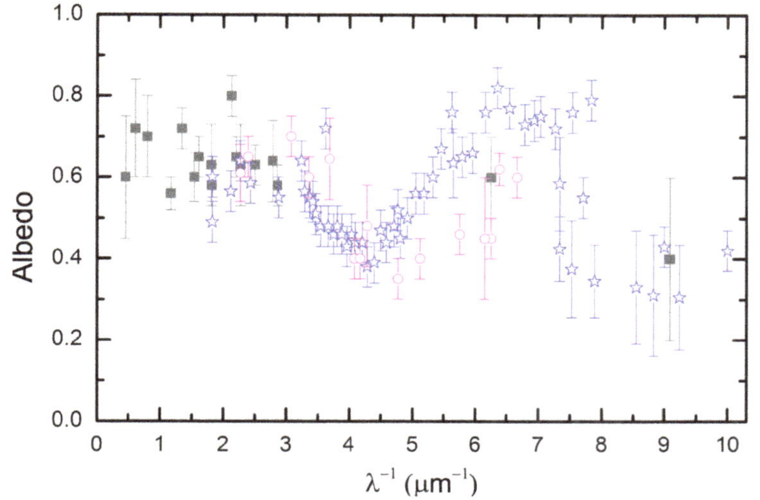

Figure 3.3. The spectral dependence of the albedo. Results are based on observations of the diffuse galactic light (magenta circles), reflection nebulae (blue stars), and diffuse scattered light from dark clouds (gray squares). Data are mostly from the compilation of previous work by Gordon (2004), with additional results from Hamden et al. (2013) and Mattila et al. (2018).

thus appears to be dominated by relatively large grains ($0.1 < 2a < 1\ \mu$m) and clearly not by grains in the small-particle limit ($a \ll 0.1\ \mu$m), for which $g(\theta) \approx 0$ is predicted.

The contributions of absorption and scattering to the total extinction are denoted by the terms $(1 - \alpha)A_\lambda/A_V$ and $\alpha A_\lambda/A_V$, respectively (see Section 2.1.3, Equation (2.18)). Estimates for the diffuse ISM are shown in Figure 3.4, based on an assessment of the albedo data in Figure 3.3. A constant value of $\alpha = 0.62$ is assumed for $\lambda^{-1} < 3\ \mu\text{m}^{-1}$, and variations in the range $3 < \lambda^{-1} < 7\ \mu\text{m}^{-1}$ are represented by a smooth interpolation; values for $\lambda^{-1} > 7\ \mu\text{m}^{-1}$ are ignored. Note that this analysis is subject not only to statistical uncertainties but also to systematic errors arising from the fact that the extinction and albedo are not measured for identical samples. Apparent fluctuations in the scattering component between 3.5 and 5.5 μm^{-1}, for example, may arise because the profile of the bump feature is not ideally represented by the albedo data, which is heavily dependent on observations of reflection nebulae. The curves in Figure 3.4 should thus be regarded as guides to general behavior rather than definitive results. Nevertheless, the data seem sufficiently robust to support two important conclusions:

- The 4.6 μm^{-1} bump in the extinction curve is a pure absorption feature.
- Scattering is a major source of continuum extinction, exceeding absorption over most of the available spectral range.

Scattering halos detected around bright X-ray point-sources provide another opportunity to study dust properties, distinct from and complementary to investigations at other wavelengths. The halos are products of small-angle scattering of

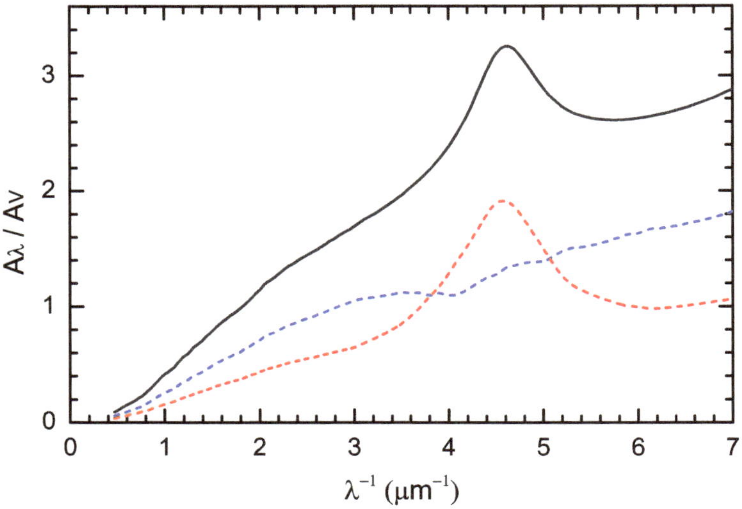

Figure 3.4. Estimated contributions of absorption (red) and scattering (blue) to the total extinction, based on an interpolation of albedo data in Figure 3.3. Extinction data (black) correspond to Figure 3.2, converted to A_λ/A_V with $R_V = 3.02$.

X-rays by dust grains along the lines of sight. In this region of the spectrum, all potential grain materials have refractive indices close to unity, i.e., very similar to that of the surrounding medium (the Rayleigh–Gans approximation; Mathis & Lee 1991; Smith & Dwek 1998). The X-ray spectrum of the dust may contain absorption edges that are diagnostic of composition (see Section 7.1.5), but the scattered component is most sensitive to the porosity and size of the particles. Additional factors that affect interpretation of results include the spatial distribution of the dust along the line of sight, which is typically unknown or poorly constrained, and the possibility of multiple scattering. The grains responsible for X-ray scattering may be much larger than the wavelengths of the scattered photons, and the observed properties of the halos are indeed most sensitive of the numbers of *large* grains in the size distribution. Observations of the halo around the well-studied X-ray source Nova Cygni 1992 have been interpreted as evidence for porous grains (>25% vacuum) with sizes extending to larger radii ($a \sim 2\ \mu$m) than those traditionally used to fit the extinction curve (Mathis et al. 1995; Witt et al. 2001). However, in an extensive survey of 35 sources, Valencic & Smith (2015) found that most are consistent with models that assume a size limit of $a \lesssim 0.4\ \mu$m, and that porosity is not a prerequisite for a satisfactory fit.

3.2.4 Power-Law Behavior in the Infrared

Earlier discussion of extinction in the infrared (Section 3.2.2) focused in its importance as a guide to calibrating the extinction curve. This spectral region also provides tools with which to explore the physical and optical properties of the dust, and thereby to discriminate between grain models. The slope of the continuum extinction at these wavelengths is governed by the size and composition of the particles, and the occurrence of absorption features identified with solid-state transitions in the grain material provides direct evidence for some of the constituents.

An alternative view of the mean diffuse–ISM extinction curve is displayed in Figure 3.5, using log–log format to afford a detailed view of its behavior in the infrared. In this format, the curve is approximately linear in the near-infrared, behavior that may be represented mathematical by a power law:

$$\frac{A_\lambda}{A_V} = k'_{\mathrm{IR}}\, \lambda^{-\beta} \tag{3.10}$$

where $k'_{\mathrm{IR}} = k_{\mathrm{IR}}/R_V$ is a scaling constant (see Equation (3.6)). A match to the data in Figure 3.5 is given by Equation (3.10) with values of $k'_{\mathrm{IR}} = 0.4$ and $\beta = 1.85$ (Martin & Whittet 1990). More recent investigations have found β values of the same order, typically in the range 1.7–2.1 (e.g., Xue et al. 2016; Wang & Chen 2019; Decleir et al. 2022).

How may this result be interpreted? Models for extinction based on Mie theory (Section 2.1) are, indeed, consistent with a transition from an approximately λ^{-1} dependence in the visible to a power law with a more negative exponent in the infrared, as the wavelength becomes large compared with the largest grains in the size distribution. If small-particle approximations come into play (Section 2.1.2), λ^{-4}

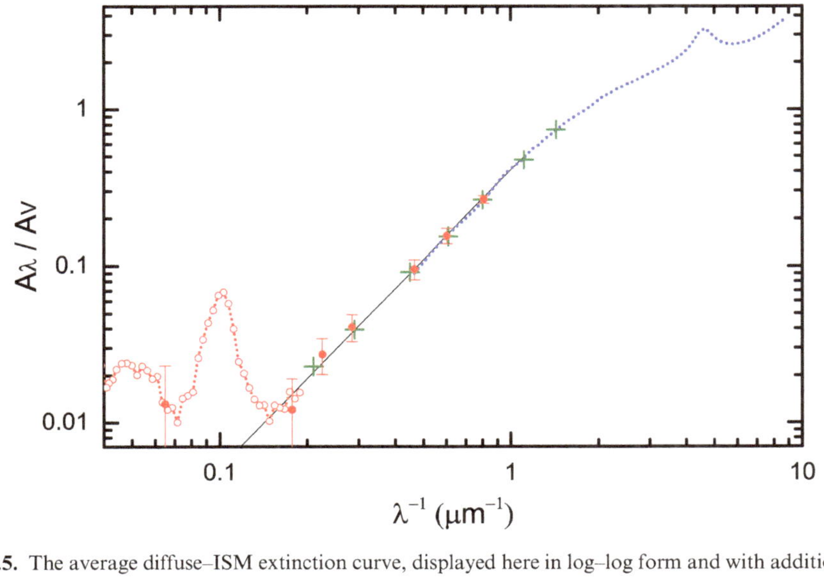

Figure 3.5. The average diffuse–ISM extinction curve, displayed here in log–log form and with additional data to illustrate behavior in the infrared. The blue dotted curve is from Fitzpatrick et al. (2019). Red filled and open circles represent photometric and spectroscopic data from Gordon et al. (2021), respectively. Green crosses are from the compilation of photometric data by Martin & Whittet (1990). The thin black line is a power law of index $\beta = 1.9$, scaled to match the data from 0.2 to 1.1 $\mu\mathrm{m}^{-1}$. Peaks centered at $\lambda^{-1} \approx 0.05$ and 0.1 $\mu\mathrm{m}^{-1}$ are attributed to absorption by silicate dust.

behavior is predicted for the scattering component of extinction by grains composed of materials with dielectric optical properties. However, Q_{sca} declines more rapidly than Q_{abs} with increasing λ, thus absorption must eventually become dominant, and the form of the extinction law then depends on the spectral dependence of the dielectric function in the infrared, i.e., on the nature of the grain material.

Spectral fingerprints of one major component of the dust are evident in Figure 3.5. Peaks centered near 10 and 20 μm are identified with absorption features superposed on the continuum extinction, arising from excitation of vibrational modes in silicates (see Chapter 5 for detailed discussions of these and other infrared absorption features attributed to dust). Because these features are broad and overlapping, it is not possible to determine the continuum extinction at wavelengths >7 μm by observations of reddened stars, and it is tempting to assume that a simple extrapolation of the power-law fit should apply (Figure 3.5). It would be surprising, however, if nature conspired to produce a transition from scattering-dominated extinction in the near-infrared (Section 3.2.3) to absorption-dominated extinction in the far-infrared consistent with a power law of constant spectral index. Nevertheless, recalling that the absorption and emission functions are equivalent according to Kirchhoff's law (Section 2.3.1), it is notable that both the near-infrared extinction and the far-infrared emissivity (Section 2.3.4) exhibit power-law behavior with broadly similar values of β.

3.2.5 Neutral Extinction

Neutral (wavelength-independent) extinction occurs when particles large compared with the wavelength of observation are present. This behavior is evident in the results of Mie theory calculations (Section 2.1.1): Q_{ext} approaches a plateau value of about 2 when $X = 2\pi a/\lambda$ becomes large (Figure 2.1). Fog in the Earth's atmosphere is a good example for visual wavelengths. Any neutral extinction produced in the interstellar medium by "giant" grains, i.e., with dimensions $\gg$1 μm, would be undetected by the observational methods described in Section 3.1; but their presence would result in additional attenuation of starlight, thereby introducing systematic errors in spectrophotometric determinations of stellar distances (Section 1.2.1). Evaluation of R_V by extrapolation of the extinction curve (Equation (3.5)) is based on the assumption that the continuum extinction becomes negligible as the wavelength becomes very large. Nevertheless, interstellar grains with sizes ~10 μm are known to exist, as revealed by observations of scattering in dark clouds (Pagani et al. 2010) and analyses of dust entering the solar system from the local ISM (Section 1.3.4), and there is no a priori reason to exclude still larger particles. The question is whether they are sufficiently numerous to contribute appreciable extinction.

Independent methods of evaluating R_V that include any contribution from neutral extinction are available to address this question. Trumpler's technique for determining the diameters of open star clusters in our Galaxy (Section 1.2.1) is an early example. The average value of the total (neutral plus wavelength-dependent) visual extinction for a star cluster may be stated as

$$\langle A_V \rangle = \langle V - M_V \rangle - \{5 \log(D/\theta) - 5\} \tag{3.11}$$

where θ is the angular diameter of the cluster in radians, and D is the linear diameter in parsecs deduced from Trumpler's morphological classification technique; thus, D/θ is the geometric distance (independent of extinction). $\langle V - M_V \rangle$ is determined by photometry and spectral classification of individual cluster members, from which the mean reddening $\langle E_{B-V} \rangle$ is also deduced. A detailed analysis of data for many clusters yields a value of $R_V = 3.15 \pm 0.20$ (Harris 1973). Another method, introduced by Sandage (1975, 1976), is based on a finding that the brightest members in clusters of external galaxies have a low dispersion in their luminosities. This enables estimates of total extinction (again including any neutral component) introduced by foreground dust in the Milky Way from observed displacements in the Hubble diagram (redshift versus magnitude). Results yield $R_V = 3.35 \pm 0.30$ for a favorably placed example close to the galactic disk.

The consistency of these results with the value $R_V = 3.05 \pm 0.10$ deduced from the mean extinction curve (Section 3.2.2) suggests that any neutral extinction must be small, at least in the diffuse ISM. This conclusion concurs with evidence arising from the abundances and depletions of condensible elements such as C, Mg, and Si (Chapter 7), inferred to reside mostly in carbon and silicate grains less than 1 μm in size to account for the λ-dependent extinction and associated spectral features, with little left to form much larger ones. But this does not, of course, exclude the possibility of larger grains. The availability of astrometric data from the Gaia mission prompts a reassessment. Gaia enables stellar parallaxes to be measured with unprecedented precision, and hence to greater distances than was previously possible (e.g., Lindegren et al. 2021). Neutral extinction introduces systematic offsets between distances measured by parallax and by spectrophotometry. Preliminary results suggest such discrepancies do indeed occur in some lines of sight, predominantly toward cold clouds (Siebenmorgen et al. 2020). A shift in the grain size distribution toward larger particles is, indeed, expected in colder, denser clouds, as confirmed by observations of wavelength-dependent extinction (Section 3.3.1), polarization (Section 4.2.6), and scattering (Pagani et al. 2010).

3.2.6 The Dust-to-Gas Ratio

Gas and dust appear to be generally well-mixed in the ISM, such that the total column density of gas scales linearly with reddening, and hence with total extinction. This correlation is confirmed by observations of absorption lines of hydrogen, the primary constituent of the gas, in the spectra of reddened stars with measured extinction (Bohlin et al. 1978; Rachford et al. 2009). Data for diffuse lines of sight toward stars within a few kiloparsecs of the Sun in the disk of our Galaxy show that

$$\left\langle \frac{N_{\rm H}}{E_{B-V}} \right\rangle \approx 5.8 \times 10^{25}\ {\rm m^{-2}\ mag^{-1}} \tag{3.12}$$

where $N_{\rm H} = N({\rm H}) + 2N({\rm H_2})$ is the column density of H nucleons summed over atomic and molecular phases. With $R_V = 3.05$ (Equation (3.7)) and assuming an additional ~10% contribution to N_H from ionic gas (Spitzer 1978, p. 156), this result may be expressed as a ratio with respect to total extinction:

$$\left\langle \frac{N_H}{A_V} \right\rangle \approx 2.1 \times 10^{25} \text{ m}^{-2} \text{ mag}^{-1} \tag{3.13}$$

in close agreement with an independent evaluation by Zhu et al. (2017). The mean accumulation rate of extinction per unit pathlength toward stars in the Bohlin et al. study is

$$\left\langle \frac{A_V}{L} \right\rangle \approx 1.0 \text{ mag kpc}^{-1}. \tag{3.14}$$

An estimate of the amount of grain material required to produce the observed extinction in the galactic disk may be deduced from these results using general principles described by Purcell (1969). The integral of the extinction efficiency factor Q_{ext} (Section 2.1.1) over all wavelengths may be obtained from the Kramers–Krönig relationship

$$\int_0^\infty Q_{\text{ext}} \mathrm{d}\lambda = 4\pi^2 a \left\{ \frac{m^2 - 1}{m^2 + 2} \right\} \tag{3.15}$$

for grains of radius a and complex refractive index m. The contribution of dust to the total density of interstellar material in a column of length L and unit cross-section is

$$\rho_d = \frac{N_d m_d}{L} \tag{3.16}$$

where N_d is the column density of the dust (i.e., the number of dust grains in the column), and $m_d = \frac{4}{3}\pi a^3 \rho_s$ is the mass of an individual grain composed of solids of mean density ρ_s. Using Equation (2.7) to replace Q_{ext} with A_λ in Equation (3.15), and substituting to eliminate N_d and m_d in Equation (3.16), we have

$$\rho_d \propto \rho_s \left\{ \frac{m^2 + 2}{m^2 - 1} \right\} \int_0^\infty \frac{A_\lambda}{L} \mathrm{d}\lambda. \tag{3.17}$$

From a knowledge of the observed mean extinction curve, ρ_d may be approximated as an average relative to $\langle A_V/L \rangle$ in mag kpc^{-1} (e.g., Spitzer 1978, p. 162):

$$\langle \rho_d \rangle \approx 1.3 \times 10^{-27} \rho_s \left\{ \frac{m^2 + 2}{m^2 - 1} \right\} \left\langle \frac{A_V}{L} \right\rangle. \tag{3.18}$$

Estimates of ρ_d are thus dependent on the assumed grain composition as well as on the observed rate at which extinction increases with distance in the galactic disk. If, for example, we adopt values of $m = 1.50 - 0i$ and $\rho_s = 2500$ kg m^{-3}, appropriate to amorphous silicates, the value of $\langle A_V/L \rangle$ from Equation (3.14) above yields a mean dust density

$$\langle \rho_d \rangle \approx 1.1 \times 10^{-23}\ \text{kg m}^{-3} \tag{3.19}$$

for the same sample of stars used to determine $\langle N_H/A_V \rangle$. It should be emphasized that this result is an average for many sight-lines that include a range of environments. Much larger values of A_V/L, and hence of ρ_d, are expected to occur locally within individual clouds. Equations (3.13) and (3.14) may be combined to eliminate A_V, resulting in an estimate of the mean hydrogen density toward the sample:

$$\langle \rho_H \rangle = m_H \left\langle \frac{N_H}{L} \right\rangle \approx 1.1 \times 10^{-21}\ \text{kg m}^{-3} \tag{3.20}$$

where L is now expressed in m. Finally, allowing for the contributions of other elements (principally helium) to the gas, the mass ratio of dust to gas is

$$Z_d = 0.71 \frac{\rho_d}{\rho_H} \approx 0.007 \tag{3.21}$$

or a little under 1%. This result is consistent with independent estimates of Z_d based on the abundances and depletions of the condensible elements (Chapter 7), evaluated toward a similar sample of stars in the Milky Way (Section 7.2.2).

Z_d is likely to vary with locus in the Galaxy. Toward the galactic center, for example, observations of sources located in the nuclear region yield values of $\langle A_V/L \rangle$ higher by a factor of about four compared with estimates for the solar neighborhood, implying a proportionate increase in $\langle \rho_d \rangle$. This difference may be attributed in large part to greater concentrations of interstellar material, both gas and dust, but may also reflect an underlying trend in metallicity with galactocentric radius (Giannetti et al. 2017). The abundance of dust is naturally sensitive to the rate at which the ISM becomes enriched with dust-forming chemical elements, and this depends on the number density and evolutionary state of the stellar population.

3.2.7 Mapping the Galactic Distribution

The availability of large stellar photometric databases enables detailed 3D mapping of the distribution of dust in the Milky Way (Green et al. 2018, 2019; Lallement et al. 2019; Guo et al. 2021). The ability to map in three dimensions arises from the fact that, with a large enough sample, a small area of sky contains multiple sources with a distribution of measured distances and reddenings. Extinction-based surveys are complementary to those based on thermal dust emission in the infrared (Schlegel et al. 1998; Planck Collaboration 2014). Infrared maps are effectively two-dimensional, being derived from the observed flux within the beam of the telescope, with no direct means of distinguishing between contributions from different regions along the line of sight. However, they do have the advantage of uniform angular resolution, capable of resolving fine structure, unbiased by the distribution of background stars or our ability to infer their intrinsic properties. Comparisons between dust maps based on extinction and on thermal emission are especially effective at intermediate and high galactic latitudes, as there is usually no ambiguity in the location of the emitting dust.

3.3 Spatial Variations in the Extinction Curve

The previous section reviews what can be learned from studies of dust in diffuse regions of the ISM, characterized by the mean extinction curve presented in Figure 3.2. We next examine significant departures from this general behavior, evident in certain lines of sight, and consider how they relate to differences in environment. The focus here is on the continuum extinction; variations specific to the bump feature are discussed further in Section 3.4.

Evidence for variations in the optical properties of the dust were first discussed by Baade & Minkowski (1937). Extinction curves for stars in the Orion Nebula (Messier 42) were found to differ from the mean curve for more typical regions of the Milky Way, in a manner consistent with selective removal of small particles from the size distribution. This insightful interpretation has stood the test of time: the variations observed in Orion and many other regions are indeed attributed, in large part, to changes in the grain size distribution in response to environment. These modifications may be imposed by a number of physical processes, discussed in detail in Chapter 9, including growth by coagulation, collisional fragmentation, and size-dependent acceleration by radiation pressure in stellar winds.

3.3.1 Size-dependent Behavior

The extinction ratio $R_V = A_V/E_{B-V}$ (Section 3.2.2) provides a means of probing variations in the grain size distribution. R_V is naturally most sensitive to the nature of the particles responsible for extinction in the blue-visible region of the spectrum. Calculations for particles with dielectric optical properties, implicated by the large scattering component (Section 3.2.3), predict that R_V varies systematically with size. The extinction efficiency (Section 2.1.1) increases most rapidly with the parameter $X = 2\pi a/\lambda$ in the range $0 < X < 3$. Thus, the extinction at visible wavelengths is especially sensitive to the relative numbers of particles with radii $a \sim 0.1$ μm. If larger grains become more dominant, E_{B-V} is reduced (because extinction at B becomes relatively less efficient compared to that at V), and hence R_V is increased. Conversely, a systematic bias toward smaller particles causes a reduction in R_V.

Representative extinction curves are shown in Figure 3.6 for three regions of the Milky Way exhibiting R_V values that differ significantly from the average for the diffuse ISM. All are associated with current or recent star formation, but with contrasting morphologies and phases of evolution. The Scorpius-Ophiuchus region, part of Gould's Belt (Section 1.3.4), includes both young stars and dark clouds; NGC 1977, adjacent to the Orion Nebula, is a more compact structure containing newborn OB stars that illuminate local reflection nebulae and H II regions; Trumpler 37 is a star cluster that recently emerged from its natal cloud. Their extinction curves display systematic behavior across the spectrum: the slope in the knee region ($\lambda^{-1} \sim 3$ μm^{-1}) and the amplitude of the extinction between the bump and the FUV rise ($\lambda^{-1} \sim 6$ μm^{-1}) vary inversely with R_V. Such behavior may be understood if the shift toward larger grains implied by a higher R_V value results in flatter (grayer) continuum extinction in the ultraviolet, arising from greater contributions by particles much larger than the wavelength. The bump feature,

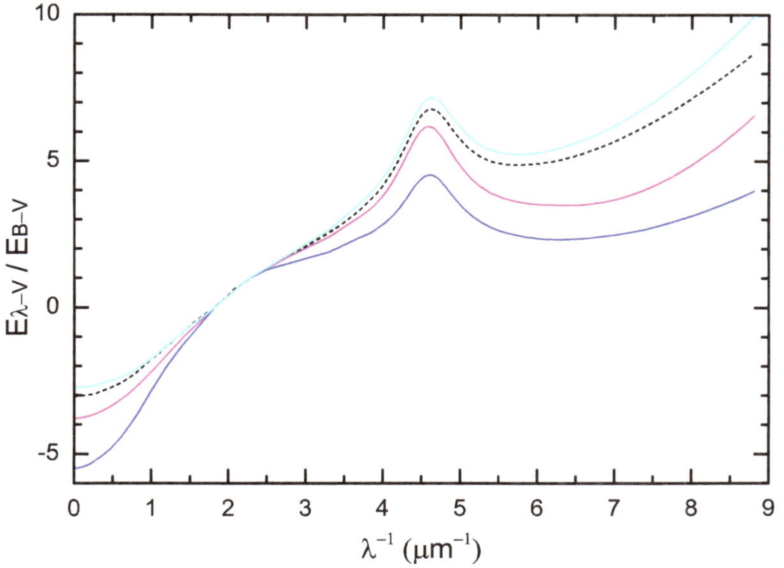

Figure 3.6. Mean extinction curves for three contrasting environments with different R_V values. The curves are empirical fits to observational data (Fitzpatrick & Massa 2007) for stars toward Trumpler 37 (cyan), the Sco-Oph dark clouds (magenta), and NGC 1977 (blue). R_V values determined from the intercepts (Equation (3.5)) are 2.75, 3.8, and 5.5, respectively. The mean diffuse–ISM curve is also shown for comparison (dashed black line; R_V = 3.02).

superposed on this continuum, shows some variation in amplitude in these lines of sight, most evident toward NGC 1977 where it is relatively weak.

R_V values significantly greater than the diffuse–ISM average are commonly associated with relatively dense environments, in which particle growth is attributed to coagulation resulting from frequent, low-impact, grain–grain collisions (Section 9.1.1). The Scorpius-Ophiuchus region includes a well-studied example (the ρ Oph dark cloud; Vrba et al. 1993). Growth by coagulation may also be a contributing factor in NGC 1977, and the effect of radiation pressure imposed by embedded luminous stars may further modify the size distribution by selectively expelling smaller grains. In contrast, dust in the vicinity of Trumpler 37 may have been modified by a supernova shock wave (Clayton & Fitzpatrick 1987; Valencic et al. 2003), causing selective destruction of larger grains.

Histograms provide a useful means of comparing R_V distributions for larger samples of stars, using spectrophotometric data without a need to construct complete extinction curves (Section 3.2.2). Some examples are illustrated in Figure 3.7, covering environments ranging from diffuse high-latitude clouds to star-forming regions in Taurus, Scorpius-Ophiuchus, and Orion. The high-latitude clouds (yellow) included in the survey by Larson & Whittet (2005) are translucent ($A_V \lesssim 1$), starless, and generally remote from any individual sources of luminosity. As such, they might be expected to exhibit typical diffuse–ISM extinction; however, the overall distribution is somewhat skewed toward lower R_V values, consistent with

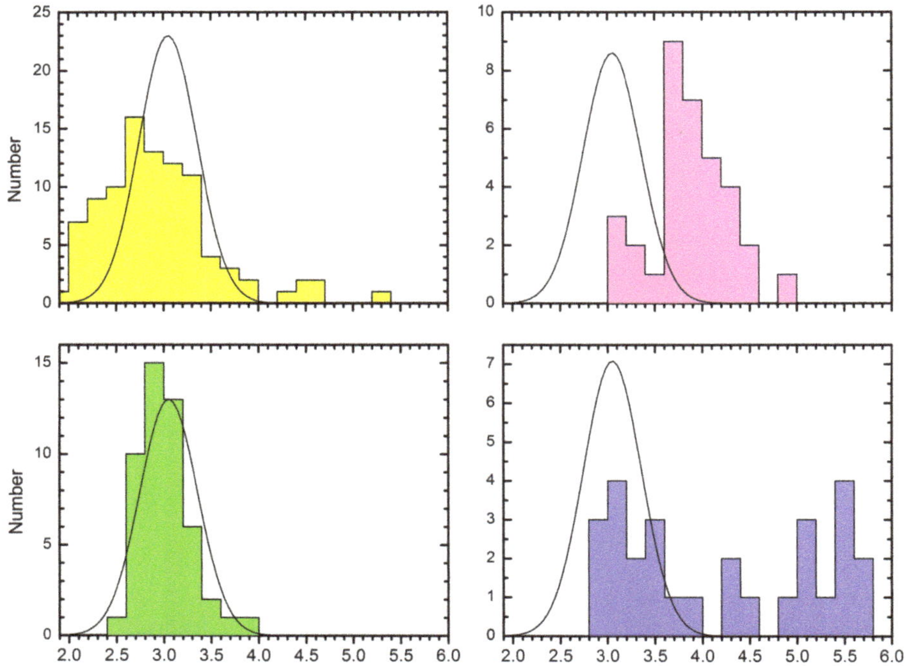

Figure 3.7. Histograms illustrating the distribution of R_V values in four diverse samples: high-latitude clouds (yellow), the Taurus dark cloud (green), the Sco-Oph region (magenta), and the Orion Nebula region (blue). Vertical axes indicate frequencies in bins of width 0.2. The black curve in each frame is a Gaussian representation of the distribution in the diffuse ISM, peaking at $R_V = 3.05$, with amplitude scaled to match the sample size.

enhanced abundances of relatively small grains in a significant fraction of the sample. The Taurus (green) and Sco-Oph (magenta) distributions make an interesting comparison. Whereas grain growth appears pervasive in Sco-Oph, it seems lacking in Taurus, the distribution closely resembling that of the diffuse ISM. The available data for the Taurus region (Whittet et al. 2001) sample dust predominantly in the diffuse outer layers of the cloud ($A_V \lesssim 3$), which appear to have an unmodified size distribution. Lastly, the apparently bimodal distribution of R_V values in Orion (blue) may be explained by inclusion of stars reddened by diffuse foreground material in the sample, as well as stars embedded in dense nebular regions such as NGC 1977, where the grains are being strongly modified.

3.3.2 Empirical Formulations

Evidence for systematic behavior described above suggests that entire extinction curves may be represented empirically by a simple mathematical formula in which R_V is a free parameter. Known as the CCM method, developed by Cardelli et al. (1989) and updated by Valencic et al. (2004), the extinction curve in its absolute normalized form is represented by the formula

$$\frac{A_\lambda}{A_V} = a(x) + \frac{b(x)}{R_V} \tag{3.22}$$

where $a(x)$ and $b(x)$ are coefficients that have unique values at a given wavenumber $x = \lambda^{-1}$. The coefficients are determined empirically from the slope and intercept of the correlation between A_λ/A_V and $R_V{}^{-1}$ for each selected value of x across the spectrum. The potential importance of such a formulation is that it allows the extinction curve to be predicted for lines of sight with measured values of R_V but otherwise unknown extinction properties. However, the accuracy of such predictions is limited because stars with similar R_V values can exhibit extinction curves with significantly different morphologies in the UV.

An analytical means of characterizing the UV portion of the extinction curve has been developed in a series of papers by Fitzpatrick & Massa (1986, 1988, 1990, 2007; the FM method). Extinction at wavenumbers $x \geqslant 3.5\ \mu\text{m}^{-1}$ is specified as the sum of three components (see Figure 3.2):

$$\frac{E_{\lambda-V}}{E_{B-V}} = (c_1 + c_2 x) + c_3 D(x) + c_4 F(x) \tag{3.23}$$

where c_1, c_2, c_3, and c_4 are constants for a given line of sight. The components are: (i) a linear term, $c_1 + c_2 x$, representing the underlying continuum extinction; (ii) a "bump" term of amplitude c_3, where $D(x)$ is a mathematical representation of its absorption profile (see Section 3.4.1); and (iii) a quadratic term of amplitude c_4 representing the FUV rise.[2] The latter is non-zero only above a certain cut-off wavenumber, x_c, such that

$$\begin{aligned} F(x) &= (x - x_c)^2 \qquad [x > x_c] \\ &= 0 \qquad\qquad\quad\ [x \leqslant x_c] \end{aligned} \tag{3.24}$$

where x_c is typically of order 6 μm^{-1}. Each term in Equations (3.23) and (3.24) is treated as a free parameter when fitting individual extinction curves.

Note that these formulae are applicable only in the ultraviolet; however, an extension of the parameter set to allow empirical representations of entire extinction curves is proposed by Fitzpatrick & Massa (2007). A power-law form of constant index $\beta = 1.84$ is assumed for the infrared segment ($x < 1\ \mu\text{m}^{-1}$), with k_{IR} and R_V treated as free parameters (Equation (3.6)). The remainder of the curve is represented by a cubic spline function, fit to a set of "anchor points," that connects the ultraviolet segment to the infrared.

Comparing results for many lines of sight, it is found that the slope of the linear component of the UV extinction exhibits an imperfect but indicative correlation with $R_V{}^{-1}$ (Fitzpatrick & Massa 2007):

$$R_V{}^{-1} \simeq 0.17 + 0.20 c_2. \tag{3.25}$$

[2] A polynomial form for $F(x)$, with an invariant cut-off at $x = 5.9\ \mu\text{m}^{-1}$, was assumed in earlier work (Fitzpatrick & Massa 1988). The revised version improves an already excellent fit.

This correlation elucidates the nature of R_V-dependent behavior in the UV extinction curve (e.g., Figure 3.6): for the most part, it stems from associated changes in the underlying level of UV continuum extinction, represented by the linear term. Support for a relationship between R_V and other components of the UV extinction is less evident. It is notable that the function $F(x)$ provides an adequate representation of the shape of the FUV rise, regardless of environmental factors and independent of the morphology of the extinction curve at longer wavelengths (Fitzpatrick & Massa 1988). This suggests that it is a distinct optical property of the dust arising in an independent small-grain population, variations in its amplitude simply reflecting fluctuations in the abundance of the carrier grains. Similar comments apply to the bump, discussed further in Section 3.4.

3.3.3 Extinction in Other Galaxies

The observed properties of extinction by dust in the Milky Way provide a standard for comparison with other galaxies. Variations from one galaxy to another may depend on several interrelated factors, such as systematic differences in the abundances of the chemical elements, in the dust-to-gas ratio, in the star formation rate, and in the range of physical conditions sampled by the observations. If the impact these factors have on our observations can be unraveled, such comparisons may advance our understanding of dust properties within as well as beyond the Milky Way.

The most accessible targets for study by the methods described in Section 3.1 are nearby galaxies in our Local Group, such as the Magellanic Clouds and the Andromeda Galaxy (M31). Foreground extinction by dust in the Milky Way in these lines of sight is generally small and can be accounted for. Results for dust in the Large and Small Magellanic Clouds (LMC and SMC) are illustrated in Figure 3.8. Each set of data points is an average for several lines of sight (Gordon et al. 2003). Estimated mean R_V values extend from 2.7 (SMC bar) to 3.4 (LMC average), well within the observed range for the Milky Way. The general average extinction for the LMC (red) is rather similar to the Milky Way average (blue), the main difference being a reduction in the amplitude of the bump. The "supershell" associated with the 30 Doradus region (orange) exhibits somewhat enhanced UV continuum extinction compared with the LMC average, but is otherwise comparable. Observations of extinction in M31 (Clayton et al. 2015) yield results that span a similar range, including examples that resemble the mean curve for the Milky Way and that of the 30 Doradus region. Far more extreme differences are seen in the results for the SMC bar (green points in Figure 3.8), which exhibit a remarkably steep UV slope and a very weak or absent bump.

A number of factors must be considered in attempting to understand these differences. Compared to the Milky Way, the Magellanic Clouds are gas-rich, metal-poor, and have more vigorous rates of star formation, the SMC being the more extreme of the two (Westerlund 1997). M31 bears a closer resemblance to the Milky Way but appears to be in a phase of lower star formation activity (van den Bergh 1999). Current star formation in the Magellanic Clouds may have been

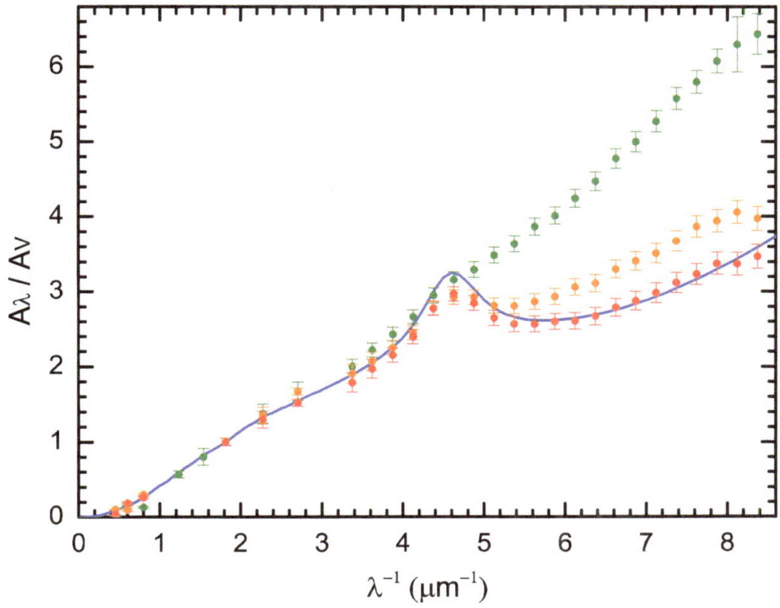

Figure 3.8. Extinction in the Magellanic Clouds (colored points) compared with the average for the Milky Way (blue curve). Results for the SMC bar region are in green. Two distinct data sets are shown for the LMC: a general average (red) and the LMC2 "supershell" region near 30 Doradus (orange). Data are from Gordon et al. (2003).

triggered in part by tidal interactions with the Milky Way, and these dwarf galaxies may thus exhibit a limited form of starburst activity (Section 1.4.6). The SMC bar and the 30 Doradus region in the LMC are, indeed, sites of rapid, high-mass star formation. Another factor that must be considered in this context, however, is observational selection. Because of their greater distances, the stellar samples used to probe extinction in other galaxies naturally tend to favor the most luminous candidates, including a higher proportion of supergiants than is the norm for studies of extinction in the Milky Way. Because they evolve more rapidly, the most luminous stars trace regions of recent star formation more closely than do stars of lesser luminosity, so these regions may tend to be over-represented in the samples. The detection of extinction curves closely *similar* to the Milky Way average in some locations in both M31 and the LMC therefore seems significant and perhaps more generally representative than the differences seen elsewhere.

Star formation regions in the Magellanic Clouds may evolve differently compared with their counterparts of similar mass in the Milky Way. Dust-to-gas ratios are lower, so there is less attenuation of the radiation fields generated by newborn stars. Energetic irradiation and shock waves driven by star formation and subsequent supernova explosions may profoundly influence the optical properties of the dust. Shock waves, in particular, are likely to destroy or modify grains by sputtering and shock-induced grain–grain collisions (Section 9.1.2), potentially increasing the relative numbers of smaller particles that produce extinction most efficiently in the UV. SMC-like extinction is inferred in starburst and ultraluminous infrared galaxies by modeling their spectral energy distributions (Gordon et al. 1997; Shimizu et al. 2011), and also seen in a region of the Milky Way where there is evidence for shock-processing of dust in a region of recent star formation (Valencic et al. 2003). The SMC bar may thus represent a prototypical environment in which this form of processing is dominant.

3.4 The "Bump" Feature

The peak centered near 4.6 μm^{-1} (217.5 nm) in the extinction curve is a conspicuous and pervasive feature of dusty environments, in our own Galaxy and in others. The presence of a corresponding minimum in the albedo confirms that it is an absorption feature, as discussed in Section 3.2.3. This section reviews the observed properties of the bump and assesses their implications for the identity of the absorber.

3.4.1 Overview of Observed Properties

The essential properties of the bump are summarized as follows (e.g., Fitzpatrick & Massa 1986, 2007; Valencic et al. 2004):

- Its *position* is remarkably stable. Star-to-star deviations from the mean wavenumber of the peak, $\langle x_0 \rangle = 4.59 \pm 0.02\ \mu m^{-1}$, do occur but tend to be small, rarely exceeding the estimated uncertainties.
- Its *profile* has a characteristic shape, closely represented by a Drude function (Equation (3.26)).
- Its *width* exhibits significant dispersion about a mean value of $\langle \gamma \rangle = 0.92 \pm 0.03\ \mu m^{-1}$ (full-width half-maximum). Examples are shown in Figure 3.9. Exceptionally broad bumps are more common than exceptionally narrow ones, and broader bumps are often associated with denser environments. There is no evidence of a correlation between position and width.
- Its *strength* is variable, the most extreme examples occurring in regions where it is weak or absent (e.g., Figures 3.8 and 3.10).

An accurate mathematical representation of the profile is useful as an analytical tool and may also provide physical insight. A formulation proposed by Fitzpatrick & Massa (1986), based on the Drude model for the energy states of electrons in metals (Bohren & Huffman 1983, p. 251), has become the standard. The Drude profile is defined in terms of $x = \lambda^{-1}$ as

$$D(x) = \frac{x^2}{(x^2 - x_0{}^2)^2 + \gamma^2 x^2} \tag{3.26}$$

where x_0 and γ specify the peak position and width in wavenumber units. A measure of the strength of the feature per unit E_{B-V} is given by the area under the profile in the normalized extinction curve:

$$A_{\rm bump} = c_3 \int_0^\infty D(x)\, dx = \frac{\pi c_3}{2\gamma} \tag{3.27}$$

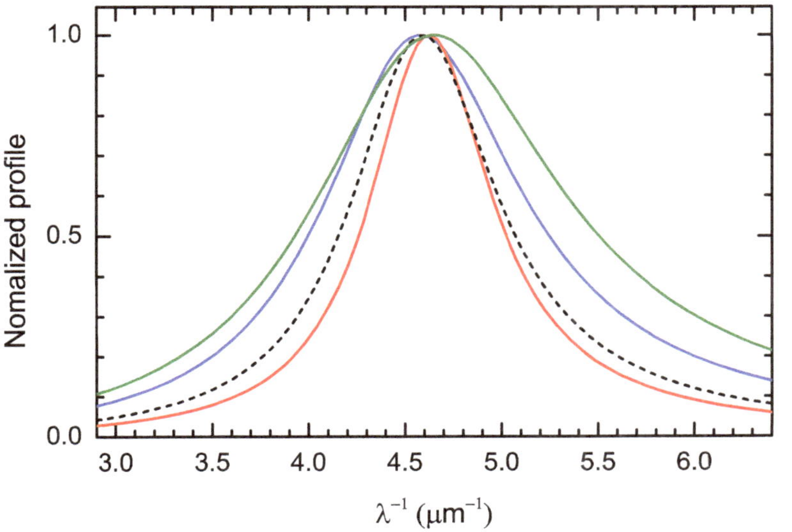

Figure 3.9. Examples of bump profiles toward individual stars exhibiting differences in width. The curves are Drude functions (Equation (3.26)) based on fits to observational data and normalized to unity at the peak. HD 93028 (red, $\gamma = 0.77\ \mu m^{-1}$), a distant star in the Milky Way toward Carina, has one of the narrowest bumps known. Broader bumps are seen toward ζ Oph (blue, $\gamma = 1.25\ \mu m^{-1}$) and especially HD 29647 (green, $\gamma = 1.58\ \mu m^{-1}$), stars associated with local clouds in the Sco-Oph and Taurus regions, respectively. The diffuse–ISM mean profile (dashed curve, $\gamma = 0.92\ \mu m^{-1}$) is shown for comparison.

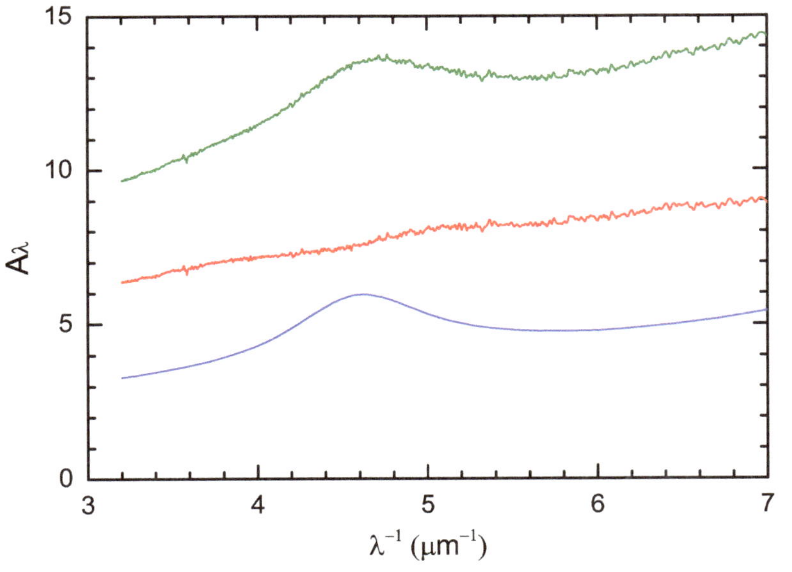

Figure 3.10. Separation of diffuse (blue) and dense (red) components of the total extinction (green) toward the star HD 283809, which lies behind the molecular core TMC–1 in Taurus, illustrating suppression of the bump in the dense core. Data are from Whittet et al. (2004).

where c_3 is the bump amplitude term in the FM formulation (Equation (3.23)). The parameters x_0, γ, and c_3 have been evaluated from empirical fits to observational data for large samples of stars covering a wide range of environments, both in the Milky Way (e.g., Valencic et al. 2004; Whittet et al. 2004; Fitzpatrick & Massa 2007; Gordon et al. 2009) and in local-group galaxies (e.g., Gordon et al. 2003; Cartledge et al. 2005; Clayton et al. 2015).

3.4.2 A Constraint on Particle Size

A remarkable property of the bump is the occurrence of variations in width, unaccompanied by variations in position. This result issues a challenge to any hypothesis for the origin of the feature. With regard to size, specifically, it strongly suggests that the absorption occurs in the small-particle limit, i.e., in particles small compared with the wavelength (effectively $a < 0.01$ μm). This deduction follows from modeling calculations, which show that x_0 increases systematically with size for larger particles because of a scattering contribution to extinction, predominantly on the long-wavelength side of the absorption peak. Any spatial fluctuation in mean grain size would then lead to both star-to-star changes in x_0 and a systematic trend in x_0 with γ, neither of which have been observed. Note, however, that this constraint does not rule out the possibility of larger grains containing graphitic carbon: if they exist they are presumably much less numerous than the bump grains, and make relatively minor contributions to extinction in the mid-ultraviolet (see Section 3.6).

3.4.3 Environmental Effects

Variations in strength and width that synchronize with environment seem likely to provide insight into the origin of the feature. Lines of sight through typical regions of the diffuse ISM in the disk of the Galaxy tend to have relatively strong bumps of average width. Lines of sight toward H II regions and/or OB star clusters tend to display weaker bumps and shallower FUV extinction, NGC 1977 being a typical example (Figure 3.6). In general, the amplitude (c_4) of the FUV rise is poorly correlated with the strength of the bump but does show a significant correlation with its width (see Section 3.4.4). Broader bumps tend to occur in dense environments, such as the Taurus dark cloud, but the converse is not always true: the bump in the mean extinction curve for the Sco-Oph region (Figure 3.6), for example, is similar in both amplitude and width to that of the diffuse ISM, notwithstanding the difference in R_V.

The most dramatic differences occur in situations where the bump is unusually weak, implying that the absorber is being destroyed or modified in a way that suppresses the absorption. Such behavior is seen in contrasting environments. These include not only regions of quasi-starburst activity, such as the SMC bar discussed in Section 3.3.3, but also a cold, quiescent dense cloud in the solar neighborhood. The latter case is illustrated in Figure 3.10. Extinction in the line of sight to the star HD 283809 (green curve), which lies behind the dense core TMC–1 in Taurus, occurs partly in the dense core itself, and partly in an extended envelope with optical properties that more closely resemble those of normal diffuse–ISM dust (see Figure 3.7). These components may be separated with reference to data for another star that lies adjacent to but not directly behind the core (blue curve, an empirical fit to the observations). The difference curve (red) yields a measure of extinction within the core, in which the bump is evidently suppressed. Valencic et al. (2003) used an analogous method to remove foreground extinction toward HD 204827 in the young cluster Trumpler 37, demonstrating the occurrence of SMC-like extinction in this region of the Milky Way.

3.4.4 Are the Bump and the FUV Rise Related?

The bump and the FUV rise are the only significant structures apparent in the ultraviolet extinction curve, sensitive searches for weaker features having yielded negative results (Clayton et al. 2003; Gordon et al. 2009). Are these landmark features related? The observed correlation between the *width* of the bump and the *amplitude* of the FUV rise (Fitzpatrick & Massa 1988) may seem surprising, but it hints at an intriguing possibility: the FUV rise might be the wing of a second absorption feature centered in the extreme ultraviolet (EUV; $\lambda < 91.2$ nm), hidden by the opacity of interstellar hydrogen beyond the Lyman limit (Section 3.1). If the bump and the putative EUV feature are produced by similar mechanisms, correlated changes in their widths may occur (Rouleau et al. 1997). Variations in the width of the EUV feature may then result in an apparent modulation of the FUV extinction amplitude, establishing a correlation with the width of the bump.

3.4.5 Implications for the Identity of the Absorber

Primary requirements for a viable candidate may be summarized as follows:

- The absorber must be cosmically abundant and sufficiently robust to survive in diffuse phases of the ISM.
- It must reproduce the observed position and profile shape of the feature, while accommodating observed variations in width and strength.
- It must not produce significant absorptions at other wavelengths where no feature is observed.

In the decades since its discovery, by far the most widely discussed candidate for the bump has been particles containing graphitic carbon (see Stecher & Donn 1965; Hensley & Draine 2021). Indeed, this remains the leading contender, satisfying almost all of the criteria listed above. Yet, despite decades of research, the precise nature of the particles remains controversial. The remainder of this section presents an assessment of the evidence.

A limit on the *abundance* of the absorber responsible for the bump may be estimated using a method described by Draine (1989, 2003). The strength of the feature, expressed as an equivalent width in frequency units, is related to the parameter $A_{\rm bump}$ (Equation (3.27)) by

$$W_\nu = cA_{\rm bump}E_{B-V}. \tag{3.28}$$

The column density N_X of the absorber is proportional to W_ν and inversely proportional to the intrinsic strength of the feature per absorber, denoted by the oscillator strength, f. Using a value of $A_{bump} \simeq 5.1 \times 10^6$ m^{-1}, deduced from the mean diffuse–ISM extinction curve, and combining Equation (3.28) with Equation (3.12) to express the abundance of the absorber relative to hydrogen, we obtain

$$\frac{N_X}{N_H} \simeq \frac{10^{-5}}{f}. \tag{3.29}$$

The strongest permitted transitions typically have $f \lesssim 1$, and so the abundance of the carrier must be $N_X/N_H \gtrsim 10^{-5}$. With reference to standard abundances (Section 7.1) and excluding noble gases, it must therefore contain one or more elements from the set {C, N, O, Mg, Si, S, Fe}. Of these, S is excluded because it resides primarily in the gas phase and cannot be a major component of the dust. N and O are also excluded because they are electron acceptors. The set of candidates is thus reduced to just four elements: {C, Mg, Si, Fe}. Carbonaceous materials such as graphite therefore pass the abundance test, with limited alternatives. The oscillator strength per carbon atom for graphitic grains in the small-particle limit is $f \sim 0.16$ (Draine 1989); Equation (3.29) thus yields $N_C/N_H \sim 6.3 \times 10^{-5}$, which is ~20% of the available carbon.

Graphite is a familiar crystalline form of solid carbon. It is sufficiently robust to survive for long periods under interstellar conditions—graphitic particles of interstellar provenance are, indeed, found in meteorites. Its crystal structure consists of layered, ordered planar sheets (graphene), each sheet being a two-dimensional lattice of sp^2-bonded carbon atoms arranged in interconnected hexagonal rings. Each atom in a sheet is connected to its three nearest neighbors by a covalent (σ) bond; with only three of the four potential bonding sites satisfied, the fourth (π) valance electron is free to migrate in the plane, making graphite electrically conductive. Absorptions arise from photon-excitation of electrons with respect to the positive ion background of the solid to produce resonance peaks in the dielectric function, a phenomenon referred to in the bulk material as plasma oscillations. Excitation of π electrons produces absorption centered near 220 nm in the mid-ultraviolet, corresponding to the bump feature. Excitation of σ electrons produces another feature, centered near 80 nm in the EUV, the presence of which may have implications for the shape of the extinction curve in the FUV region, as discussed in Section 3.4.4. An important caveat is that hydrogenation suppresses absorption by localizing the electrons (Hecht 1986). If graphite is indeed the source of the bump, the grains responsible cannot be saturated with hydrogen.

Calculations based on Mie theory for small graphite "spheres" were typically used in early attempts to model the bump, but this approach seems unrealistic. With regard to shape, graphite particles minimize their free energy when flattened, such that their shortest axes are perpendicular to the graphene layers; thus, calculations for (e.g.) oblate spheroids are more plausible. With regard to size, the applicability of classical Mie theory as a reliable approach to modeling very small particles is called into question. The realism of calculations based on bulk dielectric functions must inevitably decline as progressively smaller particles are considered: if a solid is repeatedly subdivided, its optical properties will eventually diverge from those of the bulk material as molecular orbital theory takes over from solid-state band theory. The grains invoked to explain the bump may be at an intermediate stage where the proximity of a surface significantly modifies the behavior of internal electrons (see Bohren & Huffman 1983, p. 335). The reliability of the calculations may also be affected by errors in the dielectric function for graphite, which must be determined in planes both parallel and perpendicular to the crystal axis. Surprisingly, these are not ideally well established, with considerable discrepancies existing between data from different studies (see Huffman 1977; Draine & Lee 1984; Draine & Malhotra 1993; and Draine 2016 for detailed discussions). Furthermore, real interstellar graphite grains are unlikely to be perfectly-ordered individual crystals. An alternative approach is to measure extinction directly in the laboratory for manufactured smokes that might contain realistic analogs of interstellar particles (e.g., Stephens 1980), but difficulties are then encountered in determining and controlling the shape, size distribution and degree of crystallinity of the samples. All things considered, the discrete dipole approximation (Section 2.1.4) may provide the most reliable means of modeling the absorption produced by small graphite grains (Draine 1988), allowing the effects of shape and structure to be explored.

How well do such calculations match the observed *position* and *profile shape* of the bump? The answer is very well indeed (Figure 3.11)—provided that certain specific criteria are met. Particles small compared with the wavelength are needed to reproduce its positional stability, as discussed in Section 3.4.2. However, the

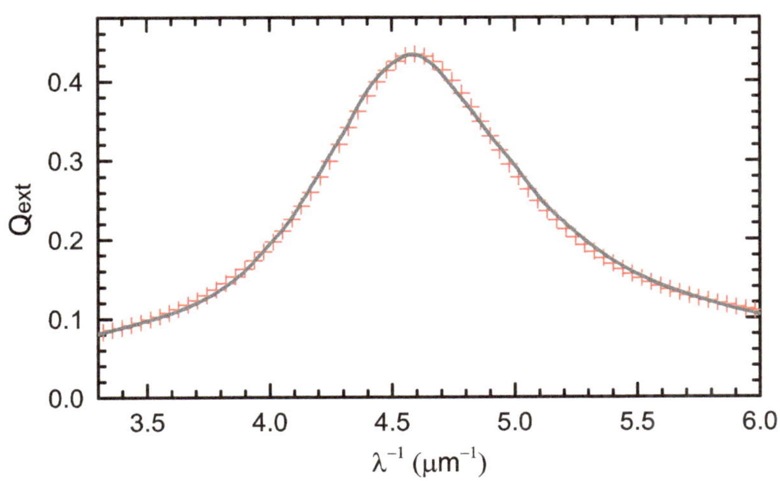

Figure 3.11. Illustration of the predicted bump feature for small graphite spheroids (gray curve) compared with a Drude profile matched to observations (red crosses). Values of the extinction efficiency factor Q_{ext} are calculated for randomly-orientated oblate spheroids of effective radius 3 nm and axial ratio 1.6 (Draine 1989). The Drude profile is superposed on a linear background and normalized to the peak of the calculated curve, its position and width corresponding to the observed mean values in the diffuse ISM.

calculations show that shape as well as size can affect the position and width of the absorption. If oblate spheroids are assumed, an excellent fit to the observed mean profile is obtained in the small-particle limit with an appropriate choice of axial ratio, as illustrated in Figure 3.11. The goodness of fit depends on both the axial ratio and the orientation of the crystal axis relative to the geometric axis, and so the properties of the particles are closely constrained—perhaps to an unrealistic degree.

Perhaps the greatest challenge to the graphite hypothesis (or indeed *any* hypothesis) for the bump is to explain the observed *variations in width* without incurring significant shifts in position. Proposed mechanisms for width variation include compositional inhomogeneities, particle clustering, and surface modifications such as mantle growth. The effects of impurities and porosity were investigated by Perrin & Sivan (1990). Although generally discouraging, their results for graphite with amorphous carbon inclusions do show that changes in width of ~10% arising from different impurity concentrations may be accompanied by changes in position of only ~0.5% in the small-particle limit. Rouleau et al. (1997) show that particle clustering can also result in width changes of appropriate magnitude, but the peak position of the observed feature is not well-matched in the small-particle limit in their calculations. Draine & Malhotra (1993) suggest that width changes may be caused by some systematic modification of the optical properties of the graphitic material itself: possibilities include variations in hydrogenation or crystallinity, or changes in electronic structure caused by adsorption and desorption of surface atoms. Deposition of surface coatings (mantles) onto the particles may broaden the feature (Mathis 1994): if narrow bumps are associated with uncoated grains, the broader bumps observed in denser regions may result from mantle growth. However, to obtain realistic fits in the small-particle limit, both the shapes of the cores and the optical properties of the mantles must be closely constrained: the mantles must be absorptive, containing (e.g.) metals or PAHs, whereas icy mantles are excluded.

The properties of carbonaceous grains in space are naturally governed by the processes that create and modify them. Likely sources include stardust condensing in the outflows of carbon-rich evolved stars (Section 8.1), and attachment of C atoms or C-bearing molecules onto existing grains in the ISM (Section 9.2.2). Grains forming in these environments are expected to be partially or predominantly amorphous (disordered) in structure rather than crystalline. Graphitization of amorphous carbon requires an input of energy, which may result from long-term exposure to ultraviolet photons in the ISM (Section 9.1.3). It is thus plausible to presume (e.g., Mennella et al. 1998) that the bump absorbers may be UV-processed, hydrogenated amorphous carbon grains. Although lacking long-range order, such particles are considered to be polycrystalline, containing a mix of hybridization states that include graphitic subunits. Spectra of laboratory analogs demonstrate the emergence of a bump-like feature in amorphous carbon with a strength that increases with the degrees of UV exposure (Gadallah et al. 2011). Variations in the width of the feature may be attributed to the degree of clustering of the particles (Schnaiter et al. 1998) or to changes in the electronic band structure in response to their previous thermal histories (Papoular & Papoular 2009).

A distinct but closely related proposal for the origin of the bump attributes it to polycyclic aromatic hydrocarbons (PAHs; e.g., Joblin et al. 1992; Duley & Seahra 1998). PAHs resemble fragments of graphene in structure, consisting of two-dimensional arrays of hexagonal rings with attached peripheral H atoms. They are known to be pervasive in the ISM from observations of their infrared emission features (Section 6.2.2). Interstellar PAHs are presumed to exist in a wide variety of shapes and sizes, but the spectra identify them as a class rather than as specific molecules. Given their structural relationship to graphite, similarities in their contributions to UV extinction are not unexpected. However, laboratory investigations show that different species have differing UV spectra that may include sharp features, in contrast to the observed smoothness of the bump profile and the general extinction curve. In addition to absorptions near 210 nm overlapping the bump, these may include absorptions in the 240–400 nm wavelength interval where no corresponding structure is observed. If PAHs are major contributors to the UV extinction, the population must be diverse, such that overlapping features from different species combine to produce smooth, quasi-continuous absorption (Steglich et al. 2011, 2012). Smoothing may be enhanced by clustering of PAHs into nanoparticles bonded by van der Waals forces. Individual PAHs, PAH clusters and graphitic dust may represent a continuous size distribution of carbonaceous nanoparticles.

Can a model based on such particles account for the observed suppression of the bump in certain environments? In the case of the SMC (Section 3.3.3) it is not clear whether the extreme weakness or absence of the bump results from differences in elemental abundances, physical conditions, or some combination. Compared with the Milky Way, the SMC is a low-metallicity system in which C is deficient (Jenkins & Wallerstein 2017), suggesting that bump suppression results from a dearth of carbonaceous grains. However, a closely similar extinction curve is observed in a region of the Milky Way disk, where elemental abundances are expected to be normal (Valencic et al. 2003). In the latter case, the dust may have been processed by shocks, but is not clear how this leads to suppression of the bump. In the case of the dense core TMC–1 (Figure 3.10), the weakness or absence of the bump is attributed to suppression of π-electron excitation in graphite by attachment of atomic hydrogen to the grains (Whittet et al. 2004). The contrast between clouds of comparable densities in Taurus, where the bump is suppressed, and in Sco-Oph, where it is nearly normal, is attributed to different levels of irradiation: embedded OB stars in the Sco-Oph region enhance the ambient radiation field, causing photodesorption of H atoms from the grains and thereby maintaining the strength of the bump.

Are there viable alternatives to a carbon-based carrier? Other possibilities have been considered, but none are compelling. Any dielectric material will absorb continuously in the UV at sufficiently high photon energies, due to excitation of electrons to the conduction band. If the onset of absorption coincides with a rapid decrease in scattering, the net effect can be to simulate a broad peak in extinction near the absorption edge. However, extreme fine-tuning of the particle size to within a specific narrow range is needed for this to occur, which seems highly implausible even if a suitable candidate material can be identified (Huffman & Stapp 1971 proposed enstatite spheres of radius $a = 0.06$ μm). The most widely discussed non-carbon-based proposal is absorption associated with OH^- ions at low-coordination sites on the surfaces of small silicate grains (Steel & Duley 1987); but no detailed laboratory measurements have been obtained to support this possibility, and an unreasonably large number of such sites may be required to account for the observed feature (Mathis 1994).

To summarize, graphitic nanoparticles are by far the most likely agent responsible for the bump. Such particles absorb by excitation of π-electrons, and reproduce the observed feature in the small-particle limit subject to certain assumptions. Observational confirmation might be provided by searches for other predicted spectral signatures of graphite. The EUV feature near 80 nm and its possible contribution to the FUV extinction was previously noted; graphite also has an infrared feature at 11.5 μm (Draine 1984), but it is intrinsically very weak and observational confirmation may not be feasible. Details regarding the nature of the absorbing particles, such as their typical structure and degree of crystallinity, remain uncertain. Further laboratory simulations and modeling calculations may be helpful, for example to test the various mechanisms proposed to explain changes in the width of the bump.

3.5 Discrete Structure in the Visible

At first glance, the extinction curve may appear smooth, exhibiting little evidence of structure between the bump in the ultraviolet and the silicate features in the infrared (e.g., Figures 3.2 and 3.5). However, sensitive spectrophotometric observations have demonstrated the occurrence of discrete absorptions and extended structure at visible and near-infrared wavelengths. It is convenient to consider these features in two groups according to width: the diffuse interstellar bands (DIBs), which have widths typically in the range 0.1–3 nm (full-width half-maximum), and broader structure, typically 30–100 nm in extent.

3.5.1 Diffuse Interstellar Bands

The presence of diffuse absorption features in the optical spectra of reddened stars has been known for more than a century. Their identity is a long-standing problem that has simultaneously fascinated and frustrated researchers for many decades (see Herbig 1995 and Krełowski 2018 for reviews). An interstellar origin was first indicated by their lack of periodic Doppler motion in the spectra of spectroscopic binary stars, and further established by correlations between their strengths and generic measures of interstellar matter, such as reddening or hydrogen column density. Several hundred DIBs have been detected, spanning wavelengths from the violet to the near-infrared (Figure 3.12).

The widths of the DIBs in comparison to typical gas-phase interstellar lines prompted proposals for a solid-state carrier in some component of the dust. For various reasons this is now considered unlikely. Features produced by relatively large grains in the size distribution should exhibit systematic changes in both profile shape and central wavelength with grain size, as discussed in the previous section

Figure 3.12. The diffuse interstellar bands. Relative intensity (white) is plotted over the color spectrum, with wavelength increasing linearly from right to left in the range 370–880 nm. Image credit: NASA/Peter Jenniskens and François-Xavier Désert, based on data from Jenniskens & Désert (1994).

with regard to the bump feature, and emission wings are also predicted; but no such effects have been observed. Searches for corresponding polarimetric structure have also yielded negative results, indicating that the absorbers cannot be common constituents of the aligned grains responsible for interstellar polarization. Absorbers hosted by grains in the small-particle limit might be consistent with these constraints. However, such particles produce extinction most efficiently at shorter wavelengths, so correlations are predicted between DIB strengths and measures of UV extinction such as the amplitudes of the bump, the UV continuum extinction, and the FUV rise. A comprehensive investigation by Xiang et al. (2017) found no evidence for any such correlation.

Although numerous, the DIBs are weak, e.g., in comparison to the bump feature, thus the absorbers need not be very abundant. The most plausible candidates appear to be large carbon-based gas-phase molecules containing ~10–100 atoms, such as PAHs, long carbon-chain molecules, or fullerenes. That PAHs are considered to be candidates for both the bump (Section 3.4.5) and the DIBs might seem suggestive, but laboratory studies as well as observational data suggest that their carriers cannot be one and the same (Steglich et al. 2011; Xiang et al. 2017). Indeed, the sheer number of known DIBs, their diversity in width, and their spectral distribution (Figure 3.12) strongly suggest that multiple carriers are required: a single species of forbidding complexity would be needed to account for all of them. Furthermore, statistical investigations have established the occurrence of DIB "families" (Krełowski 2018): the members of a family correlate closely in strength with each other but far less well with other DIBs, suggesting that each family is associated with a different absorber. A pair of adjacent features in the near-infrared has been attributed to C_{60}^{+} (ionized buckminsterfullerene; Campbell et al. 2015), but this is disputed. Krełowski (2018) concludes in his review that no proposed identification to date for any DIB can be considered secure.

3.5.2 Broadband Structure

In common with the DIBs, broadband structure represents a long-standing, unsolved problem in astronomical spectroscopy (see York 1971; van Breda &

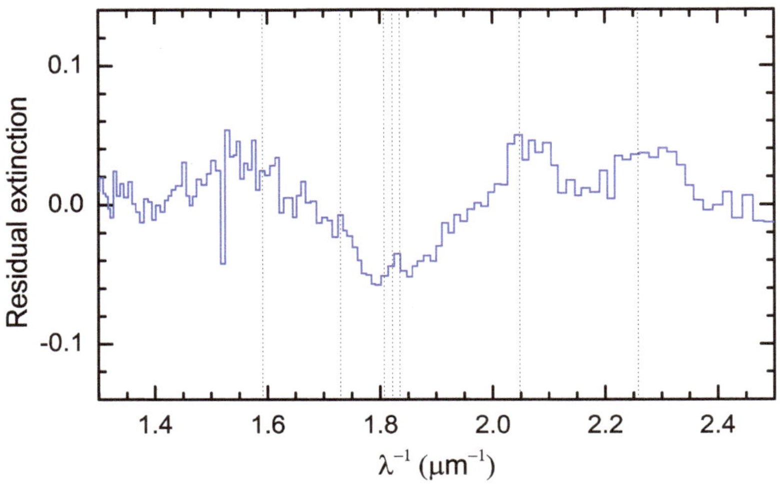

Figure 3.13. Broadband structure in the visible extinction. Residuals obtained by fitting a fourth-order polynomial to the mean $E_{\lambda-V}/E_{B-V}$ extinction curve (Massa et al. 2020) are shown in blue. The dip near 1.52 μm^{-1} corresponds to the Hα line, attributed to slight mismatches between observed and comparison spectra (Section 3.1). Vertical dotted lines indicate the positions of some of the stronger diffuse interstellar bands, centered (right to left) at wavelengths 442.8, 488.2, 545.0, 548.8, 553.5, 578.0, and 628.4 nm.

Whittet 1981; Massa et al. 2020). It appears as a shallow modulation in extinction, best displayed as residuals with respect to a smooth functional representation of the curve. Results based on a polynomial fit to the mean diffuse–ISM extinction (Figure 3.2) are shown in Figure 3.13. DIBs produce narrower sub-peaks in the data: the positions of some of the stronger ones are indicated. The overall structure of the residual curve, often referred to in the literature as "very broadband structure" (VBS), may be considered as three distinct peaks, centered near wavenumbers 1.58, 2.05 and 2.28 μm^{-1}. In this interpretation, the most prominent feature of the VBS—the dip centered near 1.82 μm—is simply the minimum between two peaks.

The shapes of the peaks in Figure 3.13 may be represented by Drude profiles (Massa et al. 2020). Their widths and strengths vary from one line of sight to another, but their positions appear to remain stable, behavior akin to that of the bump feature (Section 3.4.1). Indeed, the mean amplitude of the VBS, represented by the root-mean-square value of the residuals, correlates with the strength of the bump. It does not correlate with R_V, suggesting that the variations are independent of the size of the grains producing the underlying continuum extinction. Although these results do not establish a common origin for the VBS and the bump, they do suggest that a similar mechanism may be at work and that the VBS most likely arises in some component of the (small) dust population.

The identity of the ingredient(s) responsible for broadband structure remains unknown. It seems clear from the observations that the VBS profile is the product of (at least) three broad, shallow components, most probably absorption features. A proposal that it might be caused by an emission (luminescence) feature centered near 1.8 μm^{-1} has been ruled out (Jenniskens 1994). Structure in the dielectric functions of continuous absorbers such as magnetite (Fe_3O_4) has also been discussed (Huffman 1977), but attempts to model the profile are unconvincing (Millar 1982). Jenniskens (1994) proposed that an onset of continuous absorption in small amorphous carbon grains might account for the rise in extinction residuals at $\lambda^{-1} > 1.8\ \mu m^{-1}$, but this possibility needs to be reconsidered in the light of the more detailed observations now available.

3.6 Modeling the Extinction Curve

The general philosophy used to develop models describing the observed properties of interstellar dust is summarized in Section 1.2.7. To construct a model for extinction, specifically, one must make initial assumptions regarding the size, composition, shape and structure of the particles. Size is represented by a distribution function $n(a)$, and composition is specified by the dielectric function $\epsilon(\lambda)$ of the material. Methods described in Section 2.1 are then used to compute the extinction. Results are compared with observed extinction curves, and the initial assumptions may then be adjusted to improve and ultimately optimize the fit. Successful models must also be consistent with other evidence, such as the abundances of the condensible elements, the emissive properties of the dust, and the occurrence of spectral features arising in the particles. All things considered, it is clear that a viable model must include a minimum of two distinct constituents, of which one is presumed to be C-rich (e.g., graphite), the other O-rich (e.g., silicates).

3.6.1 The Size Distribution

The size distribution is specified in terms of an appropriate grain dimension, such as the radius of a sphere (a). The full range of particle sizes to be included in the model is sectioned into increments, and the distribution function $n(a)$ specifies the relative numbers of particles within each increment $a \rightarrow a + da$. Calculated extinction curves are particularly sensitive to the form and range of applicability assumed for $n(a)$, leading to useful constraints. Power-law representations have been found to give generally satisfactory fits, as first shown by Mathis et al. (1977) and subsequently adopted by other investigators. Known as the MRN distribution, this takes the form

$$n(a) = K\, a^{-q} \qquad [a_{\rm min} < a < a_{\rm max}] \tag{3.30}$$

where K is a constant, q is the power-law index, and $n(a) = 0$ outside of the stated limits. Using a two-component (graphite-silicate) model, Mathis et al. (1977) obtained a good fit to the observed mean extinction curve with $q \simeq 3.5$ and size ranges 0.005–1 μm for graphite and 0.025–0.25 μm for silicates. Extinction curves with higher values of R_V are generally consistent with increases in $a_{\rm min}$ and/or $a_{\rm max}$ for the silicate component (Mathis & Wallenhorst 1981). The size distribution of the graphite component is more tightly constrained by the observed properties of the bump feature, especially its positional stability.

A power-law form for $n(a)$ is physically reasonable, as it is consistent with predictions for particles subject to collisional fragmentation (Biermann & Harwit

1980; Jones et al. 1996). The assumption of hard limits at a_{min} and a_{max} is clearly unphysical, however. Mid-infrared emission from the dust (Chapter 6) implies the presence particles smaller than the a_{min} cut-off assumed in these calculations: a continuum of sizes from large molecules to small grains seems more plausible, especially for the carbonaceous component (PAHs → graphite). Similarly, a smooth decline in the numbers of larger particles is preferable to an abrupt cut-off at a_{max} (see Section 8.4.2 for related discussion). Rather that adopting a specific model for the size distribution, one may consider the modeling process as an inverse problem, in which observed extinction curves enable size distributions to be derived for grain populations of assumed composition (e.g., Kim et al. 1994; Weingartner & Draine 2001; Zubko et al. 2004).

3.6.2 Composition and Structure

Graphite and silicates are the primary constituents of most grain models for several reasons. They are composed of chemical elements from the set {C, O, Mg, Si, Fe} that make up most of the mass of dust in the ISM (Section 1.2.6). Graphitic nanoparticles or PAHs seem to provide the only viable explanation for the bump feature (Section 3.4.5), and silicates are similarly implicated by their spectral signatures (Figure 3.5). The outflows of evolved stars are likely sources of interstellar dust, and they naturally produce distinct populations of C-rich or O-rich particles according to their intrinsic chemistries (Chapter 8). Two-component models based on these materials are thus logical, and attractive for their apparent simplicity. However, they may not fully represent the structural and chemical complexities of real interstellar grains, as these materials must inevitably interact and become blended to some extent in the crucible of interstellar space. Accretion and coagulation processes may produce composite and core-mantle grains that include both carbon and silicates, on which ice mantles may accumulate inside molecular clouds, and the optical properties of all constituents evolve as the result of thermal and energetic processing.

Various approaches have been adopted in the quest for more realistic models. All have in common "bump" grains composed of graphitic nanoparticles or PAHs, but propose differing combinations and structures for grains producing the general continuum extinction. These include bare (uncoated) silicate and large carbon grains (Weingartner & Draine 2001; Zubko et al. 2004); silicate cores with carbonaceous mantles composed of organic refractory matter or amorphous carbon (Li & Greenberg 1997; Jones et al. 2013); and composite particles containing mixtures of carbon, silicates, water-ice, and voids (Zubko et al. 2004). An illustrative example is shown in Figure 3.14 (a number of other models yield broadly similar results). All make heavy demands on the reservoir of available elements, as determined by abundance and depletion data reviewed in Chapter 7.

The optical properties of silicates place common constraints on the models. As good dielectrics, silicates produce extinction primarily by scattering, and are therefore expected to contribute substantially to general extinction across the spectrum in order to explain observations of the albedo and phase function (Section 3.2.3). However, it is clear that a two-component model combining silicate grains with

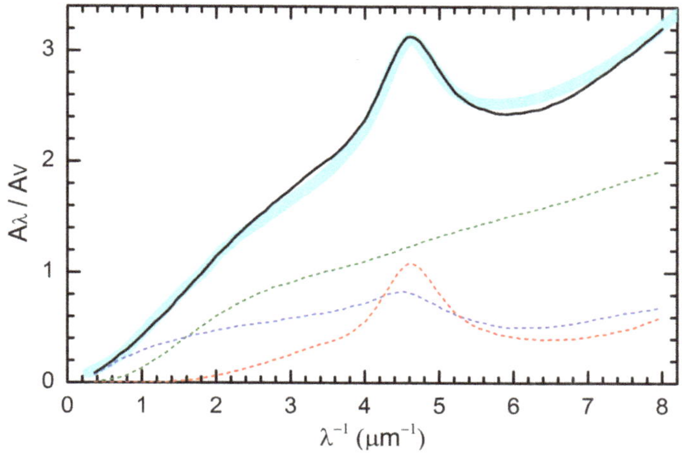

Figure 3.14. Example of a model extinction curve (black) compared with the observed diffuse–ISM mean (thick cyan curve). Individual components of the model are shown as dashed curves: silicates (green), small "bump" grains (red), and large carbonaceous grains (blue). Adapted from Weingartner & Draine (2001), based on their "Case A" model for $R_V = 3.1$ and a C/H abundance ratio of 6×10^{-5} in small grains.

small "bump" grains cannot account for the entire extinction curve. This is the case because of limits placed by both the abundance of Si and the observed silicate absorption features, which should be significantly stronger if silicates were the dominant source of general extinction. Models typically maximize the silicate extinction by assuming that all the available Si is included in compounds such as $MgSiO_3$ and $MgFeSiO_4$ within grains of appropriate size, but another component is still needed to make up the shortfall. Solids abundant enough to do so seem certain to be carbonaceous, at least in the diffuse ISM, and the various models referenced above propose different formats for this material: refractory mantles on silicates, inclusions in silicates, or a distinct population. In the example shown in Figure 3.14 it takes the form of "large" carbonaceous grains (blue curve). If these grains contain graphite, as in this example, they produce a bump feature that is broader and much shallower than that produced by the small "bump" grains. The size distribution must then be fine-tuned in order to maintain the observed stability of the bump position when extinction curves with different R_V values are considered: in effect, this constraint requires the distribution to be bimodal across all forms of solid carbon (see Weingartner & Draine 2001; Zubko et al. 2004).

The ultimate goal—to converge upon a unique model—has yet to be accomplished. Different models consistent with all known constraints are capable of matching the observations. Significant progress has been made, nevertheless: the nature of the particles responsible for the bump in the ultraviolet seem reasonably well established, and silicates implicated by absorption features in the infrared are an essential ingredient of all models. Paradoxically, it is the particles responsible for extinction in the visible, under investigation for the longest time, that remain the least well understood. Core-mantle grains, composite grains, and separate populations of carbon and silicate dust are all represented in extant models.

References

Baade, W., & Minkowski, R. 1937, ApJ, 86, 123
Biermann, P., & Harwit, M. 1980, ApJ, 241, L105
Bohlin, R. C., Savage, B. D., & Drake, J. F. 1978, ApJ, 224, 132
Bohren, C. F., & Huffman, D. R. 1983, Absorption and Scattering of Light by Small Particles (New York: Wiley)
Campbell, E. K., Holz, M., Gerlich, D., & Maier, J. P. 2015, Natur, 523, 322
Cardelli, J. A., Clayton, G. C., & Mathis, J. S. 1989, ApJ, 345, 245
Cartledge, S. I. B., Clayton, G. C., Gordon, K. D., et al. 2005, ApJ, 630, 355
Clayton, G. C., & Fitzpatrick, E. L. 1987, AJ, 93, 157
Clayton, G. C., Gordon, K. D., Salama, F., et al. 2003, ApJ, 592, 947
Clayton, G. C., Gordon, K. D., Bianchi, L. C., et al. 2015, ApJ, 815, 14
Decleir, M., Gordon, K. D., Andrews, J. E., et al. 2022, ApJ, 930, 15
Draine, B. T. 1984, ApJ, 277, L71
Draine, B. T. 1988, ApJ, 333, 848
Draine, B. T. 1989, in IAU Symp. 135, Interstellar Dust, ed. L. J. Allamandola, & A. G. G. M. Tielens (Dordrecht: Kluwer), 313
Draine, B. T. 2003, ARA&A, 41, 241
Draine, B. T. 2016, ApJ, 831, 109
Draine, B. T., & Lee, H. M. 1984, ApJ, 285, 89
Draine, B. T., & Malhotra, S. 1993, ApJ, 414, 632
Drilling, J. S., & Landolt, A. U. 2000, in Allen's Astrophysical Quantities, ed. A. N. Cox (4th ed.; New York: Springer), 381
Ducati, J. R., Bevilacqua, C. M., Rembold, S. B., & Ribeiro, D. 2001, ApJ, 558, 309
Duley, W. W., & Seahra, S. 1998, ApJ, 507, 874
Fitzpatrick, E. L., & Massa, D. 1986, ApJ, 307, 286
Fitzpatrick, E. L., & Massa, D. 1988, ApJ, 328, 734
Fitzpatrick, E. L., & Massa, D. 1990, ApJS, 72, 163
Fitzpatrick, E. L., & Massa, D. 2005, AJ, 130, 1127
Fitzpatrick, E. L., & Massa, D. 2007, ApJ, 663, 320
Fitzpatrick, E. L., & Massa, D. 2009, ApJ, 699, 1209
Fitzpatrick, E. L., Massa, D., Gordon, K. D., Bohlin, R., & Clayton, G. C. 2019, ApJ, 886, 108
Gadallah, K. A. K., Mutschke, H., & Jäger, C. 2011, A&A, 528, A56
Giannetti, A., Leurini, S., Konig, C., et al. 2017, A&A, 606, L12
Gordon, K. D. 2004, in ASP Conf. Ser. 309, Astrophysics of Dust, ed. A. N. Witt, G. C. Clayton, & B. T. Draine (San Francisco, CA: ASP), 77
Gordon, K. D., Calzetti, D., & Witt, A. N. 1997, ApJ, 487, 625
Gordon, K. D., Clayton, G. C., Misselt, K. A., Landolt, A. U., & Wolff, M. J. 2003, ApJ, 594, 279
Gordon, K. D., Cartledge, S., & Clayton, G. C. 2009, ApJ, 705, 1320
Gordon, K. D., Misselt, K. A., Bouwman, J., et al. 2021, ApJ, 916, 33
Green, G. M., Schlafly, E. F., Finkbeiner, D., et al. 2018, MNRAS, 478, 651
Green, G. M., Schlafly, E., Zucker, C., Speagle, J. S., & Finkbeiner, D. 2019, ApJ, 887, 93
Guo, H.-L., Chen, B.-Q., Yuan, H.-B., et al. 2021, ApJ, 906, 47
Hamden, E. T., Schiminovich, D., & Seibert, M. 2013, ApJ, 779, 180

Harris, D. H. 1973, in IAU Symp. 52, Interstellar Dust and Related Topics, ed. J. M. Greenberg, & H. C. van de Hulst (Dordrecht: Reidel), 31
He, L., Whittet, D. C. B., Kilkenny, D., & Spencer Jones, J. H. 1995, ApJS, 101, 335
Hecht, J. H. 1986, ApJ, 305, 817
Heck, A., Egret, D., Jaschek, M., & Jaschek, C. 1984, IUE Low-Dispersion Spectra Reference Atlas–Part 1: Normal Stars (ESA publication SP-1052)
Hensley, B. S., & Draine, B. T. 2021, ApJ, 906, 73
Herbig, G. H. 1995, ARA&A, 33, 19
Huffman, D. R. 1977, AdPhy, 26, 129
Huffman, D. R., & Stapp, J. L. 1971, NPhS, 229, 45
Indebetouw, R., Mathis, J. S., Babler, B. L., et al. 2005, ApJ, 619, 931
Jenkins, E. B., & Wallerstein, G. 2017, ApJ, 838, 85
Jenniskens, P. 1994, A&A, 284, 227
Jenniskens, P., & Désert, F.-X. 1994, A&AS, 106, 39
Joblin, C., Léger, A., & Martin, P. 1992, ApJ, 393, L79
Jones, A. P., Tielens, A. G. G. M., & Hollenbach, D. J. 1996, ApJ, 469, 740
Jones, A. P., Fanciullo, L., Köhler, M., et al. 2013, A&A, 558, A62
Kim, S. H., Martin, P. G., & Hendry, P. D. 1994, ApJ, 422, 164
Krełowski, J. 2018, PASP, 130, 071001
Lallement, R., Babusiaux, C., Vergely, J. L., et al. 2019, A&A, 625, A135
Larson, K. A., & Whittet, D. C. B. 2005, ApJ, 623, 897
Li, A., & Greenberg, J. M. 1997, A&A, 323, 566
Lindegren, L., Bastian, U., Biermann, M., et al. 2021, A&A, 649, A4
Majewski, S. R., Zasowski, G., & Nidever, D. L. 2011, ApJ, 739, 25
Martin, P. G., & Whittet, D. C. B. 1990, ApJ, 357, 113
Massa, D., Fitzpatrick, E. L., & Gordon, K. D. 2020, ApJ, 891, 67
Mathis, J. S. 1994, ApJ, 422, 176
Mathis, J. S., & Lee, C. W. 1991, ApJ, 376, 490
Mathis, J. S., & Wallenhorst, S. G. 1981, ApJ, 244, 483
Mathis, J. S., Cohen, D., Finley, J. P., & Krautter, J. 1995, ApJ, 449, 320
Mathis, J. S., Rumpl, W., & Nordsieck, K. H. 1977, ApJ, 217, 425
Mattila, K., Haas, M., Haikala, L. K., et al. 2018, A&A, 617, A42
Mennella, V., Colangeli, L., Bussoletti, E., Palumbo, P., & Rotundi, A. 1998, ApJ, 507, L177
Millar, T. J. 1982, MNRAS, 200, 527
Pagani, L., Steinacker, J., Bacmann, A., & Stutz, A. 2010, Sci, 329, 1622
Papoular, R. J., & Papoular, R. 2009, MNRAS, 394, 2175
Perrin, J. M., & Sivan, J. P. 1990, A&A, 228, 238
Planck Collaboration, 2014, A&A, 571, A11
Purcell, E. M. 1969, ApJ, 158, 433
Rachford, B. L., Snow, T. P., Destree, J. D., et al. 2009, ApJS, 180, 125
Rouleau, F., Henning, Th., & Stognienko, R. 1997, A&A, 322, 633
Saajasto, M., Juvela, M., & Malinen, J. 2018, A&A, 614, A95
Sandage, A. 1975, PASP, 87, 853
Sandage, A. 1976, PASP, 88, 367
Schlafly, E. F., Meisner, A. M., Stutz, A. M., et al. 2016, ApJ, 821, 78
Schlegel, D. J., Finkbeiner, D. P., & Davis, M. 1998, ApJ, 500, 525

Schnaiter, M., Mutschke, H., Dorschner, J., Henning, Th., & Salama, F. 1998, ApJ, 498, 486
Shimizu, T., Kawara, K., Sameshima, H., et al. 2011, MNRAS, 418, 625
Siebenmorgen, R., Krełowski, J., Smoker, J., Galazutdinov, G., & Bagnulo, S. 2020, A&A, 641, A35
Smith, R. K., & Dwek, E. 1998, ApJ, 503, 831
Spitzer, L. 1978, Physical Processes in the Interstellar Medium (New York: Wiley)
Stecher, T. P., & Donn, B. 1965, ApJ, 142, 1681
Steel, T. M., & Duley, W. W. 1987, ApJ, 315, 337
Steglich, M., Bouwman, J., & Huisken, F. 2011, ApJ, 742, 2
Steglich, M., Carpentier, Y., Jäger, C., et al. 2012, A&A, 540, A110
Stephens, J. R. 1980, ApJ, 237, 450
Valencic, L. A., & Smith, R. K. 2015, ApJ, 809, 66
Valencic, L. A., Clayton, G. C., Gordon, K. D., & Smith, T. L. 2003, ApJ, 598, 369
Valencic, L. A., Clayton, G. C., & Gordon, K. D. 2004, ApJ, 616, 912
van Breda, I. G., & Whittet, D. C. B. 1981, MNRAS, 195, 79
van den Bergh, S. 1999, A&ARv, 9, 273
Vrba, F. J., Coyne, G. V., & Tapia, S. 1993, AJ, 105, 1010
Wang, S., & Chen, X. 2019, ApJ, 877, 116
Weingartner, J. C., & Draine, B. T. 2001, ApJ, 548, 296
Westerlund, B. E. 1997, The Magellanic Clouds (Cambridge: Cambridge Univ. Press)
Whittet, D. C. B., Gerakines, P. A., Hough, J. H., & Shenoy, S. S. 2001, ApJ, 547, 872
Whittet, D. C. B., Shenoy, S. S., Clayton, G. C., & Gordon, K. D. 2004, ApJ, 602, 291
Witt, A. N. 1988, in Dust in the Universe, ed. M. E. Bailey, & D. A. Williams (Cambridge: Cambridge Univ. Press), 1
Witt, A. N. 1989, in IAU Symp. 135, Interstellar Dust, ed. L. J. Allamandola, & A. G. G. M. Tielens (Dordrecht: Kluwer)
Witt, A. N., Smith, R. K., & Dwek, E. 2001, ApJ, 550, L201
York, D. G. 1971, ApJ, 166, 65
Xiang, F. Y., Li, A., & Zhong, J. X. 2017, ApJ, 835, 107
Xue, M., Jiang, B. W., Gao, J., et al. 2016, ApJS, 224, 23
Zasowski, G., Majewski, S. R., Indebetouw, R., et al. 2009, ApJ, 707, 510
Zhu, H., Tian, W., Li, A., & Zhang, M. 2017, MNRAS, 471, 3494
Zubko, V., Dwek, E., & Arendt, R. G. 2004, ApJS, 152, 211

AAS | IOP Astronomy

Dust in the Galactic Environment (Third Edition)

D C B Whittet

Chapter 4

The Observed Properties of Dust—II. Polarization

The observed linear polarization of starlight demonstrates that the interstellar medium (ISM) is optically *dichroic*, i.e., the amplitude of extinction varies with the orientation of the electric vector in the transmitted beam. This result implies that the dust grains responsible for at least one component of the extinction have anisotropic optical properties and a preferred orientation in space. Thermal radiation emitted by these grains is similarly polarized, but in the orthogonal direction (see Section 1.2.4 for an introductory overview and Figure 2.4 for a simple schematic). If the direction of alignment changes along the line of sight, the net polarization may also include a circular component. The physical basis for interstellar polarization and the parameters used to quantify it are discussed in Section 2.2 (anisotropic extinction) and Section 2.3.5 (polarized emission). The present chapter focuses on observational results and their implications. These studies are important for several reasons. They provide information on the galactic magnetic fields that sculpt patterns of alignment, and they shed further light on the properties of the grains themselves. Furthermore, polarizing dust in our Galaxy is the screen through which we observe polarization in the cosmic microwave background (e.g., Mangilli et al. 2021), a topic beyond the scope of this book.

4.1 Polarization and Magnetic Fields

Spiral galaxies are permeated by magnetic fields (see Beck 2016 for a review). Although weak by terrestrial standards, interstellar magnetic fields have a major impact on galactic evolution.[1] Their origins remain uncertain, but they are thought to be sustained by a dynamo mechanism arising from interactions between

[1] The mean flux density of the interstellar magnetic field in the solar neighborhood is $B \simeq 5 \times 10^{-10}$ T. Note that B is usually expressed in the cgs unit μG (micro-Gauss) in the astronomical literature (1 $\mu G = 10^{-10}$ T).

doi:10.1088/2514-3433/ac7204ch4

differential galactic rotation, local turbulence, and expanding gas flows driven by supernova explosions. This process leads to complex field patterns that may be modeled as the summation of a uniform component, with long-range order, and a random component, attributed to local disturbances on size scales ~10–100 pc. Observations of polarization by dust represent a major means of elucidating this structure, both on macroscopic scales and within individual clouds and protostellar regions.

Some of the relevant physics is not yet fully understood, but it has long been accepted that grain alignment in the ISM is largely governed by magnetic fields. A spinning grain of asymmetric shape tends to rotate about its principal axis of inertia (Section 4.3.1), its angular momentum perpendicular to its longest dimension: so a long cylinder spins end-over-end, an oblate spheroid spins like a top, and a disk spins like a record on a turntable. Interactions between the internal magnetism of the particle and the external magnetic field, **B**, then cause its spin axis to align such that the longest dimension is on average perpendicular to **B** (Figure 2.4). Polarization measurements thus trace the magnetic field. In this section it is assumed that the observed polarization is parallel to **B** for dichroic extinction and perpendicular to **B** for thermal emission. Independent confirmation that this is correct arises from observations of synchrotron radiation (e.g., Andersson et al. 2015; Jones et al. 2020). A more detailed discussion of the alignment process is deferred to the final section.

4.1.1 Polarization Efficiency and Depolarization

Polarimetry provides a valuable tool for mapping magnetic fields, but it is important to bear in mind that a typical line of sight may sample a wide range of environments and morphologies, and this may impact the results. Whereas extinction is a scalar quantity for which the total is the sum of all incremental contributions, polarization is effectively a vector yielding a sum that depends on geometric factors as well as on the properties of the grains and their degree of alignment. A consequence of this difference may be seen in the observed relationship between the amplitudes of polarization and extinction, shown in Figure 4.1 for representative samples of reddened stars in the Milky Way. The scatter is far greater than can be accounted for by observational error; but it is consistent with expected variations in polarization efficiency, specified by the ratio p/A (Section 2.2.3). The vacant area toward the upper left in Figure 4.1 suggests an upper boundary to the distribution, approximated by the diagonal line. This limit may correspond to the ideal situation in which two circumstances combine: (i) physical conditions are conducive to optimal alignment, and (ii) the magnetic field is uniform and perpendicular to the line of sight, such that the grains align in the plane of the sky. In all other situations, the net polarization is reduced relative to extinction, as the result of poorer alignment, geometric depolarization, or most likely a combination of both.

Polarized submillimeter emission from some regions of low extinction imply somewhat higher polarization efficiencies than the "classical" limit of Serkowski et al. (1975). Observations of polarized starlight toward such regions are consistent with $p_{\max}/A_V$ values of up to ~5% mag^{-1} (Panopoulou et al. 2019; Planck

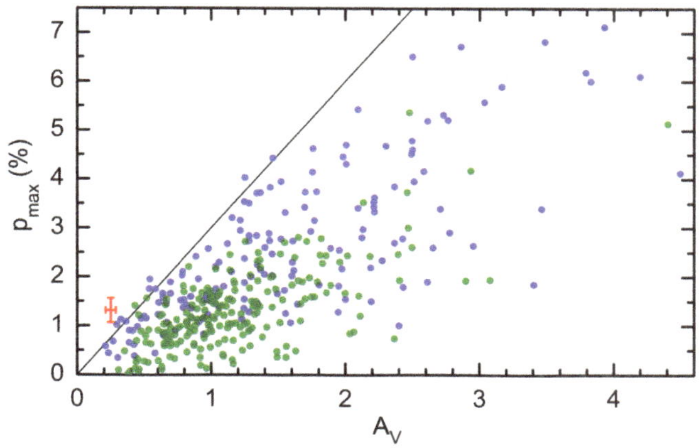

Figure 4.1. Visual extinction versus peak polarization for reddened stars in the Milky Way. Data are from (i) a general survey of the Milky Way (Serkowski et al. 1975; blue); (ii) a survey of the Coalsack region (Andersson & Potter 2005, 2007; green); and (iii) the mean result for stars in a region of exceptionally high p/A ratio identified in data from the Planck mission (Panopoulou et al. 2019; red cross). The diagonal line represents a nominal upper limit on polarization efficiency, $p_{\rm max}/A_V = 3\%\ {\rm mag}^{-1}$.

Collaboration 2020; an example is shown in Figure 4.1). Nevertheless, theoretical models can account for these results provided that the grains responsible for visual extinction and thermal emission can be closely aligned: for example, they lie well within limits suggested by calculations of dichroic extinction by perfectly aligned, highly asymmetric grains such as needle-like cylinders (Section 2.2.3).

Note, however, that the observed maximum polarization efficiency does not place a direct constraint on the *strength* of the interstellar magnetic field.[2] There are a number of reasons for this: we have no direct measure of actual typical grain shapes, and we also lack a fully quantitative understanding of the relationship between field strength and alignment efficiency. Indeed, it may be the case that only a certain threshold field strength is needed for near-perfect alignment of certain grain types. Conversely, field strengths estimated from other observations, such as synchrotron radiation, Faraday rotation and Zeeman splitting, may be used to test proposed alignment mechanisms (Section 4.3). The significance of polarization for studies of galactic magnetic fields is as a means of elucidating their *structure*.

4.1.2 Mapping Methodologies

Dichroic extinction and thermal emission provide complementary methods of investigating magnetic field structure. In either case, the component of **B** perpendicular to the line of sight is detected and mapped onto the sky. Comparisons between results for the two types of observation are generally in fair agreement (see Figure 4.2 for an example), but inconsistencies may arise from inherent differences

[2] The dispersion of polarization position angles arising from turbulence in the ISM provides an indirect measure of field strength by the Chandrasekhar–Fermi method (e.g., Crutcher et al. 2004).

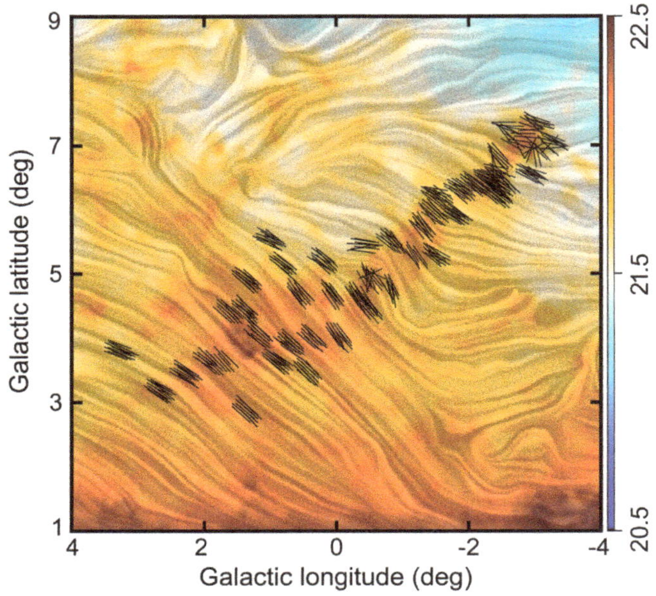

Figure 4.2. Comparison of submillimeter and visual polarization in the Pipe Nebula. A map of gas column density inferred from dust emission intensity is textured to follow magnetic field lines: the striations are orthogonal to the observed polarization at 353 GHz (850 μm) and thus parallel to **B**. The color scale indicates values of $\log N_H$ with N_H in cm^{-2}. Black lines are measures of visual polarization toward stars behind the nebula. Image credit: courtesy of Juan Soler, reproduced with permission from Astronomy & Astrophysics (Soler et al. 2016).

in sampling. Of course, observations of polarized starlight detect only matter between the star and the observer, whereas corresponding observations of emission detect all material in the line of sight within the field of view of the telescope, irrespective of distance, including any behind the star. To assess this difference, the optical depth of the emitting dust may be estimated from the observed intensity, and compared with that toward the star, estimated from its reddening. Using a sample restricted to cases where these estimates are in reasonable agreement, a tight correlation is found between optical and submillimeter measurements of position angle, with the expected 90° offset (Planck Collaboration 2015). The distribution of amplitudes is also similar, the submillimeter polarization intensity exhibiting behavior analogous to that seen in Figure 4.1 with a corresponding upper limit. The degree of depolarization tends to increase with column density, behavior attributed to greater disparity in magnetic field structure rather than systematic reductions in grain alignment (Planck Collaboration 2020).

The Pipe Nebula (Figure 4.2) provides an interesting case study. This relatively nearby filamentary dark cloud (distance ~ 150 pc) lies close to the galactic plane in the direction of the nuclear bulge. The submillimeter observations are thus subject to substantial contamination by background emission from dust extending beyond both the nebula and the stars used to measure visual polarization. The excellent agreement between the two determinations of field structure is therefore remarkable, and might conceivably be serendipitous; but a more plausible explanation is that the nebula is the dominant source of polarization in the line of sight, whereas that of the background is diminished by structural complexities leading to depolarization, as discussed above. Excellent agreement between observations of optical and submillimeter polarization is also found toward nearby clouds at higher galactic latitudes $|b|$, where background contamination is expected to be far less (Soler et al. 2016).

4.1.3 Magnetic Fields in Molecular Clouds

Studies of magnetic field structure in molecular clouds are crucial to our understanding of the star formation process (see Crutcher 2012 for a review). Magnetic fields control the flow of ions in the gas, and neutral species are collisionally coupled to the ions. Clouds thus tend to contract along magnetic field lines, as appears to have occurred in the Pipe Nebula (Figure 4.2) and other filamentary clouds (e.g., Franco & Alves 2015; Cox et al. 2016). Internal magnetic fields increase in strength with gas density ($B \propto n^{0.65}$; Crutcher 2012), and they mediate the gravitational collapse of dense cores to form new stars.

Polarimetric observations may be used to probe magnetic field structure to different optical depths within a molecular cloud, corresponding to regions of different densities and temperatures. Visual and near-infrared observations of background stars are generally limited to line-of-sight extinctions $A_V \lesssim 20$ mag; but as the dust is optically thin to far-infrared and submillimeter radiation, observations of polarized emission can be used to investigate polarization efficiency and field structure in dense cores and protostellar regions with much greater extinctions. The observations reveal a divergence in polarization efficiency according to environment, termed the "heterogeneous cloud effect" (Michail et al. 2021). Cold, dark clouds exhibit a consistent pattern of declining polarization production with increasing optical depth, whereas warmer regions subject to local irradiation and heating by embedded sources show consistently greater polarizations (e.g., Hildebrand et al. 1999; Whittet et al. 2008; Chapman et al. 2011; Alves et al. 2014; Michail et al. 2021). This trend is a potential hindrance to studies of magnetic field structure in the cold clouds (see below), but it helps to clarify the nature of the alignment process (Section 4.3). If radiative torques are implicated in the spin-up of grains that leads to alignment (Section 2.4.3), the observed polarization is predicted to depend on the ambient radiation field as well as on grain properties and the direction and uniformity of the magnetic field. Within a dark cloud lacking internal sources of luminosity and illuminated only by external radiation, the polarization efficiency is therefore expected to decline systematically with optical depth.

Because of this behavior, observations of background starlight are hindered by a sampling problem: the observed polarization is dominated by dust in the outer layers of the cloud, where alignment is driven by exposure to the external radiation field. Consequently, stars observed through relatively opaque regions of the cloud may have amplitudes of polarization little or no greater than those of stars sampling lower dust columns. If the polarization amplitudes are dominated by the outer layers, the position angles must likewise be biased, and polarization maps may thus contain little or no information on magnetic field structure inside the cloud. Indeed, it has been suggested that alignment of grains in the dense interiors of dark clouds might be negligible (e.g., Goodman et al. 1995). However, other observations clearly indicate that alignment is possible in such regions. Ices condense into mantles on dust *only* above a certain threshold extinction inside molecular clouds, and polarization has been detected in their absorption features toward background stars as well as embedded sources (Section 4.2.7). Thermal continuum emission from dust increases in proportion to the distribution of mass (Section 2.3.3), and so these observations naturally favor the dense cores of molecular clouds. Significant levels of polarization are detected in dense, starless cores (e.g., Alves et al. 2014; Liu et al. 2019) as well as in regions of current star formation where radiation from embedded sources may be enhancing alignment (e.g., Sadavoy et al. 2018).

4.1.4 Magnetic Fields in the Milky Way

We conclude this section with a big-picture view of magnetic field structure revealed by polarimetry of dust in our Galaxy. Results from observations of starlight are shown in Figure 4.3. As expected, the most highly polarized stars generally lie within a few degrees of the galactic equator (the horizontal central band). The most prominent exceptions are seen toward nearby dense clouds and star formation regions associated with Gould's Belt (Section 1.3.4), which lie ~16° north of the midplane toward the center (the Sco-Oph region), and a similar angular distance south of the midplane toward the anti-center (Taurus and Perseus to the left, Orion

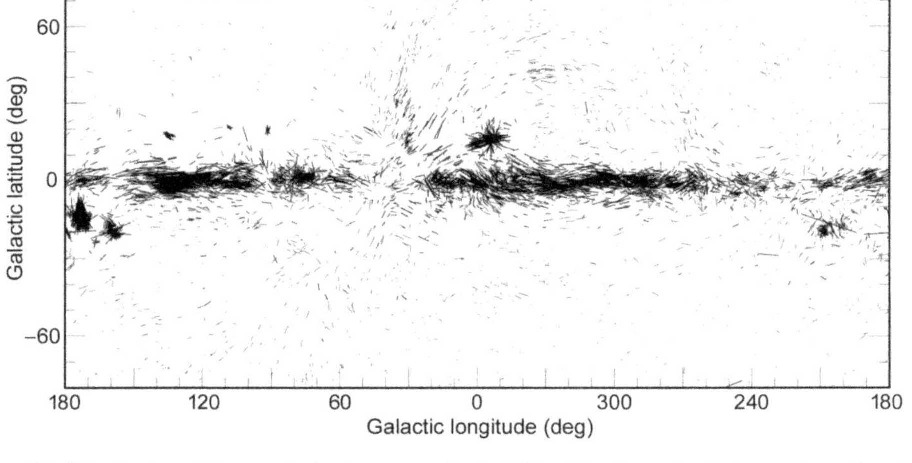

Figure 4.3. Distribution of linear polarization vectors in the Milky Way from visual observations of starlight, displayed in galactic coordinates. The galactic nucleus is toward the center of the frame and the galactic plane is horizontal through the center. The length and orientation of each line represent the amplitude and position angle of polarization detected in a star at that locus. Position angles measure the magnetic field orientation. Based on a compilation of data for 9286 stars by Heiles (2000).

to the right). Structure is also evident at higher latitudes, notably a loop extending from the Milky Way at $\ell \sim 30°$ toward the North Galactic Pole, thought to be a local disturbance caused by supernovae and/or stellar winds originating in Gould's Belt.

Observations of stars near the galactic equator provide the clearest evidence of long-range order, and they show that the **B** field is on average parallel to the galactic disk. However, exceptions are evident in certain longitude zones, where position angles appear to be randomized. This behavior is most clearly illustrated by variations in the quantity

$$q = \cos 2(\theta_G - 90°) \quad (4.1)$$

i.e., the angle-dependent component of the Stokes Q parameter (Equation (2.28)). This is shown in Figure 4.4, with the data from Figure 4.3 averaged into 10° longitude intervals (Fosalba et al. 2002). Values of q are expected to be unity for perfect alignment parallel to the galactic plane ($\theta_G = 90°$), and zero for a random distribution. The observed variations approximate to a sinusoidal form within these limits, as shown in Figure 4.4, with maxima approaching unity at $l \approx 135°$ and $-45°$, and minima close to zero at $l \approx 45°$ and $-135°$. A randomized distribution is expected for lines of sight parallel to **B**, so the positions of these minima indicate the approximate mean direction of the uniform magnetic field component in the galactic disk near the Sun. Of course, this deduction assumes that the field is linear in the solar neighborhood, which is unlikely to be strictly true: its mean direction is roughly tangential to the local Orion arm (Figure 1.7), suggesting that it traces spiral structure. Such behavior would be consistent with data for other spiral galaxies, which typically exhibit a concurrence of field direction and spiral structure (e.g., Beck 2016).

An all-sky map of polarized thermal dust emission is shown in Figure 4.5. As in Figure 4.2, the map is textured to render the magnetic field direction measured at a

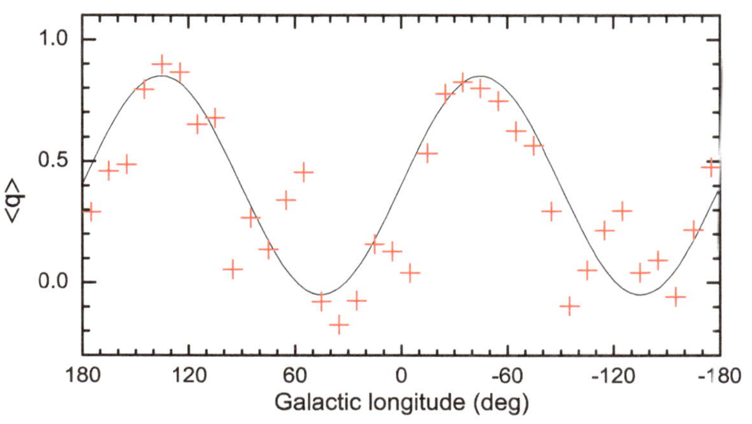

Figure 4.4. The dependence of the position-angle distribution on galactic longitude for the data in Figure 4.3. Red crosses are mean value of q (Equation (4.1)) for stars in 10° longitude bins (Fosalba et al. 2002). The curve is a sinusoidal fit with a 180° periodicity.

Figure 4.5. All-sky map in Mollweide projection, illustrating the distribution of polarized thermal dust emission and inferred magnetic field structure. The map is orientated relative to the disk of the Milky Way (red horizontal band) with the nucleus toward the center. The intensity of emission is color-coded from dark red to dark blue (greatest to least). The map is textured to render the magnetic field direction, based on measurements of polarization at 353 GHz. Image credit: European Space Agency; Planck Collaboration; CC-BY-SA-3.0.

frequency of 353 GHz. As expected, the overall pattern is consistent with the results from starlight polarization, confirming that **B** is predominantly parallel to the galactic disk at low galactic latitudes; but it also reveals much additional detail, especially in directions away from the galactic disk that are poorly sampled by observations of starlight. Whereas the visual observations present a scattershot view limited by the availability of stars with measurable interstellar polarizations, the submillimeter observations sample the entire sky at a nominal angular resolution of $\sim 5'$. The large-scale patterns seen at higher latitudes trace the local magnetic field in the solar neighborhood, which may be sculpted by the winds and shock waves thought to have formed the Local Bubble (Section 1.3.4). See Alves et al. (2018) and Planck Collaboration (2020) for detailed discussion.

4.2 Spectral Dependence and Environmental Sensitivity

4.2.1 Polarization and Extinction Curves Compared

As polarization of starlight is a form of extinction, its spectral dependence carries analogous information on the properties of dust grains and their response to environment (Chapter 3), subject to the additional constraint that only aligned grains bearing optical asymmetries are detected by the observations. The spectral dependence of linear polarization, or polarization curve, is usually displayed in the same way as extinction, i.e., as a plot of amplitude (in this case p_λ) versus λ^{-1}. A typical example is shown in Figure 4.6 and compared with extinction data for the same line of sight, presented here in log–log format for convenient display.

The polarization curve exhibits a broad peak in the visible region of the spectrum. Its slope in the infrared ($\lambda^{-1} < 1\,\mu\text{m}^{-1}$) is consistent with power-law behavior analogous to that of extinction (Martin & Whittet 1990). In contrast, a striking divergence between p_λ and A_λ is apparent in the blue through ultraviolet ($\lambda^{-1} > 2\ \mu\text{m}^{-1}$).

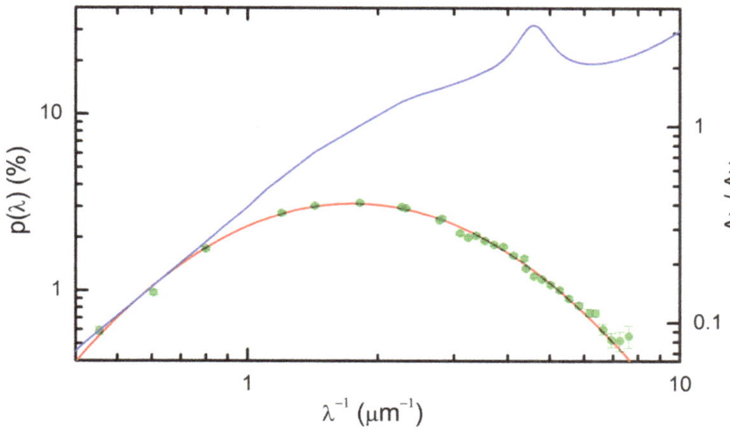

Figure 4.6. Comparison of interstellar polarization and extinction curves toward a star with typical diffuse-ISM extinction (HD 99872; $R_V \simeq 3.0$). Polarization data (green circles, left axis) are fitted by the Serkowski formula (Equation (4.2); red curve) with $\lambda_{\max} \simeq 0.57\ \mu$m. The extinction curve (blue, right axis) is an empirical fit to original data (Fitzpatrick & Massa 2007). Scaling is chosen to illustrate convergence in the infrared ($\lambda^{-1} < 1\ \mu m^{-1}$).

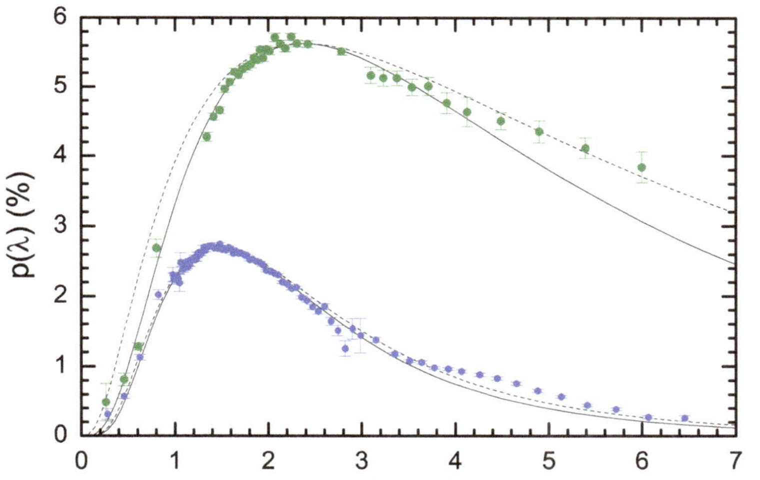

Figure 4.7. Comparison of interstellar polarization curves in contrasting environments: HD 204827 in Trumpler 37 ($R_V \simeq 2.6$; green) and ρ Oph ($R_V \simeq 4.3$; blue). Curves are empirical fits (Equation (4.2)) to data in the range $0 < \lambda^{-1} < 3\ \mu m^{-1}$, yielding $\lambda_{\max}$ values of 0.42 and 0.68 μm, respectively. Solid curves: K treated as a free parameter; dashed curves: K evaluated from Equation (4.3) using UV-optimized coefficients (Martin et al. 1999). A polarization excess corresponding to the 4.6 μm^{-1} bump is seen toward ρ Oph (Section 4.2.4).

This implies that small grains, which contribute to extinction most efficiently in the UV, are inefficient polarizers: in general, they either lack optical asymmetry or fail to align. The absence of a polarization counterpart to the 4.6 μm^{-1} extinction bump in Figure 4.6 is particularly striking (but see Figure 4.7 for an exception).

4.2.2 The Serkowski Formula

The observed spectral dependence of linear continuum polarization in the Milky Way is generally well described by an empirical formula first introduced by Serkowski (1973):

$$p_\lambda = p_{\max} \exp\{-K \ln^2(\lambda_{\max}/\lambda)\} \tag{4.2}$$

where $p_{\max}$ and $\lambda_{\max}$ specify the amplitude and position of the peak in the curve (see Figure 4.6), and K specifies its width. Broadly similar behavior is seen in data for other galaxies, suggesting that the formula has generic applicability (Rodrigues et al. 1997; Clayton et al. 2004; see Hough 1996 for a review of earlier work). Of the three parameters that specify p_λ in Equation (4.2), $\lambda_{\max}$ is of particular interest as it is predicted to depend on the characteristic size of the polarizing grains (Section 2.2.1). It has a mean value $\lambda_{\max} \simeq 0.55$ μm and an extreme range of 0.3–1 μm in the Milky Way, consistent with an origin for interstellar polarization predominantly in the larger grains of the size distribution.

The parameter K in Equation (4.2) was initially treated as a constant and set to $K = 1.15$, a value that provides a satisfactory description of polarization across the visible spectrum (Serkowski et al. 1975). The availability of observational facilities with greater sensitivity and wider spectral coverage led to subsequent refinements. With K treated as a free parameter, least-squares fits of Equation (4.2) to visual and near-infrared data for lines of sight with different $\lambda_{\max}$ values show that K correlates approximately linearly with $\lambda_{\max}$:

$$K = c_1\lambda_{\max} + c_2 \tag{4.3}$$

where c_1 and c_2 are (ideally!) constants (Wilking et al. 1980, 1982; Whittet et al. 1992). This dependence of K on $\lambda_{\max}$ has a positive slope ($c_1 > 0$), implying a systematic decrease in the width of the polarization curve with increasing $\lambda_{\max}$ (see Figure 4.7 for examples). However, the optimal values of c_1 and c_2 differ according to the spectral range used to evaluate them. They were initially determined from fits to data in the range $0.4 < \lambda^{-1} < 3$ μm^{-1}, yielding values of $c_1 = 1.66 \pm 0.09$ and $c_2 = 0.01 \pm 0.05$ (Whittet et al. 1992). With this constraint on K, the Serkowski formula yields excellent fits to data in that spectral range; but it tends to under-estimate the degree of polarization in the ultraviolet for stars with low $\lambda_{\max}$ ("super-Serkowski behavior"; Clayton et al. 1995). Data in the blue through mid-ultraviolet ($2.2 < \lambda^{-1} < 8$ μm^{-1}) are better matched with values of $c_1 = 2.56 \pm 0.38$ and $c_2 = -0.59 \pm 0.21$ (Martin et al. 1999). This difference is illustrated in Figure 4.7 for two stars with contrasted $\lambda_{\max}$ values. The improvement in the UV fit for UV-optimized coefficients is only significant in lines of sight with shortest $\lambda_{\max}$ (<0.5 μm), illustrated by the upper curve in Figure 4.7; for longer $\lambda_{\max}$ the fits are almost the same. Be this as it may, the overall agreement remains excellent at wavelengths near the peak. The Serkowski formula provides closely similar estimates of $p_{\max}$ and $\lambda_{\max}$, regardless of whether K is treated as a free parameter or constrained by an appropriate version of Equation (4.3).

4.2.3 Power-Law Behavior in the Infrared

The Serkowski formula provides a reliable representation of polarization into the infrared only for $\lambda^{-1} \gtrsim 0.4\ \mu\text{m}^{-1}$ ($\lambda \lesssim 2.5\ \mu$m); in practical terms, this limit corresponds to observations in the standard K passband centered at $\lambda \sim 2.2\ \mu$m, the longest used in the analyses described above. Beyond this, the behavior of p_λ is best described by a power law of the form

$$p_\lambda = p_{\lambda_0}(\lambda_0/\lambda)^\beta \tag{4.4}$$

where λ_0 is a reference wavelength used for normalization. Comparisons of results for different lines of sight indicate that the index β is typically in the range 1.6–2.0, closely similar to the corresponding index for extinction (Section 3.2.4), and is uncorrelated with λ_{max} (Martin et al. 1992). Equation (4.4) is appropriate for data in the spectral range from near to mid-infrared, and it provides a means of estimating the continuum underlying infrared spectral features detected in polarimetric data (Section 4.2.7).

4.2.4 Spectral Variations in Position Angle

At a given wavelength, the observed position angle of polarization is an average for the line of sight, projected onto the sky, as discussed in Section 4.1. Values measured at different wavelengths are expected to yield the same average (to within observational error) provided that the size distribution of the polarizing grains does not change significantly in different regions along the line of sight. This condition appears to hold true in the majority of cases (e.g., Whittet et al. 1992, 2001; Andersson & Potter 2007). But if, say, two clouds with systematically different size distributions are present in a line of sight, their mean values of λ_{max} will also differ. With reference to Figure 4.7, it follows that their relative contributions to the observed polarization vary with wavelength (the long-λ_{max} cloud is more dominant at longer wavelengths and conversely). So if the two clouds have different net directions of alignment, the observed position angle also varies systematically with wavelength.

In cases with relatively simple geometries and a serendipitous distribution of background stars, it is possible to separate such contributions by analysis of Stokes parameters. The Taurus dark cloud, for example, may be modeled as the sum of a diffuse screen and a dense core. Comparing data for stars behind screen alone and screen plus core, Messinger et al. (1997) showed that the screen component has $\lambda_{\text{max}} \simeq 0.55\ \mu$m (close to the diffuse-ISM average) and the core has higher values, ranging up to 0.73 μm. The analysis simultaneously yields a measure of internal magnetic field orientation, separated from that of the screen.

4.2.5 Circular Polarization

Small but measurable degrees of circular polarization are predicted when the direction of grain alignment changes systematically along the line of sight (Section 2.2.4). This has been confirmed by observations (Avery et al. 1975; Martin & Angel

1976 and references therein), but spectral coverage comparable with that of linear polarization is available for only a few stars. In these cases, the amplitude *V*/*I* of circular polarization is found to vary strongly with wavelength, generally exhibiting opposite handedness in the blue and red spectral regions. An example is shown in Figure 4.8, which compares observations of linear and circular polarization for *o* Sco. *V*/*I* changes sign at a distinct wavelength λ_c (the cross-over wavelength), which corresponds closely to the wavelength of the peak in linear polarization toward *o* Sco. Data for several lines of sight suggest this to be a general result: $\lambda_{max} \simeq \lambda_c$ to within observational error (Martin & Angel 1976; McMillan & Tapia 1977); but see also Michalsky & Schuster (1979) for an exception.

In combination, observations of linear and circular polarization place a useful constraint on the optical properties of the dust. The linear birefringence that gives rise to circular polarization is uniquely related to the linear dichroism at all frequencies by a Kramers–Krönig integral relation (Martin 1974; Chlewicki & Greenberg 1990), and this leads to a prediction that the ratio λ_{max}/λ_c is sensitive to *k*, the absorptive component of the complex refractive index. A value of unity ($\lambda_{max} = \lambda_c$) corresponds to $k = 0$, as illustrated by the model in Figure 4.8. For increasing absorption ($k > 0$), λ_c is predicted to increase relative to λ_{max}: the precise relation depends on the real as well as the imaginary part of the refractive index, but, typically, λ_{max}/λ_c is reduced to ~0.9 for $k = 0.05$ and to ~0.7 for $k = 0.3$ (see Aannestad & Greenberg 1983). The observations indicate $\lambda_{max}/\lambda_c = 1.00 \pm 0.05$, consistent with $k \lesssim 0.03$; i.e., the polarizing grains appear to be good dielectrics.

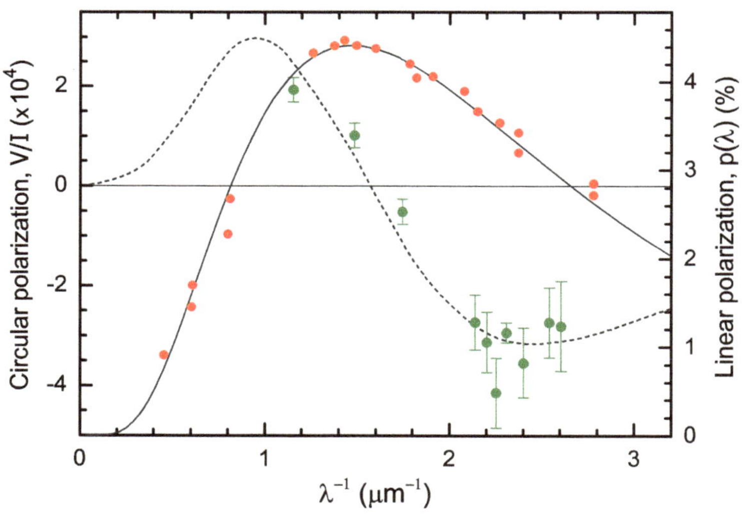

Figure 4.8. Linear and circular polarization curves for *o* Scorpii. Linear polarization data (red points, right-hand scale) are from Martin et al. (1992) and references therein, circular polarization data (green points, left-hand scale) are from Kemp & Wolstencroft (1972) and Martin (1974). The continuous curve is a fit of the Serkowski formula to the linear data. The dashed curve is a model for circular polarization by grains with $m = 1.50 - 0i$ (Martin 1974). The horizontal line indicates zero *V*/*I*. Note that the wavelength at which *V*/*I* changes sign is similar to the wavelength of maximum linear polarization ($\lambda_c \approx \lambda_{max} \approx 0.65\ \mu$m).

No significant new observations of the spectral dependence of interstellar circular polarization have been published since the pioneering work of the 1970s summarized above. Although regrettable, this hiatus is not the severe impediment to progress that it might seem. The nature of the relationship between birefringence and dichroism is such that any diagnostic properties of circular polarization should be retrievable from the linear polarization alone, provided it is known with sufficient precision over a sufficiently wide spectral range (Martin 1989). The dielectric nature of the polarizing dust is, indeed, indicated by models for the spectral dependence of linear continuum polarization, discussed in Section 4.2.8.

4.2.6 Effect of Particle Size

As previously noted, there is a theoretical basis for supposing that spatial variations in λ_{max} are related to changes in the size distribution of the aligned grains, such that larger λ_{max} values imply larger mean sizes. A correlation is thus predicted between λ_{max} and R_V, the latter being sensitive to the size distribution of particles responsible for extinction over a similar spectral range (Section 3.3.1). The predicted correlation was confirmed by Serkowski et al. (1975), and subsequently reassessed and refined as more data became available (e.g., Whittet & van Breda 1978; Clayton & Mathis 1988; Vrba et al. 1993). The general trend is shown in Figure 4.9 and compared with data for some individual stars and prototypical regions. The correlation is by no means perfect. This might be attributed in part to inclusion of stars with anomalies influencing the data, such as excess infrared emission or polarization caused by scattering in a circumstellar shell. Such objects can usually be identified and omitted

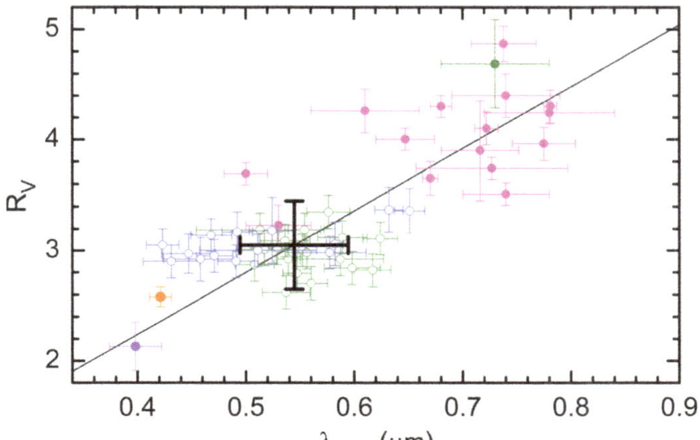

Figure 4.9. Correlation of R_V with λ_{max}. The black cross denotes mean values for the diffuse-ISM, its dimensions scaled to match the full-width half maxima of the distributions. Colored points represent individual lines of sight. Two regions with R_V values close to the mean are compared (open circles): Cygnus OB2 (blue) and the diffuse screen of the Taurus dark cloud (green). Also shown (solid circles) are data for the Taurus TMC–1 core component toward HD 283809 (green), the ρ Oph dark cloud (magenta), and two stars with unusually low R_V: HD 204827 (orange) and HD 210121 (violet). The diagonal line is the relation $R_V = 5.6\lambda_{max}$.

from the sample (as was done for the data in Figure 4.9), but dispersion about the optimal linear fit remains greater than observational error.

A probable cause of this dispersion is the sensitivity of λ_{max} to the anisotropy and alignment efficiency of the grains as well as to their size. For example, an exceptionally low λ_{max} value might occur because small grains are (i) especially abundant or (ii) optimally aligned; but of these possibilities, only (i) affects R_V. This difference might explain the relative distributions of the data in Figure 4.9 for two disparate regions that nevertheless have similar R_V values. Cygnus OB2 is a dust-enshrouded association of highly luminous young stars (McMillan & Tapia 1977; Whittet 2015); the Taurus screen is the diffuse outer envelope of a system of cold molecular clouds (Whittet et al. 2001). In both cases, the distribution of R_V is within the normal range for the diffuse-ISM; but significantly greater divergence is evident in the λ_{max} data, consistent with more efficient polarization production by small grains toward Cygnus OB2.

These factors apart, the overall trend is consistent with a simple linear correlation:

$$R_V = (5.6 \pm 0.3)\lambda_{max} \tag{4.5}$$

(e.g., Whittet & van Breda 1978), with λ_{max} expressed in micrometers. Results that fall at the high or low ends of this trend in Figure 4.9 are attributed to real, systematic shifts in size distribution relative to the mean. High values are typically seen toward dense clouds such as ρ Oph (Vrba et al. 1993) and interpreted as evidence for grain growth by coagulation (Wurm & Schnaiter 2002). Lines of sight with unusually low values of λ_{max} and R_V, signaling greater numbers of small grains, are less common but no less interesting. Two examples appear in Figure 4.9. One might expect similar processes to be at work in these line of sight, but this appears not to be the case. Extinction and polarization toward HD 210121 arises in an isolated, high-latitude dark cloud (Larson et al. 1996), apparently quiescent and remote from disruptive influences such as intense radiation fields, energetic stellar winds and shocks. The size distribution thus appears to be intrinsic to the environment, undisturbed by any external influence. In contrast, HD 204827, a member of the young cluster Trumpler 37, displays "SMC-like" extinction (Section 3.3.3) that may be attributed to shock-processing of the dust (Valencic et al. 2003).

4.2.7 Polarized Absorption Features

If the aligned grains responsible for continuum polarization contain a material that absorbs at a specific wavelength, a counterpart to the absorption feature should be seen in the polarization curve. Studies of such counterparts provide potentially valuable diagnostics of grain properties, and sensitive upper limits are also informative (see Whittet 2011 for a review). All of the primary classes of material thought to make up the bulk of the grain mass (carbonaceous matter, silicates and ices) exhibit absorption features that are candidates for spectropolarimetric detection.

By far the most prominent absorption feature attributed to carbonaceous particles is the 4.6 μm^{-1} bump, and in this case the results are somewhat equivocal. In most lines of sight, the polarimetric observations reveal no detectable counterpart and set tight upper limits (the example in Figure 4.6 is typical); but two exceptions have been found, representing about 6% of the sample (Wolff et al. 1997). One of these, ρ Oph, is shown in Figure 4.7 (the other is HD 197770). The polarization curve for ρ Oph exhibits a broad, shallow enhancement corresponding to the bump. However, in all cases, including these detections, the ratio of excess fractional polarization (or its upper limit) to optical depth within the feature is very low ($\Delta p/\Delta\tau \lesssim 0.001$; Martin et al. 1999). Thus, it may be concluded quite generally that the bump grains are very inefficient polarizers. Models that attribute the bump to small graphite grains yield the most satisfactory fits if the particles are flattened (e.g., oblate spheroids; Section 3.4.5). If so, the polarization data imply that they are at best very poorly aligned.

Other absorptions associated with carbonaceous carriers tell a similar story. Candidates include the 3.4 μm hydrocarbon feature (Section 5.2.3), and the diffuse interstellar bands (Section 3.5.1). All searches for polarimetric counterparts to these features have yielded negative results (Chiar et al. 2006, Cox et al. 2011, and references therein). Non-detections imply that the absorbing particles cannot contribute much to the continuum polarization either. This finding is most readily explained if the absorbers are very small grains (or large molecules) with dimensions $\lesssim$10 nm, presumed to be poorly aligned.

In contrast to the situation for carbonaceous grains, studies of absorption features identified with silicates and ices show a consistent pattern of associated polarization structure. Results for a well-studied line of sight toward a molecular cloud are shown in Figure 4.10. Peaks centered at $\lambda^{-1} \approx 0.05$ and 0.1 μm^{-1} are attributed to

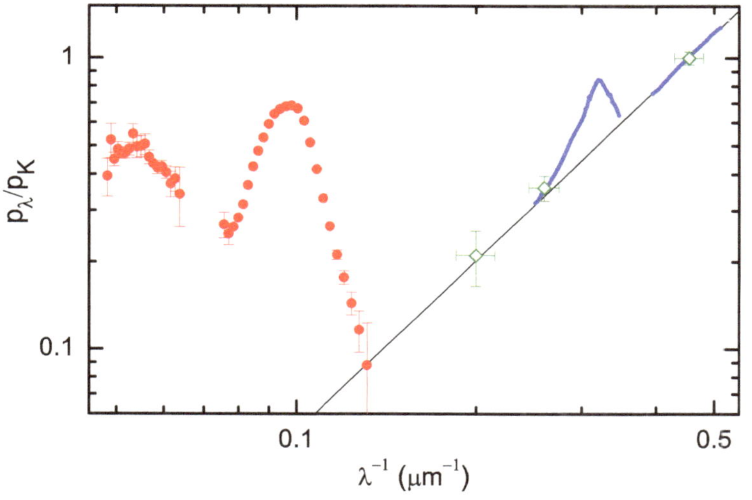

Figure 4.10. Infrared spectropolarimetry of the Becklin–Neugebauer (BN) object in the Orion OMC–1 molecular cloud, illustrating polarization peaks associated with absorption features of silicates (Smith et al. 2000, red) and H_2O-ice (Holloway et al. 2002, blue). Mean values of broadband data are also shown (green), and all data are normalized relative to the K band at 2.2 μm. The diagonal line is a power-law representation of the continuum ($p_\lambda \propto \lambda^{-1.95}$).

vibrational modes in silicates (see Figure 3.5). Silicates appear to be ubiquitous in the ISM and they are evidently major contributors to the observed polarization. Also evident is a peak centered at $\lambda^{-1} \approx 0.32\ \mu m^{-1}$, corresponding to the strong 3 μm absorption feature of solid H_2O. Ices are generally detectable only in molecular clouds, whereas silicates are present in all phases of the ISM. Nevertheless, in cases where both silicate and ice features are observed in the same line of sight, the position angles are often closely similar, indicating a consistent pattern of alignment for both absorbers. This result supports a core-mantle model for grain growth in molecular clouds, the silicates providing substrates for ice formation and accumulation. If the cores are asymmetric and aligned, then so are the mantles. Polarized ice features have been observed in both young stellar objects such as BN and background stars toward molecular clouds (e.g., Hough et al. 1988, 2008; Chrysostomou et al. 1996; Holloway et al. 2002). The results include a detection in solid CO, of particular significance as a constraint on grain alignment (Section 4.3.7) because CO freezes onto dust only in the coldest, densest regions. See Chapter 5 for full discussion of the origin and significance of these spectral features.

4.2.8 Modeling the Polarization Curve

Observations of interstellar polarization place useful constraints on grain models. A successful model for extinction (Section 3.6) should be capable of describing the spectral dependence of polarization as well, subject to additional assumptions concerning the shape and degree of alignment of the particles. The observed linear continuum polarization, p_λ (Figures 4.6–4.8), with its smooth peak in the visible and monotonic decline on either side, is consistent with production by particles that have the following general properties:

- Their dimensions are of the same order as the wavelength of the peak, i.e., toward the larger end of the size distribution.
- They have predominantly dielectric optical properties, i.e., they are weak absorbers that produce extinction primarily by scattering.

The requirement for dielectric particles arises from the smoothness of the p_λ curve, which discriminates against absorbers such as metals and graphite with dielectric functions that vary strongly with wavelength, a constraint also supported by observations of circular polarization (Section 4.2.5). The detection of polarization in the silicate absorption features (Section 4.2.7) requires that models for the continuum polarization include silicates, and indeed silicates behave as good dielectrics over a wide spectral range that spans the observed range in λ_{max}. In contrast, graphite and other carbonaceous particles—essential components of all models for extinction—appear to make little or no contribution to polarization. Consequently, it is often assumed that *all* polarization is produced by silicate dust in the diffuse-ISM, with ices also contributing in dense molecular clouds.

Models for polarization naturally depend on assumptions regarding the shapes of the particles. The range of possible grain morphologies is unbounded and almost unconstrained by available data, so "modelers are forced to limit consideration to some subset of idealized shapes" (quoted from Draine & Hensley 2021). To facilitate the calculations, simple axially-symmetric shapes such as spheroids or cylinders are often assumed. Modern computational methods are capable of simulating particles of arbitrary shape (Section 2.1.4), but there is no apparent advantage in adopting shapes more complex than the simplest needed to give good fits to the data. Particles with modest axial ratios seem more physically reasonable than those with extreme shapes: indeed, spheroids with axial ratios as low as 1.4 are sufficient to account for the observed polarization if alignment is perfect (Kim & Martin 1995). There is some evidence that flattened (oblate) spheroids give better fits to both the continuum polarization and the silicate features than elongated (prolate) ones (Kim & Martin 1995; Wright et al. 2016). This is not a definitive conclusion, however, and models developed by Guillet et al. (2018) that take polarized emission into account tend to favor prolate grains.

Examples of fits to the polarization curve are shown in Figure 4.11, based on calculations for aligned oblate spheroids with a fixed 2:1 axial ratio and composed of silicate. Results are shown for two contrasting values of $\lambda_{\rm max}$. The fits are generally excellent (a mismatch near 6.5 μm^{-1}, most noticeable in the long-$\lambda_{\rm max}$ case, may be an artifact of the chosen dielectric function; Kim & Martin 1995). The inset graph in Figure 4.11 shows mass distributions corresponding to the same models, each normalized to unity at the peak. The mass distribution function is defined by analogy with the size distribution function (Section 3.6.1) as the mass of dust residing in particles with dimensions in the range a to $a + da$. It is notable that the two

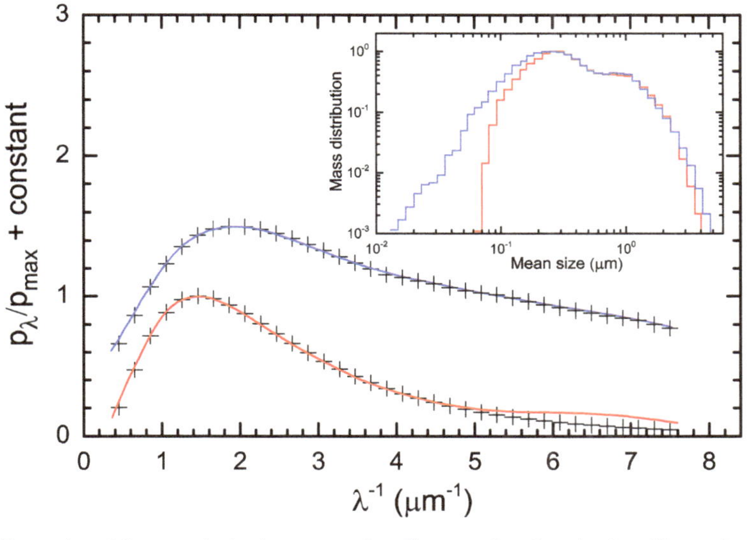

Figure 4.11. Examples of linear polarization curves for silicate grains. Results for oblate spheroids of axial ratio 2:1 composed of "modified astronomical silicate" are taken from the models of Kim & Martin (1995). Blue and red curves correspond to $\lambda_{\rm max} = 0.52$ μm and 0.68 μm, respectively, and are compared with representative observational data (crosses). Polarization values are normalized to unity at the peak (the upper curve is displaced for display). The inset graph shows grain mass distributions corresponding to the same models. Steps in the distribution functions correspond to individual size bins.

distributions almost overlap for larger grains, and diverge for smaller ones. The models thus elucidate the origin of observed spatial variations in λ_{max} (Section 4.2.6): longer λ_{max} values correspond to a reduced contribution from smaller (0.02–0.1 μm) grains, and conversely, whereas the distribution appears to be relatively stable for larger grains.

4.3 Grain Alignment

Results reviewed in the preceding sections of this chapter provide a firm basis for concluding that interstellar polarization is produced by aligned, asymmetric grains that represent a subset of all those responsible for extinction and thermal emission. The physical processes that lead to alignment are constrained by the observations in several ways:

- The distribution of position angles on the sky is highly consistent with a mechanism that imposes alignment with respect to magnetic field lines, independently confirmed by other observations such as radio synchrotron emission.
- Larger grains are aligned far more efficiently than smaller ones.
- Silicates (with or without ice mantles) are confirmed contributors to the observed polarization, whereas the contributions of carbonaceous solids appear to be minimal.
- Polarization is produced more efficiently in regions exposed to starlight than in cold, dark clouds.

This section summarizes the general principles of alignment theory and assesses the proposed mechanisms (Table 4.1) in the light of these observational constraints. References cited in the table highlight the original source and a few subsequent major developments or reassessments in each case. Further discussion and more

Table 4.1. Summary of Grain Alignment Mechanisms.

Mechanism	Description	References
Davis–Greenstein (DG)	Paramagnetic relaxation of thermally-spinning grains	Davis & Greenstein (1951) Purcell (1979) Roberge & Lazarian (1999)
Superparamagnetic (SPM) alignment	Superparamagnetic relaxation of thermally-spinning grains with iron inclusions	Jones & Spitzer (1967) Mathis (1986) Hoang & Lazarian (2016)
Purcell mechanism	Paramagnetic relaxation of suprathermally spinning grains (spin-up by H_2 formation)	Purcell (1979) Spitzer & McGlynn (1979) Lazarian (1995)
Radiative torques (RATs)	Paramagnetic relaxation of suprathermally spinning grains (spin-up by radiative torques)	Dolginov & Mytrophanov (1976) Draine & Weingartner (1996, 1997) Lazarian & Hoang (2007, 2019)
Gold mechanism	Mechanical alignment in supersonic flows	Gold (1952) Lazarian (1997)

extensive guides to previous literature may be found in the reviews by Andersson et al. (2015) and Lazarian et al. (2015).

4.3.1 General Principles of Magnetic Alignment

All proposed magnetic alignment mechanisms are related by their dependence on interactions between the rotational motion of a grain and the ambient magnetic field, **B**. Alignment results from two distinct processes that operate in tandem: internal alignment between the orientation of the principal geometric axis of the grain and its angular momentum vector, **J**, by rotational dissipation, and alignment of **J** with respect to **B** by magnetic relaxation. We consider these processes in turn.

Several factors may influence the rotational motion of a dust grain (Section 2.4.3). Grains are said to be spinning *thermally* if their rotational motion is the result of random elastic collisions with gas particles, such that their mean rotation temperature approximates to the kinetic temperature of the gas. The equilibrium rms angular speed for a particle of radius a is then predicted to vary in proportion to $a^{-2.5}$ (Equation (2.56)), i.e., smaller grains rotate on average far more rapidly than larger ones. Grains are said to be spinning *suprathermally* if their angular speeds are systematically higher than expected from thermal collisions. Processes that may cause suprathermal spin include impulsive torques resulting from release of binding energy at sites of molecule formation on a grain surface, and radiative torques arising from any inherent chirality of a grain's extinction cross-section.

Consider a rapidly-spinning grain of arbitrary shape. If the grain reaches a state of collisional equilibrium with the gas, the rotational kinetic energy associated with spin about each of the three principal axes of inertia become equalized. As $E_{\rm rot} = \frac{1}{2}I\omega^2 = \frac{1}{2}J^2/I$ for each axis, the angular momentum must then be greatest along the axis of maximum inertia. However, energy equipartition may be disturbed if rotational energy is dissipated by internal processes (Purcell 1979). Several factors may come into play that damp the grain's rotation. Frictional stresses dissipate rotational energy as heat, but thermal fluctuations may oppose this, converting vibrational energy into rotational energy. However, in the presence of a magnetic field, dissipation is greatly augmented by the Barnett effect, which causes spontaneous internal magnetization by aligning the spins of unpaired electrons and nuclear particles parallel to the axis of rotation. Because these processes are internal, angular momentum must be conserved, and the grain is driven toward a state of minimum rotational energy with spin about the principal axis of inertia. Elongated grains such as cylinders or prolate spheroids therefore spin end-over-end, with their rotation axes perpendicular to their axes of symmetry, and flattened grains such as disks or oblate spheroids spin like wheels, with their rotation axes parallel to their axes of symmetry. The timescale for this situation to be reached is ~ 1 yr for thermally-spinning grains in the size range needed to explain the observed polarization (Hildebrand 1988), and much less if the spin is suprathermal. However, suprathermally spinning grains may be subject to periods of "cross-over" (Spitzer & McGlynn 1979; Lazarian & Draine 1997) in which the direction of the external torque is reversed, causing a grain to spin down and potentially lose the internal alignment between J and its axis of maximum inertia. Larger grains have greater immunity to this randomization process than smaller ones.

If internal alignment has occurred, external alignment consistent with the observed polarization will be achieved for either flattened or elongated particles if their angular momenta become orientated parallel to the magnetic field. In reality, their angular momenta are expected to *precess* about the magnetic field rather than align in perfect picket-fence fashion, independent of the alignment mechanism (see Figure 2.4). If a spinning grain bears electrical charge, it is endowed with a magnetic moment that leads to precession; furthermore, the magnetic moment induced by the Barnett effect will impose precession even for an uncharged grain (Dolginov & Mytrophanov 1976; Purcell 1979). The precession period is expected to be short compared with timescales for other dynamical effects acting on the grain, including randomizing collisions and the alignment mechanism itself. On a time-average, the distribution of spin axes will thus be symmetric about the magnetic field.

4.3.2 Paramagnetic Relaxation: The DG Mechanism

The classical theory of alignment by paramagnetic relaxation was formulated by Davis & Greenstein (1951) soon after the discovery of interstellar polarization in 1949. It was quickly realized that grains do not behave ferromagnetically (like compass needles). Even pure iron grains cannot maintain ferromagnetic alignment when subjected to collisional torques (Spitzer & Tukey 1951), and in any case ferromagnetism would orientate the grains with their long-axes parallel to the field direction, orthogonal to that implied by the observations. The DG mechanism predicts alignment with the correct orientation by paramagnetic dissipation of rotational kinetic energy in thermally-spinning grains.

The presence of an external magnetic field of flux density **B** causes the induction of an internal field within the grain, the strength of which depends on the magnetic susceptibility of the material. In a static situation, the internal and external fields would be perfectly aligned. However, for a spinning grain, the internal field cannot adjust instantaneously to the direction of the external field, and so there is always a slight misalignment. This results in a dissipative torque that slowly removes components of rotation perpendicular to **B**, tending to bring the angular momentum into alignment with **B**. Alignment is opposed by gas-grain collisions, which tend to restore random orientation. A measure of the efficiency with which grains can be aligned by the DG mechanism is thus given by the ratio

$$\delta = t_c / t_r \tag{4.6}$$

where t_c is the collisional damping time, defined as the time taken by a grain to collide with a mass of gas equal to its own mass, and t_r is the timescale for paramagnetic relaxation. To achieve significant net alignment, we require $\delta \gtrsim 1$.

Another condition that must be satisfied arises from the fact that the mechanism is analogous to a heat engine operating between gas and dust (Jones & Spitzer 1967): if the temperatures of these reservoirs are the same, no work can be done. The gas is generally much warmer than the dust in diffuse regions of the ISM. This is because gas-grain collisions are unimportant as a source of heat exchange at such low densities, and so gas and dust are each in independent thermal equilibrium with the interstellar radiation field. Equilibrium temperatures $T_d \sim 15$ K are predicted for the dust (Section 2.3.1), compared with $\gtrsim 50$ K for the gas (Section 1.3.1). The DG mechanism, and, indeed, any mechanism that operates in an analogous way, depends on this temperature difference to achieve alignment. If the gas and dust temperatures equilibriate, the mechanism becomes inoperative.

The paramagnetic relaxation time, t_r, is determined by the ratio of the angular momentum about an axis perpendicular to **B** to the magnitude of the dissipative torque (e.g., Spitzer 1978, p. 187–189):

$$t_r \propto \frac{I\omega}{V\chi'' B^2 \sin\theta}$$

where V is the volume of the grain, θ is the angle between **B** and ω, and χ'' is the imaginary part of the magnetic susceptibility ($\chi = \chi' + i\chi''$). The latter is related to the angular speed, the properties of the grain material, and its temperature, T_d. Assuming grains of effective radius a (the radius of an equivalent sphere) composed of typical paramagnetic solids of density ρ_s, and setting $I = \frac{2}{5} m_d a^2$, we have

$$t_r \approx 1.6 \times 10^4 \left(\frac{a^2 \rho_s T_d}{B^2 \sin\theta} \right) \tag{4.7}$$

in SI units. The collisional damping time is given by

$$t_c = \frac{4a\rho_s}{3(1.2 n_H m_H \langle v_H \rangle)} \tag{4.8}$$

where

$$\langle v_H \rangle = \left(\frac{8kT_{gas}}{\pi m_H} \right)^{\frac{1}{2}} \tag{4.9}$$

is the mean speed of H atoms, and the factor 1.2 in Equation (4.8) allows for the fact that ~10% of the impinging atoms are in fact He, with mass $4m_H$ and mean speed $0.5\langle v_H \rangle$. Combining Equations (4.6)–(4.9) yields an expression for δ:

$$\delta \approx \frac{4.3 \times 10^{-5}\ B^2 \sin\theta}{n_H a T_d (k T_{gas} m_H)^{\frac{1}{2}}}. \tag{4.10}$$

Equation (4.10) may be used to test the DG mechanism under typical conditions. The observations imply efficient alignment of relatively large grains ($a \sim 0.2 \times 10^{-6}$ m, $T_d \sim 15$ K) in the diffuse-ISM. Let us assume they are immersed in atomic gas of number density $n_H \sim 3 \times 10^7$ m^{-3} and temperature $T_{gas} \sim 80$ K (Table 1.1), and subject to a magnetic field of mean flux density $B \sim 5 \times 10^{-10}$ T orientated such that $\sin\theta \sim 0.5$. With these values, Equation (4.10) yields $\delta \sim 0.04$, implying that alignment is randomized by collisions on timescales much shorter than those needed to establish it.

This calculation illustrates a general conclusion: the DG mechanism fails quantitatively to predict significant alignment by paramagnetic relaxation of grains in the size range needed to account for the observed polarization under typical interstellar conditions. Furthermore, small grains are predicted to be aligned more efficiently than large ones (δ is inversely proportional to a in Equation (4.10)), contrary to observational evidence. Nevertheless, the DG mechanism stands as a major milestone in our quest to understand grain alignment. It correctly predicts the *geometric* properties of the observed polarization, suggesting that some analogous process may be operating on more favorable timescales. This outcome prompted a focus on models that may be considered as enhancements of DG alignment.

4.3.3 Superparamagnetic Alignment

Iron is one of the "big five" chemical elements that contribute most of the dust mass (Section 1.2.6), and so its presence in polarizing grains is to be expected. As previously noted, the grains cannot be ferromagnetic. But if a paramagnetic grain contains small clusters of ferromagnetic atoms or molecules (e.g., metallic Fe, Fe_3O_4, or other oxides or sulfides of iron, with $\sim$100 Fe atoms per cluster), the magnitude of the magnetic susceptibility component χ'' may be enhanced by factors up to $\sim 10^6$ over values typical of paramagnetic materials (Jones & Spitzer 1967). This effect is termed superparamagnetism (SPM), and the clusters are referred to as SPM inclusions. Alignment may then proceeds in a manner exactly analogous to the DG mechanism described above, but the relaxation time t_r (Equation (4.7)) is reduced in proportion to the increase in χ''. Hence, alignment is inherently more efficient and more stable, with $\delta \gg 1$ for any reasonable choice of the other relevant parameters, provided that the temperature difference between gas and dust is maintained. SPM alignment is thus "robust" in the sense that the alignment process is not marginal over a range of physical conditions; only if gas and dust are in a state of thermal equilibrium will it fail.

If the polarizing grains are composite particles, formed by coagulation of smaller units (Wurm & Schnaiter 2002), the number of SPM inclusions in each grain is expected to be proportional to its volume: a large grain may contain many, a small grain may contain few or none at all. This suggests a physical explanation for the observation that only the larger grains tend to be aligned. Mathis (1986) postulates that a grain is aligned if, and only if, it contains at least one SPM inclusion: a critical size limit of $a \gtrsim 0.09$ μm for SPM alignment predicts the observed mean polarization curve for silicate grains following the MRN size distribution (Section 3.6.1). The wavelength of maximum polarization is then determined by the average size of silicate grains containing SPM inclusions.

A caveat arises from the observational constraint that carbonaceous grains make little or no contribution to the observed polarization. It is therefore necessary to assume that SPM inclusions are common in silicates but generally lacking in carbonaceous grains, irrespective of their size, unless they are isotropic in shape. This assumption might be justified in terms of their differing origins. At least some of the carbonaceous dust in the ISM forms in C-rich stellar atmospheres, whereas silicate stardust forms in O-rich atmospheres that may promote the simultaneous growth of iron oxides or other Fe-rich solids (Chapter 8). Any differences in the drift speeds of silicate and Fe-rich particles in the outflows of such stars will cause them to collide and potentially coagulate before they merge into the ISM.

Indirect support for the SPM hypothesis arises from studies of meteorites and interplanetary dust. Some components of this material appear to be relatively pristine remnants of the solar nebula, composed of elements in proportions closely similar to those predicted for interstellar dust (Chapter 7). For example, GEMS (glasses with embedded metal and sulfides) contain Fe-rich clusters within a silicate matrix (Bradley 1994), and the spatial frequency of the clusters is consistent with the requirements of the SPM model (Goodman & Whittet 1995). Although GEMS are not necessarily unmodified interstellar grains, they may have formed under similar conditions.

4.3.4 Suprathermal Spin: The Purcell Mechanism

The Purcell (1979) mechanism postulates alignment of suprathermally spinning grains by paramagnetic relaxation. Real interstellar grains are unlikely to have smooth, uniform surfaces, and collisions between gas and grains are unlikely to be elastic. In particular, a hydrogen atom colliding with a grain may stick, and subsequently migrate across the surface until it combines with another to form H_2 (Section 9.1.4). The release of binding energy ejects the molecule from the surface and simultaneously imparts angular momentum to the grain. Molecule formation is likely to occur preferentially at active sites associated with defects or impurity centers on the surface. A migrating H atom, which would be held only by weak van der Waals forces elsewhere, may become trapped at an active site until recombination with another migrating atom occurs. The distribution and orientation of active sites over the surface influences the rotational dynamics of the particle. Systematic contributions to angular momentum arising from a sequence of recombination events at a limited number of active sites with a favorable geometry may then result in a "pinwheel" effect, spinning the particle to angular speeds exceeding those predicted by random elastic collisions (Section 2.4.3) by factors of up to $\sim 10^4$.

The Purcell mechanism differs from classical DG alignment in two important respects. First, because the rotational energy of a suprathermally spinning grain greatly exceeds kT, where T is any kinetic temperature in the system, the mechanism does not require a temperature difference between gas and dust: the heat engine obtains its "fuel" from the binding energy released when H_2 is formed. However, the mechanism becomes inactive if the grains are too warm to retain surface H, or if they are immersed in a cloud that is fully molecular. Second, a suprathermally spinning grain is far less vulnerable to disruption of its orientation by random collisions, because the energy imparted by the collisions is small compared with its rotational energy. The timescale for paramagnetic relaxation is independent of the angular speed (Equation (4.7)). The time available for alignment is therefore limited not by gas damping but by the timescale on which the grain maintains spin-up in the same direction. This depends on the survival time of the surface properties that lead to suprathermal spin.

Active sites that promote H_2 formation may be "poisoned" by attachment of atoms other than H, especially oxygen (Lazarian 1995). Depending on the number and distribution of affected sites, this may lead to cross-over periods (Section 4.3.1) in which the pinwheel effect is lost or even reversed, causing the grain to spin down and potentially lose internal alignment. The timescale for this to occur is uncertain. One approach is to consider the growth time of a thin surface coating or mantle (Section 9.2.2), given by

$$t_{\rm m} = \frac{2.5\rho_{\rm s}\Delta a}{\xi n(kT_{\rm gas}m)^{\frac{1}{2}}} \tag{4.11}$$

where Δa is the thickness of the surface layer formed by accretion of atoms of mass m and number density n, and ξ is the sticking coefficient (also known as the accommodation coefficient), i.e., the probability that impinging atoms stick to the surface ($0 \leqslant \xi \leqslant 1$). If Δa is set to typical molecular dimensions (≈ 0.4 nm) in Equation (4.11), $t_{\rm m}$ becomes the time to accrete a *monolayer*. We may then assess the mechanism by reformulating the calculation in Section 4.3.2 in terms of $\delta = t_{\rm m}/t_{\rm r}$ instead of $t_{\rm c}/t_{\rm r}$, with $t_{\rm m}$ estimated for atomic O (using the appropriate abundance ratio of O to H and allowing for prior depletion of O into silicates; see Chapter 7), and assuming that the resulting mantle is composed of H_2O-ice ($\rho_{\rm s} \simeq 1000$ kg m^{-3}). Such calculations suggest that the mechanism is stable in H I clouds if the sticking coefficient for oxygen is much less than unity.

Can Purcell alignment account for the preferential alignment of large grains? The formulation presented above suggests size selectivity of the wrong sense, as $t_{\rm r}$ depends on a^2 and $t_{\rm m}$ is independent of a (Equations (4.7) and (4.11)). However, the number of active sites on a grain is obviously related to its surface area, and a critical number may be necessary to maintain suprathermal spin. The accreting O atoms are somewhat mobile and tend to seek out and poison active sites. Whereas a large grain with many active sites might maintain spin-up for times that approach the monolayer accretion time, poisoning may dominate for smaller grains, causing spin-up to be short-lived.

4.3.5 Radiative Alignment Torques

The discussion in Section 4.3.4 above assumes that H_2 recombination is the dominant process by which a grain acquires suprathermal spin. Purcell (1979) considered two other process (surface variations in the sticking coefficient and photoelectric emission), but found them to be relatively insignificant. Another possibility that has become a major focus of discussion in more recent years is the rotational motion imparted to a grain by irradiation.

A grain experiences radiative torques if angular momentum is transferred from photons in the ambient radiation field (see Section 2.4.3). For this to occur, the structure of the particle must include some degree of inherent chirality, meaning that the extinction cross-sections for left-handed and right-handed circular polarizations are not identical. Furthermore, the radiation field must be anisotropic, i.e., incident photons must be arriving predominantly from a preferred direction, otherwise the torques will tend to balance out. Anisotropic radiation fields are expected to be common occurrences in the ISM, as the result of (e.g.) the proximity of a luminous star or the inherent inhomogeneity of the medium. A grain in the outer layers of a dark cloud, for example, is exposed to stellar radiation preferentially from the direction opposite to that of the cloud.

In a detailed analysis of the mechanism, Draine & Weingartner (1996, 1997) show that spin-up by radiative torques is effective for grains in the size range needed to account for the observed polarization spectrum, potentially exceeding that from H_2 formation. Because radiative torques depend on the structure of the grain rather than on surface conditions, the resulting suprathermal rotation is expected to be long-lived, allowing ample time for alignment by paramagnetic relaxation. The orientation of the spinning grain is then constrained primarily by the magnetic field, as in other forms of magnetic alignment, but the efficiency of alignment is sensitive to the angle between **B** and the direction of incident radiation (see Andersson et al. 2015 for detailed discussion). The spectrum of the interstellar radiation field, with its inherent cut-off at the Lyman limit (Section 2.4.3) renders radiative torques ineffective for smaller grains ($a \lesssim 0.05\ \mu$m), consistent with observational constraints.

4.3.6 Mechanical Alignment: The Gold Mechanism

An elongated grain in relative motion through a gaseous medium tends to align with its spin axis perpendicular to the direction of flow (Gold 1952). Alignment is efficient only if the flow is supersonic, i.e., the drift speed exceeds the thermal speed of the gas atoms (Lazarian 1997). Examples of processes that may cause such streaming include cloud–cloud collisions, turbulence, the transverse motion of Alfvén waves, the orbital motion of charged particles around magnetic field lines, and differential acceleration of gas and dust by radiation pressure in stellar outflows.

The Gold mechanism does not depend on magnetism to produce alignment, but the dynamics of spinning grains are nevertheless highly constrained by magnetic fields. As interstellar grains tend to acquire charge by the photoelectric effect, significant drift speeds are generally reached only in directions parallel to **B**. Streaming would then tend to produce polarization with position angle orthogonal to that predicted by magnetic alignment, which is incompatible with the observed distribution of polarization. Thus, the Gold mechanism cannot account for the macroscopic pattern of alignment in the Milky Way (Section 4.1.4). Nevertheless, streaming is undoubtedly important in some situations. Mechanical alignment may become significant during cross-over periods when suprathermal spin is inoperative. In weakly ionized clouds, charged grains tend to drift through the gas in a direction *normal* to **B** due to ambipolar diffusion, at speeds that may be sufficient to produce significant mechanical alignment with orientation consistent with observations (Roberge et al. 1995).

4.3.7 Alignment in Dense Clouds

The discussion so far in this section focuses primarily on alignment in diffuse regions of the ISM. In such environments the observed polarization efficiency p/A is often high, the temperature difference between gas and dust is maintained, and the grains are expected to lack volatile surface coatings. Magnetic alignment appears to satisfy major observational constraints under such condition (geometric pattern, size selectivity, maximum p/A ratio) for realistic values of the field strength, provided that the grains are superparamagnetic and/or subject to suprathermal enhancement of their spin. There is ample availability of both photons to impart radiative torques and atomic hydrogen to feed the Purcell mechanism.

Dense clouds present a significantly different "laboratory" in which to test alignment mechanisms. The observed decline in p/A with increasing optical depth in cold, quiescent regions of dense clouds (Section 4.1.3) is broadly consistent with all models. The efficiency of SPM alignment is expected to vary with $(1 - T_d/T_{gas})$, becoming ineffective as gas and dust temperatures converge with increasing density. The Purcell mechanism fails if surface coatings suppress active sites of H_2 formation, or if $H \rightarrow H_2$ conversion is complete. And shielding by the outer layers of a dark cloud naturally attenuates the internal radiation field, reducing the efficiency of spin-up by radiative torques, especially for the smaller polarizing grains that interact most strongly with photons in the visible and UV.

A greater challenge is to explain how alignment consistent with observed levels of polarization in molecular clouds is, indeed, possible. That significant alignment does occur is demonstrated by both polarized continuum emission from cloud cores and polarized absorption corresponding to spectral feature of ices. The solid CO feature at 4.67 μm places the strongest constraint on physical conditions in which alignment may occur. CO is an abundant constituent of molecular gas, and with a sublimation temperature $\sim$17 K it is the most volatile common constituent of interstellar ices. The absorption and resultant polarization must therefore occur exclusively in cold, dense environments, where we expect gas and dust temperatures to be closely coupled and grains to be fully mantled. Neither superparamagnetic nor Purcell mechanisms seem viable in such regions, but alignment driven by radiative torques might still be effective for larger grains ($a \sim 0.5$ μm; Hough et al. 2008). Large grains may acquire suprathermal spin from infrared photons, which penetrate far deeper into a cloud than those of shorter wavelength. The mean extinction curve (Section 3.2.1) predicts that extinction in the K (2.2 μm) band, for example, is an order of magnitude less than that in the visible ($A_K/A_V \sim 0.1$). Another possibility is that cosmic rays might enhance the rotational energies of grains in molecular clouds, but a quantitative evaluation by Lazarian & Roberge (1997) suggests that the resulting torques are insignificant.

A related challenge is to explain the heterogeneous cloud effect described in Section 4.1.3. Systematically enhanced levels of polarization observed in regions of active star formation signify a notable departure from the systematic decline in cold, dark regions of molecular clouds. The implication is that embedded young stars can impose alignment on dust in the surrounding medium. This might arise in different ways. Mechanical alignment may be induced by supersonic flows associated with winds from a luminous star, or by streaming driven by radiation pressure on the dust; but it is not clear that such processes, which depend on circumstances specific to each source, can account for the observed systematic heterogeneity. A process common to all dust-enshrouded stars, however, is the production of copious infrared (re)radiation, and this may result in widespread alignment of large grains by radiative torques. Observations of polarized CO ice absorption again provide insight. Detection toward the luminous young stellar object W33A (Chrysostomou et al. 1996) implies alignment of grains in a cold region sufficiently remote from the source to retain CO on the grains, inconsistent with a circumstellar origin but consistent with radiative alignment in the molecular cloud if dust temperatures remain low enough to retain CO. The extent to which radiative torques can induce alignment without significant warming is thus an important constraint on the mechanism that might be tested by observations of polarized far-infrared and submillimeter emission. See Michail et al. (2021) and Planck Collaboration (2020) for examples (and differing interpretations) of such data.

To summarize, several alignment mechanisms may be effective in diffuse regions of the ISM—it would indeed be surprising if grains are entirely devoid of superparamagnetic inclusions or surface properties that may promote suprathermal spin—but radiative torques seem most likely to account for polarization in dense clouds. Radiation fields are expected to be anisotropic regardless of whether the sources of luminosity are external or internal, and alignment of large grains will follow if there is sufficient penetration of infrared photons.

References

Aannestad, P. A., & Greenberg, J. M. 1983, ApJ, 272, 551
Alves, F. O., Frau, P., Girart, J. M., et al. 2014, A&A, 569, L1
Alves, M. I. R., Boulanger, F., Ferrière, K., & Montier, L. 2018, A&A, 611, L5
Andersson, B.-G., & Potter, S. B. 2005, MNRAS, 356, 1088
Andersson, B.-G., & Potter, S. B. 2007, ApJ, 665, 369
Andersson, B.-G., Lazarian, A., & Vaillancourt, J. E. 2015, ARA&A, 53, 501
Avery, R. W., Stokes, R. A., Michalsky, J. J., & Ekstrom, P. A. 1975, AJ, 80, 1026
Beck, R. 2016, A&ARv, 24, 4
Bradley, J. P. 1994, Sci, 265, 925
Chapman, N. L., Goldsmith, P. F., Pineda, J. L., et al. 2011, ApJ, 741, 21
Chiar, J. E., Adamson, A. J., Whittet, D. C. B., et al. 2006, ApJ, 651, 268
Chlewicki, G., & Greenberg, J. M. 1990, ApJ, 365, 230
Chrysostomou, A., Hough, J. H., Whittet, D. C. B., et al. 1996, ApJ, 465, L61
Clayton, G. C., & Mathis, J. S. 1988, ApJ, 327, 911
Clayton, G. C., Wolff, M. J., Allen, R. G., & Lupie, O. L. 1995, ApJ, 445, 947
Clayton, G. C., Wolff, M. J., Gordon, K. D., et al. 2004, AJ, 127, 3382
Cox, N. L. J., Ehrenfreund, P., Foing, B. H., et al. 2011, A&A, 531, A25
Cox, N. L. J., Arzoumanian, D., André, Ph., et al. 2016, A&A, 590, A110
Crutcher, R. M. 2012, ARA&A, 50, 29
Crutcher, R. M., Nutter, D. J., Ward-Thompson, D., & Kirk, J. M. 2004, ApJ, 600, 279
Davis, L., & Greenstein, J. L. 1951, ApJ, 114, 206
Dolginov, A. Z., & Mytrophanov, I. G. 1976, Ap&SS, 43, 291
Draine, B. T., & Hensley, B. S. 2021, ApJ, 910, 47
Draine, B. T., & Weingartner, J. C. 1996, ApJ, 470, 551

Draine, B. T., & Weingartner, J. C. 1997, ApJ, 480, 633
Fitzpatrick, E. L., & Massa, D. 2007, ApJ, 663, 320
Fosalba, P., Lazarian, A., Prunet, S., & Tauber, J. 2002, ApJ, 564, 762
Franco, G. A. P., & Alves, F. O. 2015, ApJ, 807, 5
Gold, T. 1952, MNRAS, 112, 215
Goodman, A. A., & Whittet, D. C. B. 1995, ApJ, 455, L181
Goodman, A. A., Jones, T. J., Lada, E. A., & Myers, P. C. 1995, ApJ, 448, 748
Guillet, V., Fanciullo, L., Verstraete, L., et al. 2018, A&A, 610, A16
Heiles, C. 2000, AJ, 119, 923
Hildebrand, R. H. 1988, QJRAS, 29, 327
Hildebrand, R. H., Dotson, J. L., Dowell, C. D., Schleuning, D. A., & Vaillancourt, J. E. 1999, ApJ, 516, 834
Hoang, T., & Lazarian, A. 2016, ApJ, 831, 159
Holloway, R. P., Chrysostomou, A., Aitken, D. K., Hough, J. H., & McCall, A. 2002, MNRAS, 336, 425
Hough, J. H. 1996, in ASP Conf. Ser. 97, Polarimetry of the Interstellar Medium, ed. W. G. Roberge, & D. C. B. Whittet (San Francisco, CA: ASP), 569
Hough, J. H., Sato, S., Tamura, M., et al. 1988, MNRAS, 230, 107
Hough, J. H., Aiken, D. K., Whittet, D. C. B., Adamson, A. J., & Chrysostomou, A. 2008, MNRAS, 387, 797
Jones, R. V., & Spitzer, L. 1967, ApJ, 147, 943
Jones, T. J., Kim, J.-A., Dowell, C. D., et al. 2020, AJ, 160, 167
Kemp, J. C., & Wolstencroft, R. D. 1972, ApJ, 176, L115
Kim, S. H., & Martin, P. G. 1995, ApJ, 444, 293
Larson, K. A., Whittet, D. C. B., & Hough, J. H. 1996, ApJ, 472, 755
Lazarian, A. 1995, MNRAS, 274, 679
Lazarian, A. 1997, ApJ, 483, 296
Lazarian, A., & Draine, B. T. 1997, ApJ, 487, 248
Lazarian, A., & Hoang, T. 2007, MNRAS, 378, 910
Lazarian, A., & Hoang, T. 2019, ApJ, 883, 122
Lazarian, A., & Roberge, W. G. 1997, MNRAS, 287, 941
Lazarian, A., Andersson, B.-G., & Hoang, T. 2015, in Polarimetry of Stars and Planetary Systems, ed. L. Kolokolova, et al. (Cambridge: Cambridge Univ. Press), 81
Liu, J., Qiu, K., Berry, D., et al. 2019, ApJ, 877, 43
Mangilli, A., Aumont, J., Rotti, A., et al. 2021, A&A, 647, A52
Martin, P. G. 1974, ApJ, 187, 461
Martin, P. G. 1989, in IAU Symp. 135, Interstellar Dust, ed. L. J. Allamandola, & A. G. G. M. Tielens (Dordrecht: Kluwer), 55
Martin, P. G., & Angel, J. R. P. 1976, ApJ, 207, 126
Martin, P. G., & Whittet, D. C. B. 1990, ApJ, 357, 113
Martin, P. G., Adamson, A. J., Whittet, D. C. B., et al. 1992, ApJ, 392, 691
Martin, P. G., Clayton, G. C., & Wolff, M. J. 1999, ApJ, 510, 905
Mathis, J. S. 1986, ApJ, 308, 281
McMillan, R. S., & Tapia, S. 1977, ApJ, 212, 714
Messinger, D. W., Whittet, D. C. B., & Roberge, W. G. 1997, ApJ, 487, 314
Michail, J. M., Ashton, P. C., Berthoud, M. G., et al. 2021, ApJ, 907, 46

Michalsky, J. J., & Schuster, G. J. 1979, ApJ, 231, 73
Panopoulou, G. V., Hensley, B. S., Skalidis, R., Blinov, D., & Tassis, K. 2019, A&A, 624, L8
Planck Collaboration, 2015, A&A, 576, A106
Planck Collaboration, 2020, A&A, 641, A12
Purcell, E. M. 1979, ApJ, 231, 404
Roberge, W. G., & Lazarian, A. 1999, MNRAS, 305, 615
Roberge, W. G., Hanany, S., & Messinger, D. W. 1995, ApJ, 453, 238
Rodrigues, C. V., Magalhães, A. M., Coyne, G. V., & Piirola, V. 1997, ApJ, 485, 618
Sadavoy, S., Myers, P. C., Stephens, I. W., et al. 2018, ApJ, 859, 165
Serkowski, K. 1973, in IAU Symp. 52, Interstellar Dust and Related Topics, ed. J. M. Greenberg, & H. C. van de Hulst (Dordrecht: Reidel), 145
Serkowski, K., Mathewson, D. S., & Ford, V. L. 1975, ApJ, 196, 261
Smith, C. H., Wright, C. M., Aitken, D. K., Roche, P. F., & Hough, J. H. 2000, MNRAS, 312, 327
Soler, J. D., Alves, F., Boulanger, F., et al. 2016, A&A, 596, A93
Spitzer, L. 1978, Physical Processes in the Interstellar Medium (New York: Wiley)
Spitzer, L., & McGlynn, T. A. 1979, ApJ, 231, 417
Spitzer, L., & Tukey, J. W. 1951, ApJ, 114, 187
Valencic, L. A., Clayton, G. C., Gordon, K. D., & Smith, T. L. 2003, ApJ, 598, 369
Vrba, F. J., Coyne, G. V., & Tapia, S. 1993, AJ, 105, 1010
Whittet, D. C. B. 2011, in ASP Conf. Ser. 449, Astronomical Polarimetry 2008: Science from Small to Large Telescopes, ed. P. Bastien, et al. (San Francisco, CA: ASP), 93
Whittet, D. C. B. 2015, ApJ, 811, 110
Whittet, D. C. B., & van Breda, I. G. 1978, A&A, 66, 57
Whittet, D. C. B., Martin, P. G., Hough, J. H., et al. 1992, ApJ, 386, 562
Whittet, D. C. B., Gerakines, P. A., Hough, J. H., & Shenoy, S. S. 2001, ApJ, 547, 872
Whittet, D. C. B., Hough, J. H., Lazarian, A., & Hoang, T. 2008, ApJ, 674, 304
Wilking, B. A., Lebofsky, M. J., Martin, P. G., Rieke, G. H., & Kemp, J. C. 1980, ApJ, 235, 905
Wilking, B. A., Lebofsky, M. J., & Rieke, G. H. 1982, AJ, 87, 695
Wolff, M. J., Clayton, G. C., Kim, S. H., Martin, P. G., & Anderson, C. M. 1997, ApJ, 478, 395
Wright, C. M., Do Duy, T., & Lawson, W. 2016, MNRAS, 457, 1593
Wurm, G., & Schnaiter, M. 2002, ApJ, 567, 370

Dust in the Galactic Environment (Third Edition)

D C B Whittet

Chapter 5

The Observed Properties of Dust—III. Infrared Absorption Features

Spectroscopy of solid-state absorption features provides a direct and powerful technique for investigating the composition and evolution of cosmic dust: see Sections 1.2.5 and 2.1.4 for introductory discussions. A prominent absorption feature attributed to dust—the ultraviolet extinction "bump" centered at 217.5 nm—is reviewed in some detail in Chapter 3. In the present chapter, we turn our attention to the infrared region of the spectrum. The vibrational resonances of virtually all molecules of astrophysical interest occur at wavelengths in the spectral range from 2 to 30 μm. We begin with a brief discussion of the principles of infrared spectroscopy and its application to studies of absorption in solid particles. Subsequent sections review the results and their implications for dust in diffuse and dense phases of the interstellar medium (ISM; Sections 5.2 and 5.3, respectively). The topic of infrared spectral emission features is addressed in Chapter 6, and some applications of infrared spectroscopy to studies of the origin and evolution of the dust are discussed in Chapters 8 and 9.

5.1 Basics of Infrared Spectroscopy

5.1.1 Vibrational Modes in Solids

Absorption features at infrared wavelengths result from molecular vibrations within the grain material. The frequency of vibration for a given molecule is determined by the masses of the vibrating atoms, the molecular geometry, and the forces holding the atoms in their equilibrium positions. Consider for simplicity a diatomic molecule containing atoms of masses m_1 and m_2. The vibrations of the molecule may be represented by a harmonic oscillator in which the masses are joined by a spring obeying Hooke's law. The fundamental frequency of vibration is then given by

doi:10.1088/2514-3433/ac7204ch5

$$\nu_F = \frac{1}{2\pi}\left(\frac{k_F}{\mu}\right)^{\frac{1}{2}} \tag{5.1}$$

where $\mu = m_1 m_2/(m_1 + m_2)$ is the reduced mass and k_F is the force constant of the chemical bond.[1] The vibrations induce optical activity centered at wavelength $\lambda_F = c/\nu_F$. For gas-phase molecules, rotational splitting of vibrational energy levels results in molecular "bands" composed of many closely-spaced lines. This is illustrated in Figure 5.1 (upper frame) for the fundamental vibrational mode of carbon monoxide: separate bands, termed P and R branches, arise from application of the rule $\Delta J = \pm 1$, where J is the rotational quantum number. All rotation is suppressed in the solid phase, however, and the P and R branches are replaced by a single, continuous spectral feature centered near λ_F, as shown in Figure 5.1 (lower frame). The solid state feature is broader than the individual gas-phase lines because of interactions between neighboring molecules in the solid, but much narrower than the entire band. Solid and gas-phase absorptions are thus easily distinguishable, even if the spectral resolution is insufficient to resolve the individual lines in the gaseous

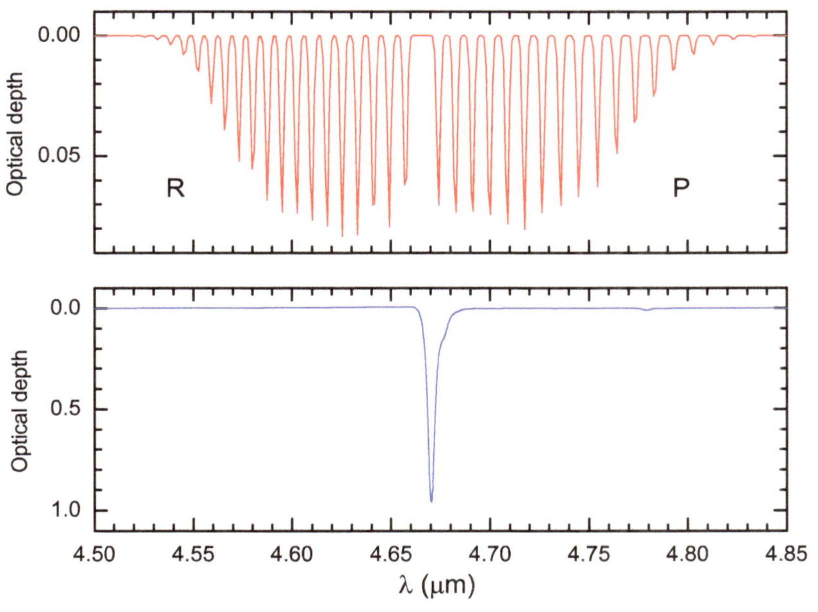

Figure 5.1. Infrared spectra of the fundamental vibrational mode of carbon monoxide in gas and solid form. Upper frame: gaseous CO at $T = 100$ K, showing the P and R branches caused by rotational splitting of the vibrational transition. Lower frame: Solid CO at $T = 10$ K, in which absorption is purely vibrational. Data courtesy of Jean Chiar.

[1] As molecules are not perfect harmonic oscillators, the actual vibrational frequency differs from ν_F by a factor that depends on the properties of the molecule and the vibrational quantum number of the energy level (e.g., Banwell & McCash 1994). At the low energy states considered here, this factor is close to unity.

spectrum. Note that vibrations are also possible at the harmonic (overtone) frequencies $2\nu_F$, $3\nu_F$, etc., but these do not generally produce observable features in interstellar dust.

Polyatomic molecules naturally have more possibilities for vibrational motion. A linear species such as CO_2 has vibrations associated with bending of the O=C=O structure in addition to longitudinal stretching of each C=O bond. For larger molecules, the complexity of the spectrum tends to increase rapidly with the number of atoms. Even a relatively simple organic molecule such as methanol (CH_3OH) exhibits distinct features arising from stretching vibrations of the C–H, C–O, and O–H bonds, structural deformation of the CH_3 unit, and weaker combination features arising when two fundamental modes are excited simultaneously. Absorptions occurring in these and other species that are detected or candidates for detection in interstellar solids are listed in Tables 5.1 (refractory solids) and 5.2 (ices).

A prerequisite for the production of a spectral feature is that the dipole moment of the molecule oscillates during the vibration. This is true in most cases of astrophysical interest, but the homonuclear molecules O_2 and N_2 are important exceptions. These species are infrared-inactive, producing no features in their pure state. Weak features may be induced when such a molecule is present as an "impurity" in a host matrix if interactions with neighboring species perturb the symmetry of the molecule (Ehrenfreund & van Dishoeck 1998). However, searches for such features in interstellar ices have so far been unsuccessful, yielding only loose upper limits on the abundances of these molecules (e.g., Boogert et al. 2015).

Comparison with laboratory data is the key to reliable interpretation of observational spectra. Assignments of solid state features to specific molecules cannot always be made simply on the basis of wavelength coincidence. Absorptions are associated in the first instance with chemical bonds or molecular subunits rather than specific molecules, so ambiguities may occur, and problems of identification are exacerbated if the observed features are weak, broad and/or overlapping, as is often the case. The availability of laboratory spectra for a wide range of materials obtained under realistic physical conditions allows the possibility to distinguish between potential candidates and gain new insight into the nature of the absorbing material.

Table 5.1. Absorptions in Some Refractory Solids of Astrophysical Interest.

Material	Mode	λ (μm)	κ_λ ($m^2\ kg^{-1}$)	References
"HAC"	aliphatic C–H str.	3.4	150	Mennella et al. (2002);
	aromatic C–C str.	6.2	30	Furton et al. (1999)
"Astrosilicate"	Si–O stretch	9.7	270	Li & Draine (2001);
	O–Si–O bend	19	110	Demyck et al. (2017a, 2017b)
Silicon carbide	Si–C stretch	11.2	660	Whittet et al. (1990)

Notes: Mean values of wavelength λ and mass absorption coefficient κ_λ correspond to peak absorption in each case. See text for discussion, caveats, and additional references.

Table 5.2. Vibrational Modes in Some Molecular Ices of Astrophysical Interest.

Molecule	Mode	λ (μm)	$\mathcal{A}$ (m/molecule)
H_2O	O–H stretch	3.05	2.0×10^{-18}
	H–O–H bend	6.0	1.1×10^{-19}
	libration	12.5	3.1×10^{-19}
	lattice mode	44	4.4×10^{-20}
NH_3	N–H stretch	2.96	2.7×10^{-19}
	deformation	6.16	5.6×10^{-20}
	inversion	9.35	1.9×10^{-19}
CH_4	C–H stretch	3.32	1.3×10^{-19}
	deformation	7.69	1.0×10^{-19}
CO	C=O stretch	4.67	1.3×10^{-19}
CO_2	C=O stretch	4.27	1.4×10^{-18}
	O=C=O bend	15.3	1.8×10^{-19}
CH_3OH	O–H stretch	3.08	1.0×10^{-18}
	C–H stretch	3.53	5.3×10^{-20}
	CH_3 deformation	6.85	1.1×10^{-19}
	CH_3 rock	8.85	1.5×10^{-20}
	C–O stretch	9.75	1.7×10^{-19}
H_2CO	C–H stretch (asym.)	3.47	2.8×10^{-20}
	C–H stretch (sym.)	3.54	3.7×10^{-20}
	C=O stretch	5.81	9.6×10^{-20}
	CH_2 scissor	6.69	3.9×10^{-20}
HCOOH	C=O stretch	5.85	6.0×10^{-19}
	CH deformation	7.25	3.0×10^{-20}
	C–O stretch	8.22	3.0×10^{-19}
C_2H_6	C–H stretch	3.36	1.6×10^{-19}
	CH_3 deformation	6.85	6.0×10^{-20}
CH_3CN	C≡N stretch	4.41	3.0×10^{-20}
OCN^-	C≡N stretch	4.62	1.0×10^{-18}
H_2S	S–H stretch	3.93	2.9×10^{-19}
OCS	O=C=S stretch	4.93	1.5×10^{-18}
SO_2	S=O stretch	7.55	3.4×10^{-19}
O_3	deformation	9.62	1.4×10^{-19}

Notes: Values of the band strength $\mathcal{A}$ are from Gerakines & Hudson (2020), Bouilloud et al. (2015), Giuliano et al. (2014), Schutte (1999), and references therein.

The spectrum of a vibrating molecule is influenced by its molecular environment. For example, absorptions near $\lambda \sim 3.4\ \mu$m attributed to the C–H bond may occur in many different compounds, but the width, shape, and peak position of the profile vary according to the structure of the molecule and the nature of the host matrix. Thus, we may distinguish between aliphatic and aromatic hydrocarbons

(Pendleton & Allamandola 2002), and between C–H-bearing ices such as CH_4 and CH_3OH (e.g., Boogert et al. 2015). Complete spectral coverage is also essential, enabling, for example, pure hydrocarbons to be distinguished from organics that contain other "CHON" elements. The absorption features of species such as CO and CO_2 exhibit different profiles depending on whether they are in pure form or a constituent of ice mixtures dominated by another molecule such as H_2O (e.g., Ehrenfreund et al. 1997). Finally, both ices and refractory solids show distinct absorption profiles in their ordered (crystalline) and disordered (amorphous) states. The crystallinity of an interstellar solid is governed by its conditions of formation, subsequent temperature changes, and exposure to energetic radiation, particle winds and cosmic rays. Thus, infrared spectroscopy offers the possibility to explore the physical as well as the chemical evolution of the dust.

5.1.2 Intrinsic Strengths

Laboratory measurements also provide information on the intrinsic strengths of the features. Such data are needed to calculate the amount of a given absorber represented by the observed strength of its absorption. For a generic refractory material such as amorphous carbon or silicate (Table 5.1), where many different molecular structures may be present, it is convenient to specify the intrinsic strength in terms of the mass absorption coefficient, defined as the absorption cross-section per unit mass:

$$\kappa_\lambda = \frac{C_{\rm abs}}{m_{\rm d}} = \frac{3Q_{\rm abs}}{4\rho_{\rm s}a} \tag{5.2}$$

where $Q_{\rm abs} = C_{\rm abs}/\pi a^2$ (Section 2.1.1), and spherical dust grains of mass $m_{\rm d}$, radius a, and density $\rho_{\rm s}$ are assumed. Note that the quantity $Q_{\rm abs}/a$ is independent of grain size in the small-particle limit (Section 2.1.2), generally a reasonable assumption at mid-infrared wavelengths. The volume-averaged density of an absorber in a column of length L and unit cross-section needed to account for the strength of a feature centered at wavelength λ with optical depth τ_λ relative to the continuum is then given by

$$\rho = \frac{\tau_\lambda}{\kappa_\lambda L}. \tag{5.3}$$

Results may be compared with equivalent data for total extinction (Section 3.2.6) to estimate the contribution of the absorber to the total mass of dust.

Mass absorption coefficients may be calculated from Mie theory if appropriate values of the complex refractive indices of the material are available, or measured directly for laboratory samples. Results are sensitive to the quality and suitability of the data used in the calculations and to methods of sample preparation, laboratory conditions and experimental procedures in empirical work. Values listed in Table 5.1 are intended only as a rough guide. Results for the C–H feature in hydrogenated amorphous carbon (HAC) naturally depend on the degree of hydrogenation of the particles: the value in Table 5.1 is appropriate to H/C ~ 0.6, as proposed by Li & Greenberg (2002) based on data from Furton et al. (1999) and Mennella et al. (2002).

Published data for amorphous silicates cover a range of stoichiometries, from the pure magnesian forms enstatite ($MgSiO_3$) and forsterite (Mg_2SiO_4) to pyroxenes and olivines with significant iron content (see Section A.3: silicates). Optical properties are expected to vary with composition, but significant dispersion is also apparent in the available data for nominally similar compositions, most likely caused by differences in methodology and laboratory practice (see Day 1981; Dorschner et al. 1988, 1995; Coupeaud et al. 2011; Demyk et al. 2017a, 2017b). Results in Table 5.1 for "astrosilicate" are based on calculations from Demyk et al. for a size distribution of generic amorphous silicates using optical data from Li & Draine (2001). Finally, silicon carbide is included in Table 5.1 for completeness, although it appears to be no more than a trace constituent of interstellar dust.

The intrinsic strength of an absorption feature attributed to a specific vibrational mode in a specific molecule may be expressed as a band strength, $\mathcal{A}$, defined as the integrated absorption cross-section per molecule. For an unsaturated absorption line, the column density N of the absorber is

$$N = \frac{\int \tau(x)\,\mathrm{d}x}{\mathcal{A}} \tag{5.4}$$

where $x = \lambda^{-1}$ and integration is over the entire profile. For features with Gaussian-like profiles, the approximation $N \simeq \gamma\ \tau_{\max}/\mathcal{A}$ may be used, where γ is the width (full width at half maximum) of the profile in wavenumber units. Representative data for band strengths in various laboratory ices are listed in Table 5.2; again, these should be regarded as guides to typical values in the literature rather than definitive results. The reviews by Schutte (1999) and Bouilloud et al. (2015) provide useful summaries of previous work and descriptions of experimental procedures.

5.1.3 Observational Approach

The apparatus used to measure infrared spectra in a terrestrial laboratory typically consists of a radiation source, a sample chamber, and an instrument to record the spectrum. These elements have equivalents in the astronomical context: the radiation source is typically a star, the sample chamber is the intervening interstellar medium, and the recording instrument is a spectrometer attached to a telescope. An additional factor for ground-based astronomy is absorption in the Earth's atmosphere, which limits data collection to certain atmospheric "windows" nested between regions of strong telluric absorption. Indeed, several of the species we may wish to study in the cosmos (H_2O, CO_2, CH_4) are also greenhouse gases responsible for infrared opacity in the atmosphere. This restriction may be alleviated by placing telescopes at high altitude: the mean scale height of water vapor ($\sim$2 km) is sufficiently low that considerable advantage is gained by observing from a mountain-top site or an airborne observatory. However, some telluric features are so strong that the spectral regions they block are accessible only to observation from space. Facilities such as the Infrared Space Observatory (ISO) and the Spitzer Space Telescope enabled these gaps to be filled, extending the inventory of major interstellar condensates available for study by infrared techniques. In addition to spectral coverage, it is important that the instruments should have sufficient resolution to detect discrete structure in the solid state features and to enable contaminating stellar lines to be separated. Resolving powers needed to do this ($\lambda/\Delta\lambda \gtrsim 1000$) are now routinely available on ground-based telescopes and in archival data from the ISO mission.

The sources selected for observation generally fall into one of two groups, as illustrated schematically in Figure 5.2. A background field star may be used to investigate absorption by foreground dust in a manner analogous to the construction of extinction curves (Section 3.1). The absorption may arise primarily in a single cloud (as depicted in Figure 5.2) or in a number of clouds distributed along the line of sight. The other possibility is that the star is *embedded* in the cloud responsible for most or all of the absorption. This is a common situation in molecular clouds, as the brightest available sources are typically young stars enclosed in dusty envelopes of protostellar material. The observed spectra are usually calibrated in units of flux density (see Section A.3). Examples are shown in Figures 5.3 and 5.4.

Each observed spectrum is a combination of the source spectrum and the extinction (both absorption and scattering) produced by intervening interstellar material. To enable quantitative analysis of the data, these components need to be separated to yield the interstellar spectrum independent of the background source. Ideally, a reliable model for the intrinsic spectrum should be constructed. In the case of a typical background star with a normal photosphere, the flux declines with

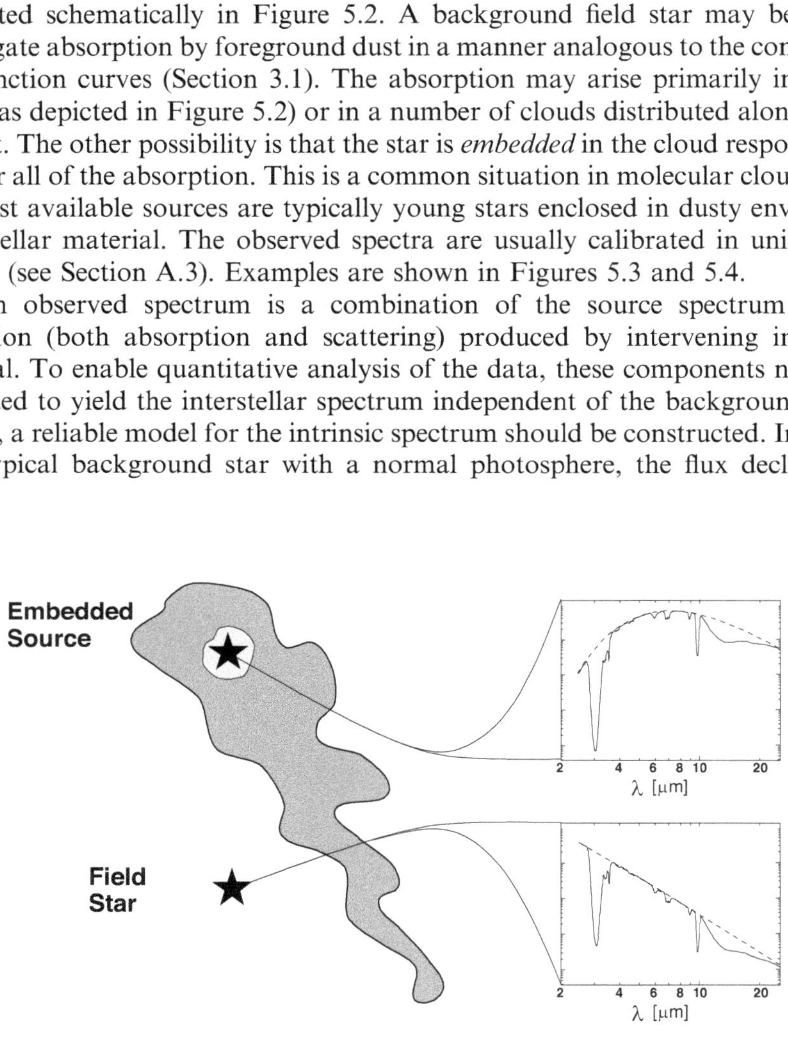

Figure 5.2. Schematic illustration of two types of infrared source used to study interstellar features in a molecular cloud. The field star is not physically associated with the cloud, displaying a photospheric spectrum with foreground absorption and reddening. Embedded stars are typically much brighter at mid-infrared wavelengths because of circumstellar emission from warm dust. Image credit: reproduced courtesy of Perry Gerakines.

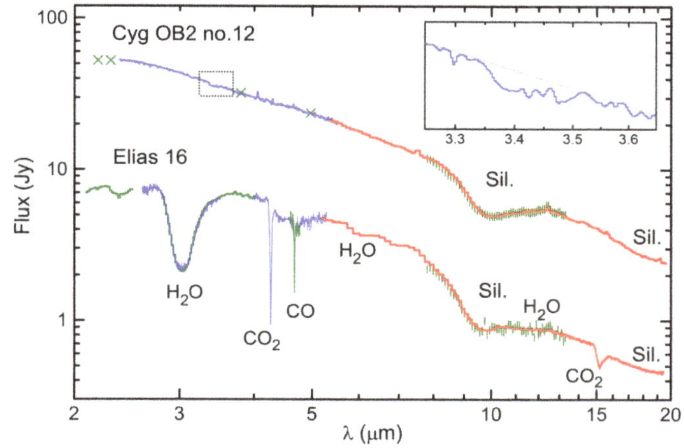

Figure 5.3. Infrared spectra of contrasting environments toward field stars. Diffuse clouds toward Cyg OB2 no. 12 produce silicate absorptions near 10 and 19 μm but no discernible ice absorption. The inset at top right is an enlargement of data in the dotted rectangle, containing the shallow 3.4 μm C–H dust feature; the gray line is a local continuum. Denser material in the Taurus dark cloud toward Elias 16 exhibits both silicate and ice absorptions. Photospheric features are also present in both spectra, notably the CO overtone band near 2.4 μm in Elias 16 (a red giant), and narrow H absorption and emission lines in Cyg OB2 no. 12 (a blue supergiant). Data are from the Infrared Space Observatory and Spitzer Space Telescope archives (blue and red, respectively), and ground-based observations (green) by Bowey et al. (1998), Chiar et al. (1995), and Smith et al. (1993).

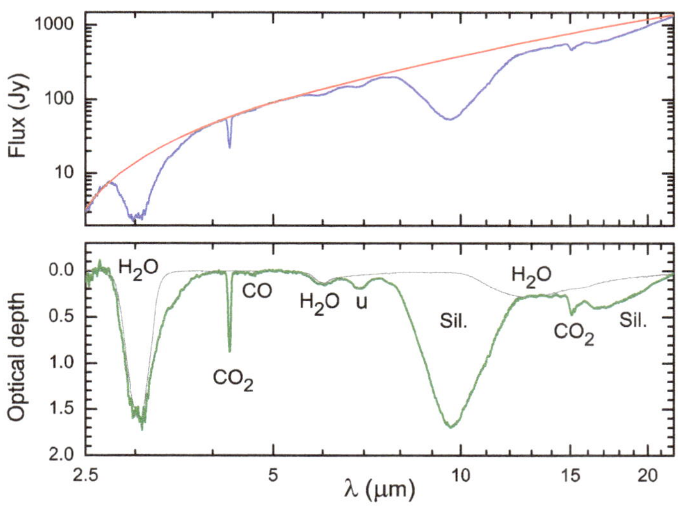

Figure 5.4. Infrared spectra of the embedded source S140, a luminous young stellar object in a molecular cloud. Upper frame: flux spectrum (ISO data, blue), and the adopted continuum (red) from 3rd-order polynomial fits to featureless regions of the spectrum (separate fits to near and mid-infrared data are joined at 5 μm). Lower frame: resultant optical depth spectrum (green). Also shown is a laboratory spectrum of pure amorphous H_2O-ice (gray curve), scaled to match the depths of the 3.0, 6.0, and 13 μm features in the observed spectrum. These and other ice features are labeled; the feature labeled "u" lacks a secure identification.

increasing wavelength and the Rayleigh–Jeans approximation (Equation (2.48)) may provide a reasonable representation of the continuum. If photospheric spectral lines are a significant source of contamination, standard spectra may be used to identify and remove them. The continuum component of the extinction, which approximates to a power law (Section 3.2.4), may also be removed to leave only absorption features. In the case of an embedded source, the underlying spectrum is typically some unknown combination of a reddened photosphere and emission from circumstellar dust over a range of temperatures. In such cases, the most realistic approach may be to determine an empirical "local" continuum for each feature or group of features. This is accomplished by fitting a mathematical function such as a polynomial to smooth regions of the spectrum adjacent to the interstellar feature(s). With a reliable continuum determined by such methods, the flux spectrum may be expressed in terms of optical depth:

$$\tau(\lambda) = \ln\left\{\frac{I_0(\lambda)}{I(\lambda)}\right\} \tag{5.5}$$

where the functions $I(\lambda)$ and $I_0(\lambda)$ represent the observed spectrum and the adopted continuum, respectively: see Figure 5.4 for an example. This format facilitates comparisons, both of one source to another and of observed sources to laboratory data.

Applying these techniques to observations of a diverse selection of targets, dust spectra may be compared over a wide range of environments. In so doing, a general pattern emerges: silicate absorption is ubiquitous, appearing in both diffuse and dense phases of the ISM whenever a sufficient dust column is present; but evidence for ices is limited to lines of sight that intercept dense molecular clouds (see Figure 5.3 for representative examples). The features arising in these distinct environments are discussed in the following sections.

5.2 The Diffuse ISM

Even the strongest infrared features observed in the diffuse ISM are weak in comparison to the continuum extinction in the visible (see Figure 3.5). Because of this, there is a sparsity of suitable sources for study of dust features in diffuse environments: OB stars with normal photospheres, traditionally used to study extinction at shorter wavelengths (Section 3.1), have steeply declining spectra in the infrared, requiring observations of extreme sensitivity to yield reliable data. Ideal targets are exceptionally luminous sources with predictable intrinsic spectra, located at distances in the galactic disk sufficient for accumulation of high extinction without intercepting a molecular cloud. Reasonable approximations to this ideal include distant supergiants and Wolf-Rayet stars, of which the well-studied B-type supergiant Cygnus OB2 no. 12 is widely used as a benchmark source. Considered to be one of the most luminous stars in the Galaxy, it remains accessible to observation in the spectral range of interest, yet with sufficient extinction ($A_V \simeq 10$) to produce measurable dust features. There is much discussion in the literature on the nature of this star and the distribution of dust in the line of sight (might some of it be circumstellar?), but the balance of evidence supports the assumption that it is an appropriate choice for studies of diffuse-ISM dust. See Whittet (2015) for a detailed analysis and review.

Other targets widely used for this purpose include members of a cluster of infrared sources located near the galactic center (GC) of the Milky Way. These sources appear to be luminous stars obscured by some 30 magnitudes of visual extinction, mostly accumulated along an 8 kpc pathlength that passes through several spiral arms (see Figure 1.7). However, some extinction toward these sources is evidently local to the GC, arising in the "central molecular zone," a disk-like structure surrounding the nucleus that contains both diffuse atomic and molecular gas (see Geballe et al. 2021 for a detailed description). Because of this, and perhaps also because of chance alignments with molecular clouds in the intervening spiral arms, the spectra of these sources are composite, including both diffuse-ISM and molecular-cloud dust signatures (see Figure 5.5 for an example). The ice features vary in depth from source to source on small angular scales (Chiar et al. 2000, 2002), consistent with spatial differences in the contributions of molecular clouds. The diffuse-ISM component seems more consistent and usually dominant, as indicated by the $\tau_{3.0}/\tau_{9.7}$ ratio (i.e., the ratio of ice to silicate, based on the amplitudes of their strongest features), which is typically ~0.15 for most sources in the GC cluster. For comparison, $\tau_{3.0}/\tau_{9.7} \sim 2$ in the molecular cloud toward Elias 16 and <0.04 in the diffuse ISM toward Cyg OB2 no. 12.

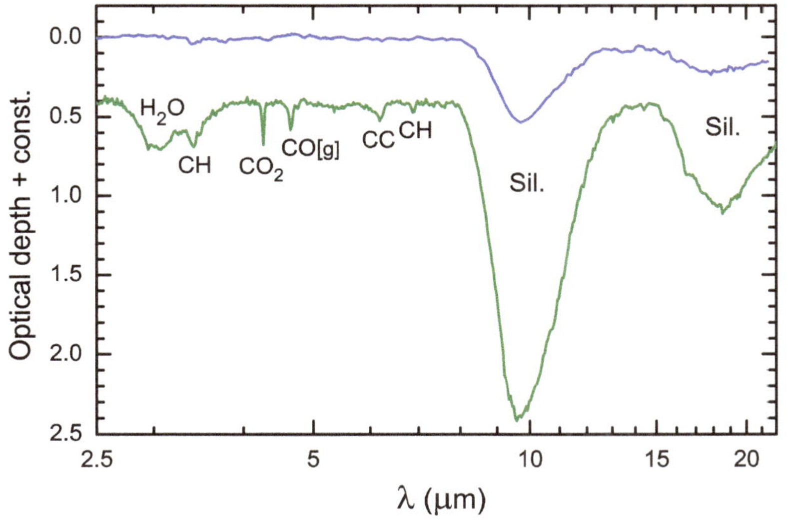

Figure 5.5. Optical-depth spectra of sources sampling (mostly) diffuse-ISM dust. Above: Cyg OB2 no. 12 (blue), based on data from the ISO and Spitzer archives (see Figure 5.3). Below: the GC source GC3 (green; ISO data, displaced downward by 0.4 for display). The feature labeled CO[g] is a blend of gas-phase CO lines; those labeled H_2O and CO_2 are ice features, and those labeled CH and CC are attributed to hydrocarbon dust.

Optical-depth spectra are compared in Figure 5.5 for Cyg OB2 no. 12 and GC3. Both display prominent silicate absorptions but only GC3 shows detectable ice features, consistent with a modest contribution to its total extinction from molecular clouds. Features at 3.4, 6.2, and 6.85 μm are attributed to hydrocarbons (CH stretch, CC stretch, and CH deformation modes, respectively): these features are clearly seen toward the GC, and also detected at much lower amplitudes in Cyg OB2 no. 12 (see Hensley & Draine 2021 for a review). Overall, the spectra provide convincing evidence for two generic types of grain material in the diffuse ISM: silicates and hydrocarbons.

5.2.1 Silicates

Silicates are ubiquitous constituents of solids in space (see Henning 2010 for a review). Their signature absorptions near 9.7 and 19.5 μm dominate the spectrum of dust in the Milky Way (Figure 5.5) and are also widely observed in the nuclei of dusty active galaxies (Alonso-Herrero et al. 2016). The 9.7 μm feature is both the strongest and the most extensively studied, in part because it is readily accessible to ground-based as well as space-based observation. The position, width, and shape of the profile are influenced by several variables that specify the nature of the absorbing particles, relating to both chemical composition (e.g., stoichiometry, Fe/Mg ratio, hydration) and physical structure (crystallinity, porosity, particle shape, and size). Studies of laboratory analogs are essential to interpretation of the observations, but a convolution of chemical and physical effects can hinder attempts to attribute spectral differences to changes in any given variable (Speck et al. 2011). This results in a level of degeneracy that prevents a truly unambiguous specification of typical "astronomical silicate." Overall, the results tend to favor Mg-rich forms with stoichiometries intermediate between olivine and pyroxene (Kemper et al. 2004; Min et al. 2007; Fogerty et al. 2016).

Progress has been made in addressing the question of crystallinity, a variable that strongly influences the profiles of the features (e.g., Fabian et al. 2000). Amorphous, disordered silicates produce broad, smooth features, whereas crystalline silicates (common in terrestrial igneous rocks) produce profiles with distinctive structure (see Figure 5.6 for an example). Spectral evidence for crystalline silicates has been found in the circumstellar environments of both old stars that inject dust into the ISM and young stars that form within dense clouds (Henning 2010); but crystallinity may be lost over time in the ISM, e.g., as the result of cosmic ray exposure. The observed interstellar features are broad and smooth with minimal structure, their profiles well matched by models based on laboratory data for amorphous silicates (Figure 5.6). It has been argued, based on such results, that silicates in the diffuse ISM are almost entirely amorphous (Kemper et al. 2004, 2005), but shallow structure centered near 11.1 μm in the wing of the 9.7 μm profile suggests a minor crystalline component ($\sim$1.5% by mass; Wright et al. 2016; Do-Duy et al. 2020). Observations of Mg and Si absorption edges at X-ray wavelengths (Section 7.1.5) have been interpreted as evidence for crystalline fractions as high as $\sim$10% toward the GC (Rogantini et al. 2020).

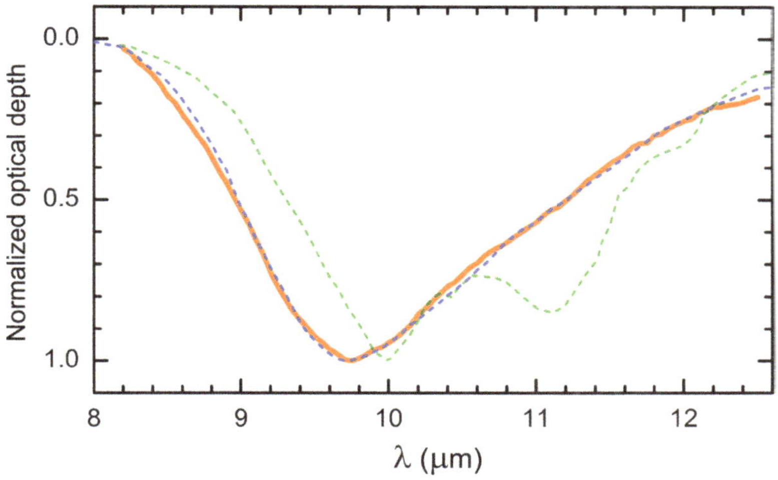

Figure 5.6. The 9.7 μm silicate profile. Observations of the GC source Sgr A* (orange) are compared with calculations for small grains composed of amorphous olivine ($MgFeSiO_4$, blue) and crystalline forsterite (Mg_2SiO_4, green), based on results from Kemper et al. (2004) and Fabian et al. (2000). Each profile is normalized to unity at peak optical depth.

Further insight into the nature of the silicates arises from spectropolarimetric observations. Detections of polarization counterparts to the absorption features confirm that silicate grains are responsive to alignment (Section 4.2.7). The theory of dichroic absorption by aligned grains (Aitken 1989) predicts that the position of peak polarization should be shifted to a somewhat longer wavelength compared with the peak in optical depth, and this is indeed observed (see Figure 5.7). The polarization profile is sensitive to the optical properties of the particles, and in particular to their crystallinity. The observations are consistent with absorption by amorphous, spheroidal olivine particles, lending further support to the conclusion that interstellar silicates are predominantly non-crystalline.

Most naturally-occurring terrestrial silicates are poor spectroscopic matches to the ISM because of their inherent crystallinity. Some members of the phyllosilicate group are potential exceptions, exhibiting smooth absorption profiles that resemble the interstellar features (Knacke & Krätschmer 1980). Phyllosilicates are characterized by a layered lattice structure. They are also hydrous, i.e., they contain integral OH units, and this property enables an observational test: hydrous silicates absorb in the 2.6–2.9 μm spectral range due to vibrational excitation of OH. An ice feature arising from OH bonds in solid H_2O overlaps this region in spectra of the GC, and because of this the spectrum of Cyg OB2 no. 12 (Figure 5.5) yields a cleaner constraint. No OH feature is discernible, setting an upper limit consistent with no more than about 2% by mass of OH in the silicates (a tentative detection based on an early release of ISO data was not confirmed in the final data set; see Whittet et al. 1997, 2001b). If this line of sight is typical, interstellar silicates must be largely anhydrous. In contrast, terrestrial and meteoritic hydrous silicates, which form in aqueous environments that presumably do not exist in the ISM, contain 5%–30% by mass of OH.

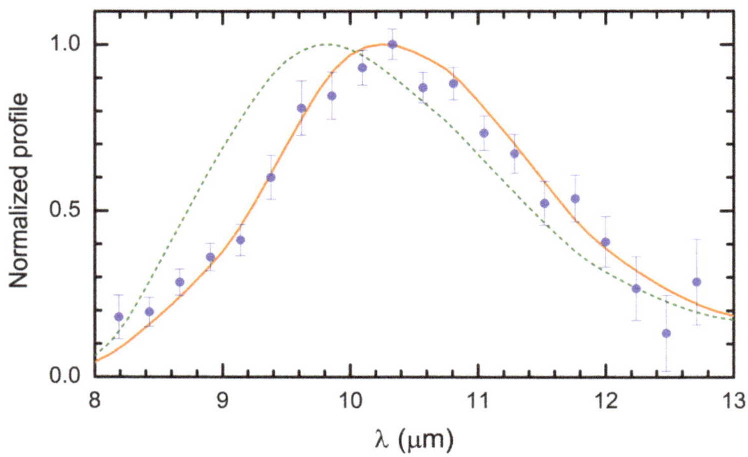

Figure 5.7. The polarization profile of the 9.7 μm silicate feature in the diffuse ISM. Blue points are mean observed values of p_λ for two early-type stars (WR 48A and GL 2104; Smith et al. 2000). The orange curve is the calculated p_λ profile for aligned oblate spheroids of size $a = 0.1$ μm and axial ratio $b/a = 2{:}1$, composed of amorphous olivine (Wright et al. 2002); the corresponding optical depth profile is also shown (dashed green curve). All profiles are normalized to unity at the peak. Data courtesy of Christopher Wright.

Silicate dust is an essential constituent of realistic models for interstellar extinction at all wavelengths (Section 3.6.2). The amplitude of its absorption is therefore predicted to increase proportionately with A_V if silicates are well-mixed with other forms of dust. This correlation is shown in Figure 5.8: the sample includes early-type (OB and Wolf-Rayet) stars in the galactic disk, mostly within about 3 kpc of the Sun, and sources in the GC region some 8 kpc distant. The early-type stars do, indeed, show a tight correlation, consistent with a linear fit through the origin:

$$A_V/\tau_{9.7} = 17.0 \pm 0.5. \tag{5.6}$$

Silicate features toward the GC sources, most remarkably Sgr A*, are deeper than predicted by extrapolation this linear fit. This result suggests systematic differences in the proportions of silicates and other dust materials along the line of sight. The 3.4 μm C–H feature exhibits similar behavior, discussed in Section 5.2.3.

A rough estimate of the silicate abundance needed to account for the observed 9.7 μm feature may be deduced from the result in Equation (5.6). Combining Equations (3.18) and (5.3), and assuming appropriate values of the refractive index and mass density of a typical grain, we may show that the density ratio of silicate dust to all dust responsible for visual extinction is

$$\frac{\rho_{\rm sil}}{\rho_{\rm d}} \approx \frac{2.9 \times 10^3}{\kappa_{9.7}} \left\{\frac{\tau_{9.7}}{A_V}\right\}. \tag{5.7}$$

Assuming $\kappa_{9.7} \approx 270$ m^2 kg^{-1} (Table 5.1), the result in Equation (5.6) suggests that silicates contribute about 60% of the mass of dust in the diffuse ISM. This estimate is

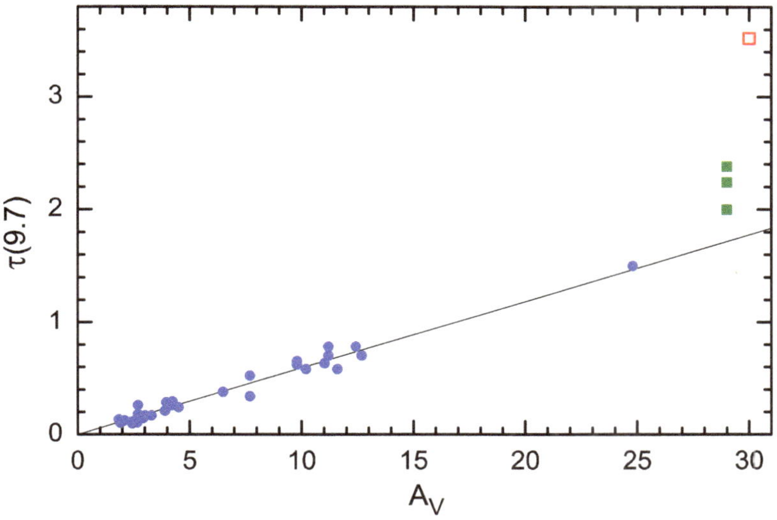

Figure 5.8. The correlation of peak optical depth in the 9.7 μm silicate feature with visual extinction. Blue circles correspond to luminous stars in the galactic disk. Infrared sources near the GC are denoted by squares in green (Quintuplet region) and red (Sgr A*). The diagonal line is the linear least-squares fit through the origin, limited to points with $A_V < 15$. Data are from Roche & Aitken (1984), Rieke & Lebofsky (1985), Chiar & Tielens (2001, 2006), van Breemen et al. (2011), and Gordon et al. (2021).

subject to considerable uncertainty but broadly consistent with predictions based on the depletions of the dust-forming chemical elements (Section 7.3).

5.2.2 Silicon Carbide?

One candidate listed in Table 5.1 that is surprisingly elusive in interstellar spectra is the silicon carbide (SiC) feature centered near 11.2 μm. This feature is often seen in the spectra of carbon-rich red giants (Section 8.2.2), and SiC is therefore evidently a constituent of circumstellar dust injected by such stars into the interstellar medium. Indeed, SiC stardust particles of extrasolar origin have been identified in meteorites (Section 8.3), providing further evidence for the existence of this compound at some level in interstellar dust, although the overall disposition of elemental Si in meteorites heavily favors silicates. The intrinsic strength of the feature associated with the Si–C bond is expected to exceed that of the corresponding Si–O feature in silicates (Table 5.1); so if similar quantities of O-bonded and C-bonded Si were present, one would expect the 11.2 μm feature to be *stronger* than the 9.7 μm feature, which is clearly not the case.

If present, SiC absorption should appear as blended structure in the 9.7 μm profile (e.g., Figure 5.6). Any such structure is very weak at best, and attempts to detect it face several challenges. Laboratory studies show that the position, width, shape, and strength of the SiC feature are sensitive to the assumed properties of the particles (size, shape, purity, crystallinity; Mutschke et al. 1999). Also, there is potential for confusion with structure centered at a similar wavelength arising in crystalline silicates, although the SiC feature is likely to be broader (Do-Duy et al. 2020). Estimated upper limits imply that no more than a few percent of the available Si is in the form of silicon carbide in our Galaxy (Whittet et al. 1990; Kemper et al. 2004; Chiar & Tielens 2006) or a well-studied external galaxy (Roche et al. 2015). This result is in accord with the requirement that essentially all of the available Si needs to be in silicates to explain the strengths of the observed 9.7 μm and 19.5 μm features (Section 7.3.2), and is also consistent with the preponderance of silicates over SiC in meteoritic presolar dust. The scarcity of SiC in the ISM is nevertheless surprising if carbon stars are a major source of interstellar grains. See Chen et al. (2022) for further discussion.

5.2.3 Hydrocarbons and Organic Refractory Matter

In addition to silicates, models consistent with the observed properties of dust in the diffuse ISM generally require two distinct populations of carbonaceous grains: (i) nanoparticles to explain the ultraviolet extinction bump and PAH emission features, and (ii) larger grains to make up the balance of the continuum extinction. The nature of the latter population is a major area of uncertainty in the modeling (Section 3.6). This section reviews the extent to which infrared spectroscopy may contribute to resolving the ambiguities. This goal is challenging, as even the strongest feature attributed to carbonaceous dust in the diffuse ISM (the aliphatic C–H stretch at 3.4 μm) is much weaker than the silicate features.

Materials considered as candidates for the 3.4 μm feature include hydrogenated amorphous carbon (HAC) and various forms of complex organic refractory matter (e.g., Duley & Williams 1981; Greenberg et al. 1995; Pendleton & Allamandola 2002). The position of peak absorption confirms that the vibrating atoms are situated in aliphatic chains rather than aromatic rings (the aromatic C–H stretch is centered at somewhat shorter wavelength, see below). Representative laboratory spectra are shown in Figure 5.9. An absorption feature centered at 3.4 μm is clearly seen in each case, but substantial differences are evident at other wavelengths. Organic kerogen extract from the Murchison meteorite (MK) lacks a hydration feature, but its spectrum is otherwise broadly similar to that of organic refractory matter (ORM) produced synthetically by UV irradiation of interstellar ice analogs: notably, both exhibit a rapid onset of absorption at $\lambda \simeq 5.8$ μm, attributed to organic carbonyl (C=O) bonds. Overlapping absorptions at longer wavelengths may include the C–C stretch at 6.2 μm and many other vibrational modes that naturally occur in complex organic matter, associated with (e.g.) C–OH and C–NH_2 functional groups and aromatic skeletal resonances (e.g., Pendleton & Allamandola 2002; Kebukawa et al. 2011). It is notable in both the MK and ORM spectra that the absorption edge near 5.8 μm is much deeper than the 3.4 μm feature, and this is inconsistent with observational data (e.g., Figure 5.5). In contrast, the HAC spectrum is relatively simple, with the C–H stretch at 3.4 μm its strongest feature, consistent with a dearth vibrational modes involving bonded O and N atoms in the material. HAC is thus a better match to the observations.

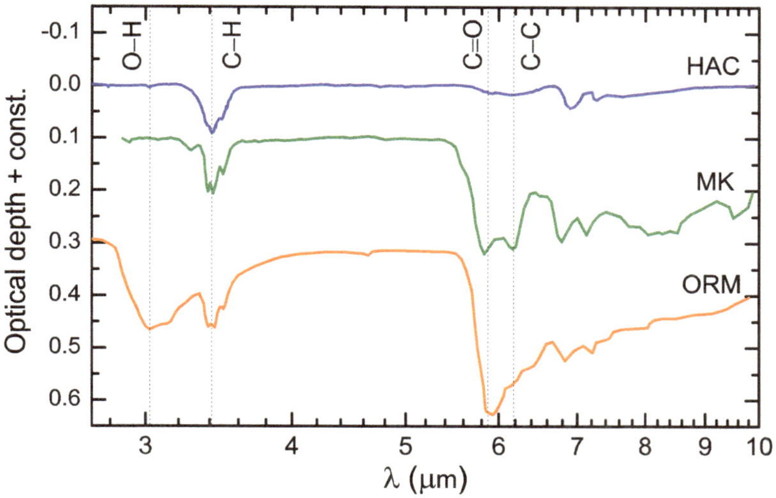

Figure 5.9. Laboratory spectra of some C-rich refractory dust candidates. From the top: hydrogenated amorphous carbon (HAC; Furton et al. 1999); meteoritic kerogen (MK) from Murchison (de Vries et al. 1993); and organic refractory matter (ORM) produced by UV photolysis of a laboratory ice mixture (Greenberg et al. 1995). Vertical dotted lines indicate the positions of some molecular resonances discussed in the text. The spectra are scaled to yield similar central depths in the 3.4 μm feature and displaced vertically for display.

It should be noted that this result does not necessarily rule out processed ice as a *source* of carbonaceous dust in the diffuse ISM. Ices are expected to form as mantles on refractory grain cores in molecular clouds (Section 9.2.2), and to undergo photochemical processing when the clouds give birth to new stars (Section 9.3.3). Following subsequent dispersal into diffuse environments, long-term exposure to energetic photons and particle winds may eventually "carbonize" the organics, reducing them (like charcoal) to amorphous carbon.

However, an important additional clue arises from spectropolarimetric observations, reviewed in Section 4.2.7. If the 3.4 μm absorber remains as a coating on grain cores that include silicates, as predicted by the formation sequence described above, a corresponding polarization signature should be detectable because silicate dust is a known source of interstellar polarization. Searches for this signature have yielded negative results (Adamson et al. 1999; Chiar et al. 2006). Models that propose silicate-carbon core-mantle or composite particles as major sources of both continuum extinction and 3.4 μm absorption thus appear to be ruled out (Li et al. 2014). The absorber must reside in particles that either lack optical anisotropy or fail to align. Potential hosts that remain viable in the light of this result are carbon nanoparticles or a distinct population of larger carbonaceous grains that do not produce significant polarization.

Insight into the complexity of the absorber is gained from detailed studies of the profiles. The 3.4 and 6.2 μm features observed toward the GC are shown in Figure 5.10 after subtraction of overlapping ice features (Chiar et al. 2013). Aliphatic hydrocarbons with the general formula CH_3–$(CH_2)_n$–CH_3 produce

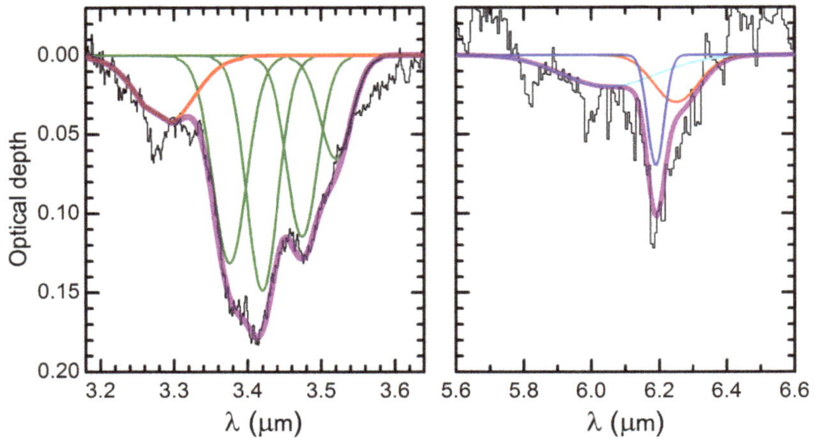

Figure 5.10. Gaussian profile fits to hydrocarbon absorptions in mean spectra for the Quintuplet sources near the GC (black). Left: the 3.4 μm (C–H stretch) feature is fitted by an aromatic component (red) and four aliphatic components (green). Right: the 6.2 μm (C–C stretch) feature is fitted by aromatic and olefinic components (red and blue) and an additional component (cyan) possibly due to carbonyl groups. The thick purple curve in each frame is the sum of components. Data courtesy of Jean Chiar, from Chiar et al. (2013).

distinct absorptions associated with the CH_2 and CH_3 groups, each of which may vibrate symmetrically or asymmetrically to produce structure in the profile (Sandford et al. 1991; Mennella et al. 1999). The 3.4 μm profile in Figure 5.10 is therefore decomposed into four Gaussian curves (green), representing (in order of increasing wavelength) the CH_3 asymmetric, CH_2 asymmetric, CH_3 symmetric, and CH_2 symmetric stretching modes. An additional component (red curve) corresponds to the *aromatic* C–H stretching mode centered near 3.3 μm. The 3.3 μm feature is significantly broader than expected for free-flying PAH molecules but consistent with an aromatic component of the hydrocarbon dust. A narrower dip centered at 3.28 μm (not included in the fitting process) is attributed to an olefinic component, i.e., hydrocarbon chain molecules that include double carbon (C=C) bonds. Analysis of the much weaker 3.4 μm feature observed in Cyg OB2 no. 12 yields broadly similar results (Hensley & Draine 2020), suggesting a degree of consistency in the nature of the hydrocarbon spectrum in different diffuse environments.

Similarly, absorptions in the spectral range that includes the 6.2 μm feature toward the GC (Figure 5.10) may be decomposed into distinct Gaussian curves. Components attributed to aromatic and olefinic C–C stretching modes match the primary feature; a broader, shallower component is tentatively attributed to carbonyl groups.

These results enable the properties of the hydrocarbons to be clarified. The relevant parameters are the abundance ratios of aromatic (sp^2-bonded) and aliphatic (sp^3-bonded) carbon atoms to each other and to hydrogen in the material. As the individual components identified in the profiles include both C–H and C–C resonances, and both aromatic and aliphatic modes, relative abundances can be estimated without a need to calibrate in terms of absolute values (the total hydrogen column density in the line of sight to the GC is poorly constrained). Results indicate that the absorbers are hydrogen-deficient (low H/C ratio) and highly aromatic. A grain structure consistent with these constraints consists of a core of aromatic amorphous carbon with a hydrogenated surface layer. See Chiar et al. (2013) for detailed discussion and an extensive guide to the literature on the properties of solid carbon.

So far we have focused on observations toward the GC because the absorption profiles are sufficiently deep and well-defined to allow detailed analysis. Toward other lines of sight with lower extinctions, data on the depth of the strongest (3.4 μm) hydrocarbon feature are available for a sample large enough to test its correlation with total extinction. The peak optical depth ($\tau_{3.4}$) of the feature tends to increase systematically with A_V, as expected if the carrier grains are well-mixed with other forms of dust in the galactic disk (see Godard et al. 2012, which includes a survey of previous work). The behavior of $\tau_{3.4}$ with respect to A_V resembles that of silicates (Figure 5.8), albeit with greater scatter arising from the comparative weakness of the 3.4 μm absorption. Similarly, the data for stars in the galactic disk are consistent with a linear correlation through the origin, represented by

$$A_V/\tau_{3.4} = 250 \pm 40. \tag{5.8}$$

An estimate based on this result, analogous to the calculation in Section 5.2.1 for silicates, suggests that the carrier grains account for $\sim 10\%$ of the dust mass if the value of $\kappa_{3.4}$ from Table 5.1 is assumed. This calculation may underestimate the total mass fraction of grains hosting 3.4 μm absorbers, however. As discussed above, the absorption is expected to occur in the hydrogenated surface layers of grains that are on average relatively H-poor. The mean H/C ratio of typical interstellar hydrocarbon dust may be significantly less than that of laboratory samples used to evaluate mass absorption coefficients, in which case the abundance of dust needed to account for the observed feature would be proportionately higher.

The 3.4 μm feature is significantly stronger per unit A_V (on average by a factor of 2) toward the GC compared with regions closer to the Sun (e.g., Adamson et al. 1990; Pendleton et al. 1994; Sandford et al. 1995; Gao et al. 2010), its behavior again resembling that of the silicates (Figure 5.8). This result is puzzling: silicates and hydrocarbons represent independent grain populations, according to the polarization data, yet they share the same anomaly. That silicate and hydrocarbon features should correlate with each other is unsurprising if the absorbers are similarly distributed, but this does not explain why A_V/τ_λ ratios are a factor $\sim$2 less for both features toward the GC compared with the results in Equations (5.6) and (5.8) for other lines of sight. Assuming that extinction estimates for the GC sources are broadly correct (they would need to be doubled to remove the anomaly!), some other effect must come into play.

Metallicity gradients are known to occur in the galactic disk (Section 7.1.2). These gradients may alter the relative rates of C-rich and O-rich stardust production (Section 8.1), in the sense that less C-rich dust is produced in higher metallicity environments. Thus, there could be a dearth of (larger) C-rich grains that contribute to A_V toward the GC. Differences in hydrogenation may also be important. Whereas most of the mass is in the large grains, most of the surface area available for H attachment is on the far more numerous nanoparticles, and they produce little continuum extinction. A reduction in the overall mass of C-rich dust could thus cause a proportionate reduction in A_V/τ_λ for both silicate and hydrocarbon features if the number of C–H absorbers on small grains holds up.

5.3 The Dense ISM

The presence of detectable ice absorption is a definitive attribute that distinguishes dense clouds from the diffuse ISM, as illustrated by the examples in Figure 5.3. The amplitude of the strongest ice feature per unit extinction ($\tau_{3.0}/A_V$) is a factor $\sim$30 times greater toward Elias 16 than its *upper limit* toward Cyg OB2 no. 12. Other ice features follow the same pattern. Differences are also evident within dense environments, comparing spectra of background stars such as Elias 16 with those of embedded young stars. For example, the solid CO feature at 4.67 μm, prominent toward Elias 16 (Figure 5.3), is weak or absent toward S140 (Figure 5.4), implying warmer temperatures in the environs of the embedded star that maintain this highly volatile species in the gas phase. These observations are qualitatively consistent with a model in which refractory cores, present in all types of cloud, act as substrates for the condensation of icy mantles in the shielded environments of dense clouds. Initially, the mantles contain a mix of species with differing volatilities; subsequent star formation causes thermal and/or photochemical evolution and sublimation of the ices.

5.3.1 An Inventory of Interstellar Ices

Table 5.3 presents a list of confirmed or probable constituents of interstellar ices based on infrared spectroscopy (data from Boogert et al. 2015 and references therein). Abundances are by number, relative to the most common constituent, H_2O. Results are included for distinct environments: quiescent dark clouds (based on observations of background stars), and two types of star formation region toward embedded stars, distinguished by mass (see Section 9.3.1). Cometary data are also included for comparison, discussion of which is deferred to Section 9.3.7. The mean abundances of most detected species lie in the 1%–30% range relative to H_2O; those with abundances < 1% and typical band strengths are generally difficult or impossible to detect with current observational facilities. In some cases, blending hinders detection and increases uncertainties in the results: NH_3 is a notable example, each of its principal absorptions (Table 5.2) overlapping a stronger feature of H_2O or silicates.

The results in Table 5.3 suggest that a typical interstellar ice mixture in a dark cloud or low-mass star formation region contains H_2O, CO, CO_2, and NH_3 roughly in the proportions 100:25:25:7, with generally smaller contributions from other species. The infrared-inactive species O_2 and N_2 might also be important but their contributions are unknown. Comparing results for low-mass and high-mass young stellar objects (YSOs), the latter exhibit generally lower solid-CO abundances,

Table 5.3. Confirmed or Probable Constituents of Interstellar and Cometary Ices.

Species	Dark Clouds	Low-mass YSOs	High-mass YSOs	Comets
H_2O	100	100	100	100
NH_3	$\lesssim$10	~6	$\lesssim$7	0.2–1.4
CO	25 (10–65)	21 (0–85)	7 (0–26)	0.4–30
CO_2	26 (14–43)	28 (12–50)	19 (11–27)	15 (4–30)
CH_3OH	8 (0–12)	6 (0–25)	9 (3–30)	0.2–7
CH_4	<3	5 (1–11)	2	0.4–1.6
H_2CO	<5	~6	~6	0.1–1
HCOOH	<2	<4	~4	~0.1
OCN^-	<0.5	0–1	0–2	—
OCS	<0.2	$\lesssim$1.6	0–0.2	~0.3

Notes: Abundance values are percentage by number relative to H_2O. Mean values, upper limits and/or ranges are given; ranges imply real spatial variations between different targets.

reflecting their greater luminosity and consequent impact on their surroundings. It is notable that the oxides of carbon (CO, CO_2) are consistently more abundant than common hydrogenated forms (CH_4, CH_3OH, H_2CO) in all environments, a finding that places important constraints on chemical models.

5.3.2 Extinction Thresholds

Quiescent dark clouds undisturbed by massive star formation represent ideal "laboratories" in which to study ices in a relatively pristine state. Insight into the conditions that lead to mantle growth on the dust may be gained by observing ice features in the spectra of field stars lying behind such clouds across a range of extinctions. Figure 5.11 plots the optical depth ($\tau_{3.0}$) of the principal H_2O–ice feature against visual extinction for stars toward a well-studied example: the Taurus dark cloud is nearby ($d \sim 140$ pc) and situated sufficiently far from the galactic disk that almost all extinction in the line of sight originates in the cloud itself. A correlation is evident between $\tau_{3.0}$ and A_V, consistent with a linear fit of the form

$$\tau_{3.0} = q\,(A_V - A_V^0) \tag{5.9}$$

where q is the slope, and the intercept A_V^0 represents the *threshold* visual extinction for detection of H_2O–ice in the cloud. This behavior stands in contrast to that of silicates in the diffuse ISM (Figure 5.8), for which a null intercept is indicated.

This difference is readily explained if the presence or absence of ice mantles on dust is governed by prevailing physical conditions. Dark clouds typically contain clumps of molecular gas embedded in an envelope of more diffuse material (Section 1.3.2). Photons from the external radiation field penetrate into the envelope, resulting in temperatures and dissociation rates similar to those in a typical diffuse cloud. Accumulation of molecular surface layers on the dust in such regions therefore tends to be limited by desorption and dissociation. A line of sight that does not intersect a dense clump thus lacks an ice feature and displays low to modest

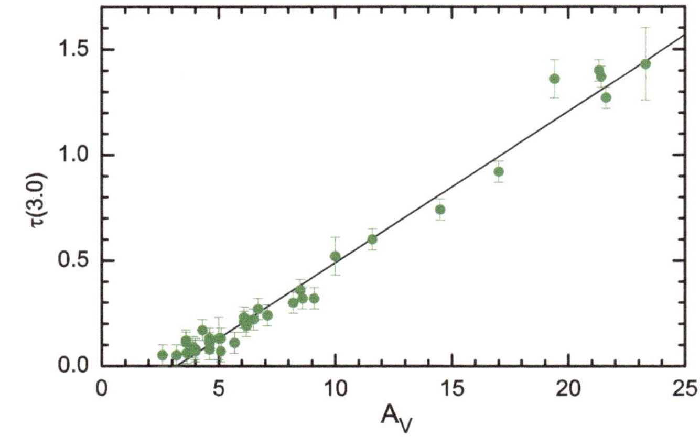

Figure 5.11. Plot of optical depth in the 3.0 μm ice feature against visual extinction for field stars located behind the Taurus Dark Cloud. The diagonal line is the linear least-squares fit (Equation (5.9)), yielding values of $q = 0.072 \pm 0.002$ and $A_V^0 = 3.2 \pm 0.1$ for the slope and intercept. Data are from Whittet et al. (2001a) and references therein.

extinction. However, the clumps are effectively shielded by the envelope from energetic external radiation, and higher densities enhance the frequency of gas-grain interaction, conditions conducive to formation and retention of ice layers on the dust. The threshold extinction thus corresponds to a critical level of shielding, sufficient to allow formation of a complete monolayer, after which growth may proceed more rapidly (Williams et al. 1992; Rawlings & Williams 2021). If the external radiation field is isotropic, the critical level is $A_V^0/2$ because the line of sight to a background star passes through diffuse material on both near and far sides of the cloud. Once H_2O–ice mantles become established, the amplitude of the 3 μm feature correlates with the column density of dust, for which A_V is a proxy, resulting in the behavior seen in Figure 5.11.

This correlation enables the abundance of H_2O molecules in the ices to be estimated. Using the Gaussian approximation in Equation (5.4) with a band strength of 2.0×10^{-18} m per molecule (Table 5.2), and expressing the width parameter $\gamma_{3.0}$ in μm^{-1}, we have

$$N(H_2O) \approx 5 \times 10^{23}\, \gamma_{3.0}\, \tau_{3.0}. \tag{5.10}$$

Assuming the mean value of N_H/A_V (Equation (3.13)) for the ISM near the Sun, the hydrogen column density for cloud material *above the ice threshold* is

$$N_H \approx 2.1 \times 10^{25}\,(A_V - A_V^0) \tag{5.11}$$

and combining Equations (5.9), (5.10), and (5.11), we have

$$\frac{N(H_2O)}{N_H} \approx 0.024\, q\, \gamma_{3.0}. \tag{5.12}$$

Inserting data for the Taurus cloud ($q \approx 0.072$; $\gamma_{3.0} \approx 0.036\ \mu m^{-1}$) into Equation (5.12) yields $N(H_2O)/N_H \approx 6 \times 10^{-5}$. This result suggests that ~15% of the elemental oxygen abundance is taken up by H_2O-ice in the cloud.

The threshold effect is a widespread phenomenon, first discovered in Taurus and subsequently confirmed in several other dark clouds that are either starless or lacking in high-mass star formation (Whittet et al. 1983, 1996, 2007, 2011, 2013; Chiar et al. 1995, 2011; Boogert et al. 2013, 2015; Chu et al. 2020). Threshold values for H_2O typically lie in the range $2 < A_V^0 < 4$ after correction for any foreground extinction. The slopes of the correlation lines for $A_V > A_V^0$ are also comparable, indicating reasonable consistency from cloud to cloud in the abundance of H_2O in solid form. Higher thresholds are suggested for clouds associated with more active sites of star formation, most likely caused by local enhancements of the ambient radiation field (Williams et al. 1992). Other constituents of the ices have been found to exhibit analogous behavior. The threshold for CO_2 is similar to that for H_2O, implying that these species form solids under similar conditions, whereas CO and CH_3OH require more shielding and/or higher densities to reach detectable abundance levels.

5.3.3 Water and Ammonia: The 3 μm Profile

The profile of the 3 μm feature is illustrated in Figure 5.12 for a typical dense-cloud environment. Ices deposited and maintained at the prevailing temperatures in such clouds ($T_d \lesssim 20$ K; Section 2.3.1) are expected to be structurally amorphous, and this is largely confirmed by the observations. There is a caveat, however: the overall shape of the absorption profile is notable for its asymmetry, a property that is inconsistent with *pure* water-ice on *small* core-mantle grains. In the small-particle

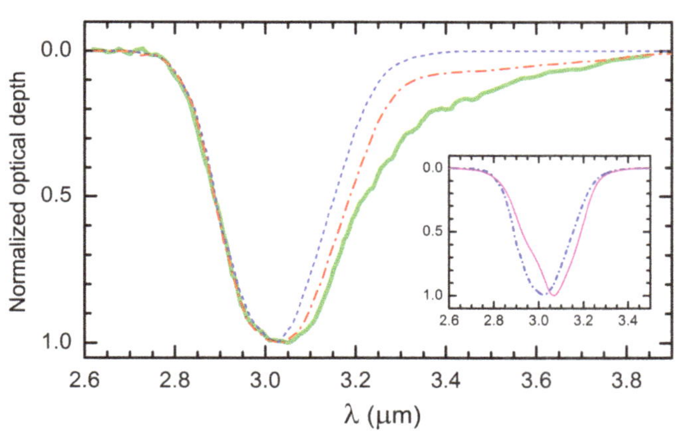

Figure 5.12. The 3 μm feature in low-temperature ices. The observed profile toward the field star Elias 16 (thick green curve, deduced from the spectrum in Figure 5.3) is compared with numerical models for a continuous distribution of ellipsoidal grain shapes in the small-particle limit, using laboratory data for amorphous ices at 10 K: pure H_2O-ice (blue) and a 10:1 H_2O:NH_3 ice mixture (red). The inset graph shows the effect of temperature, comparing pure H_2O-ice at 10 K (blue) and 80 K (magenta). All curves are normalized to unit optical depth at peak absorption.

limit, a model based on pure amorphous H_2O-ice yields a symmetric, Gaussian-like profile that matches the absorption shortward of the peak but cannot reproduce the extended long-wavelength wing (compare blue and green curves in Figure 5.12). This extended wing is ubiquitous, widely observed in the spectra of YSOs (e.g., Figure 5.4) as well as background stars, with some variation in its amplitude relative to the peak. Possible explanations include blended absorptions of other molecules in the ices, and extinction by scattering from larger grains.

Organic molecules produce generic C–H absorption features in the $3.2 < \lambda < 3.6$ μm region of the spectrum, as previously discussed, the precise position and width depending on the structure of the specific molecule and its interactions with its neighbors in the solid. Individually, such features are typically much narrower than the wing, so a blend would be required to fully account for it. However, the wing appears to be continuous. Blended features have been detected in this spectral range, including absorptions attributed to PAHs (3.25 μm), CH_4 (3.32 μm) and CH_3OH (3.53 μm), and another, centered near 3.47 μm, that lacks a secure identification; but when present, they appear merely as substructure superposed on the wing. Furthermore, C–H absorptions attributed to refractory grain materials such as hydrocarbons (Section 5.2.3) are generally lacking in lines of sight that display strong 3 μm ice features, suggesting that the presence of ice mantles suppresses the 3.4 μm feature seen in the diffuse ISM (Section 5.3.6). The amplitude, width and relative smoothness of the wing across many lines of sight is thus inconsistent with an origin in blended features in a diverse mix of organic species. See Allamandola et al. (1992), Brooke et al. (1999), Shenoy et al. (2003), Boogert et al. (2015) and Chiar et al. (2021) for further details and discussion.

Ammonia (NH_3) may be a better a candidate, based on the nature of its interaction with H_2O. Pure NH_3 produces absorption centered near 2.96 μm arising from the N–H stretching vibrational mode. However, NH_3 molecules in a water-dominated ice mixture tend to form ammonium hydrate groups ($NH_3{\cdots}OH_2$, where the dots represent the hydrogen bond). The O–H vibrations of H_2O molecules tied up in these groups are damped, resulting in a shift to longer wavelengths, thereby introducing additional absorption in the $3.0 < \lambda < 3.8$ μm spectral range: compare blue and red curves in Figure 5.12. The 10:1 H_2O:NH_3 mixture yields a closer fit to the observed profile than pure H_2O-ice. However, it does not fully account for the wing: an NH_3 concentration of 20%–30% would be required to match its average depth, resulting in enhanced N–H absorption at ~2.96 μm, inconsistent with the observed profile (e.g., Smith et al. 1989; Whittet et al. 1996; Dartois & d'Hendecourt 2001; Gibb et al. 2001). A limit of $\lesssim$10% on the ammonia concentration is also imposed by observations of its features at other wavelengths (Table 5.2; see Gibb et al. 2001, 2004; Öberg et al. 2011). Thus, some additional process must come into play.

The shape and smoothness of the wing suggest a possible origin in scattering by relatively large ice grains (Léger et al. 1983; Smith et al. 1989; Pendleton et al. 1990). Scattering becomes important when $2\pi a/\lambda \gtrsim 1$, i.e., for grain diameters $2a \gtrsim 1$ μm at $\lambda \approx 3$ μm, and its effect is to shift the extinction peak to longer wavelength. Particles in this size regime may be accommodated by a modest increase in the maximum grain radius assumed in typical models for general extinction (Section 3.6.1). A comparison of absorption profiles at 3 μm and 6 μm, both of which include blended H_2O and NH_3 ice features, led Dartois & d'Hendecourt (2001) to conclude that only the 3 μm profile is affected by scattering, suggesting that a 2-fold increase in λ is sufficient re-establish the applicability of the small-particle approximation. The observed polarization counterpart to the 3 μm absorption feature (Section 4.2.7) also addresses this question: the wavelength offset between the peaks in optical depth and polarization is sensitive to grain size and is best matched by a preponderance of small grains ($a \lesssim 0.1$ μm; Hough et al. 1988, 1996), although some larger grains may still be present. Detailed simultaneous modeling of optical depth and polarization profiles is needed to allow a fully quantitative test of the scattering model for the wing.

Laboratory studies show that warming modifies the 3 μm profile, altering its shape, width and peak position, as the ice transitions from an amorphous to a crystalline state. An example is illustrated in Figure 5.12 (inset graph). Cold, amorphous ices yield the most realistic fits to features observed in dense-cloud environments, but warmer ices across a range of temperatures are often needed to fit those of YSOs (e.g., Smith et al. 1989; Gibb et al. 2004; Boogert et al. 2015). The 3 μm profile is thus one of several spectroscopic tools that may be used to study the thermal evolution of dust in regions of recent star formation. See Section 9.3.2 for further discussion.

5.3.4 CO and CO_2: Polar and Apolar Mantles

Carbon monoxide is the most abundant gas-phase interstellar molecule after H_2, and its presence as a condensate on interstellar dust was predicted (Duley 1974) a decade prior to its first detection (Lacy et al. 1984). The only available spectral signature of the dominant isotopic form is the C–O stretch at 4.67 μm (the weaker ^{13}CO feature at 4.78 μm is also detected). Observations targeting the 4.67 μm feature in different lines of sight show that the abundance of solid CO is highly variable, more so than any other known species in the ices, with an extreme range of 0%–85% relative to H_2O (Table 5.3). This finding is readily explained by its volatility: in pure form, CO sublimes at ~17 K. Temperatures both above and below this value may often be present in different regions along the same line of sight, most notably toward YSOs, demonstrated by superposition of both gas-phase and solid CO absorptions in the same spectrum: an example is shown in Figure 5.13.

The observed profile of the CO feature provides insights into the composition and structure of the ice mantles. Laboratory spectra of ices containing CO demonstrate that the position, width, shape and amplitude of the profile are sensitive to the composition of the mixture (Sandford et al. 1988; Ehrenfreund et al. 1996, 1997). Pure solid CO produces a narrow (FWHM ~0.01 μm) feature centered near 4.674 μm. Introduction of other species may lead to broadening and shifts in position. The differences are slight if CO remains dominant, more substantial when CO is a minority constituent, and the nature of these changes depends on the dipole moment of the mixture. CO itself is only very weakly polar, and mixtures of CO with apolar

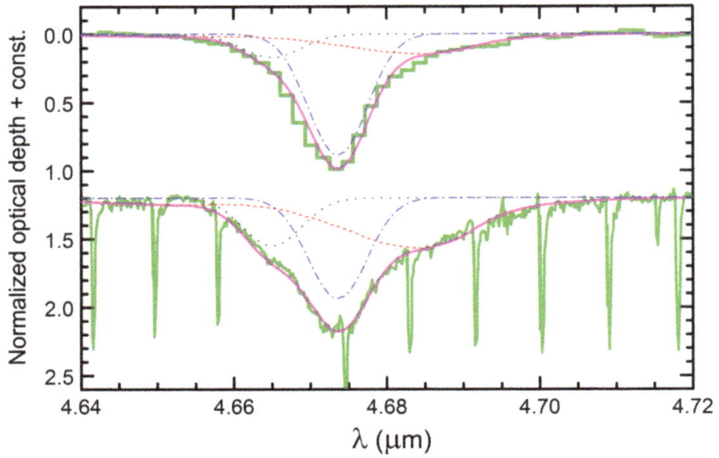

Figure 5.13. Modeling the CO profile. Green curves are observed optical-depth spectra for a field star (Elias 16, top) and a low-mass YSO (L1489 IRS), each normalized to unit optical depth at the peak of the ice feature, the latter displaced downward for display. Narrow gas-phase CO lines are also prominent in the YSO spectrum. Fits to the observed ice features (magenta curves) are summations of three components with differing amplitudes in each case: pure solid CO (blue), a 4:1 H_2O:CO mix (red), and an additional Gaussian component (gray; see text). Based on data from Boogert et al. (2015) and references therein.

molecules such as O_2 or N_2 have 4.67 μm profiles closely similar to that of pure CO. However, mixtures of CO with polar molecules such as H_2O or NH_3 exhibit more dramatic differences. All CO-bearing laboratory mixtures in which H_2O is the dominant constituent produce features shifted to significantly longer wavelength (~4.685 μm) and substantially broadened (FWHM ~0.02 μm); compare blue and red curves in Figure 5.13. Another notable property of H_2O-dominated mixtures is that they can retain CO at temperatures up to about 90 K, well above the nominal sublimation temperature of pure CO. Thus, the profile shows substantial evolution with temperature.

The absorption profiles of strong and relatively sharp solid-state features such as that of pure CO are also highly sensitive to grain shape in the small-particle limit. This dependence is caused by shape-related resonances affecting the motions of the vibrating molecules (Ehrenfreund et al. 1997). Predicted profiles therefore depend critically on the assumed grain model as well as on the optical functions of candidate solids measured in the laboratory. The shapes of real interstellar grains are, of course, poorly constrained (see Section 4.2.8). A method used to address this issue in numerical modeling is to adopt a continuous distribution of ellipsoids to represent an ensemble of particles with a wide range of morphologies and orientations (Bohren & Huffman 1983; Ehrenfreund et al. 1997; Draine & Hensley 2021). Indeed, fits to observed CO ice profiles based on this assumption are typically excellent and superior to those using spherical grains of the same composition (Pontoppidan et al. 2003). Observations of the much weaker ^{13}CO feature, which is insensitive to shape but sensitive to composition, provide further verification of the method (Boogert et al. 2002, 2015).

Examples of fits to the observed CO absorption profile are illustrated in Figure 5.13 for two contrasting lines of sight. Three distinct components are needed, their amplitudes adjusted to optimize the fit to each spectrum. Results for the dense cloud toward Elias 16 (top) show that the profile is dominated by a central "pure" component (blue), with lesser contributions from an H_2O-dominated mixture (red), and an additional Gaussian component (gray), the latter tentatively attributed to CO mixed with CO_2 (Boogert et al. 2015). The dominant component need not be literally pure CO but should have a low net dipole moment, so it might include significant quantities of CO_2 as well as O_2 and N_2. Similarly, the composition of the H_2O-dominated component is not tightly constrained and may include other species such as NH_3, CO_2, and CH_3OH. Compared to the field star, the YSO spectrum displays a broader profile, accommodated by a small reduction in the relative amplitude of the deep central component. This difference may be explained by partial sublimation of the volatile CO-dominated ices, consistent with the presence of gas-phase CO lines in the spectrum.

A schematic structure suggested by the observations is illustrated in Figure 5.14. Grains provide substrates for ice-mantle formation in regions sufficiently shielded from the external radiation field (Section 5.3.2). H_2O-rich ices are presumed to form first and to incorporate other species such as NH_3, CO, and CO_2. Production of H_2O by surface reactions requires an ample supply of atomic H, becoming less efficient in denser environments where most of the gas-phase hydrogen is molecular. Thus, a transition from H_2O-dominated to CO-dominated accumulation occurs as growth proceeds in the inner, more shielded, regions of the cloud. Formation of CO_2 by oxidation of CO continues through the transition. The resultant layers are often referred to as polar and apolar (non-polar) mantles, although in reality the "apolar" layer has a small but non-zero dipole moment. Note that the illustration in Figure 5.14 is idealized and not to scale: the ice layers are likely to be relatively thin, the interface between layers is unlikely to be sharp, other molecules such as CH_3OH are expected to be present in low to moderate concentrations, and the

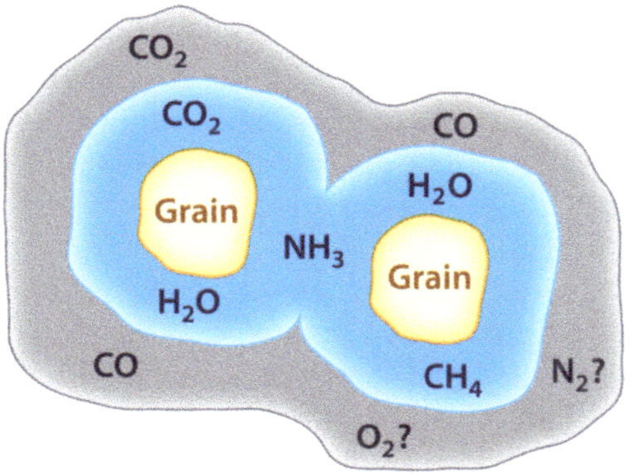

Figure 5.14. Schematic illustration of a composite particle consisting of grain cores in a common mantle composed of polar and apolar ices (blue and gray). The dominant constituents of the ice layers are H_2O and CO, respectively. Adapted from Boogert et al. (2015).

nominally apolar layer is unlikely to be entirely devoid of H_2O. The key difference is the switch between majority species. See Section 9.2.2 for extended discussion on the physics and chemistry of mantle growth.

The spectral features of CO_2 have similar diagnostic properties. Those detected include the principal stretching and bending modes of $^{12}CO_2$ at 4.27 μm and 15.3 μm, and the stretching mode of $^{13}CO_2$ at 4.38 μm (Gerakines et al. 1999; Boogert et al. 2000, 2015). Results are consistent with the general scenario outlined above, and provide further insight into the evolution of the ices. As solid CO_2 has a lower extinction threshold compared with CO, it is more widely distributed and can be detected in a wider range of molecular environments. The profile of the bending mode has proven to be a particularly powerful tracer of both the composition and thermal history of the ices, and a wealth of data is available thanks to the sensitivity of the spectrometer carried by the Spitzer Space Telescope in this region of the spectrum. Results confirm the presence of distinct polar and apolar layers in unprocessed ices (Whittet et al. 2009), and they also provide evidence for heat-induced distillation of CO_2 into a purer, more crystalline form in the environs of YSOs (Pontoppidan et al. 2008). These processes are discussed in detail in Section 9.3.2.

5.3.5 The 6.0 and 6.85 μm Features: A Potpourri

The spectral range from 5 to 8 μm is populated by vibrational modes of many candidates for detection (Table 5.2), and may therefore carry valuable information on the composition of the ices. It includes not only the bending mode of H_2O but also the deformation modes of species such as NH_3, CH_4, and CH_3OH and the carbonyl (C=O) modes of oxygen-bearing organics such as H_2CO and HCOOH. However, their contributions have proven difficult to untangle. The two principal absorption features in this region, centered near 6.0 and 6.85 μm (Figure 5.15), are both evidently blends of overlapping absorptions. The 6.0 μm feature is attributed primarily to the H_2O bending mode, modified by the presence of other species such as oxygen-bearing organics and NH_3. The 6.85 μm feature appears likewise to be a composite of several absorbers, most of which lack secure identifications. The problem is compounded by a limitation of the observational data: at the time of writing, spectra of the quality ideally needed (in terms of both spectral resolution and signal to noise ratio) to fully analyze this region of the spectrum are almost entirely limited to YSOs; the profiles of these features in unmodified regions of cold, dense clouds are less well constrained.

A representative YSO spectrum is compared with data for some candidate species in Figure 5.15. The magenta curve superposed on the observed spectrum is a notional combination of the laboratory curves, simulating an H_2O-dominated ice mixture; other species are included in proportions limited by their maximal abundances, as constrained by the amplitudes of their absorptions at all wavelengths. For example, the deformation mode of CH_3OH may contribute to the 6.85 μm feature, but it cannot be the primary absorber because its abundance is limited by observations of its features at other wavelengths, such as the C–H and C–O

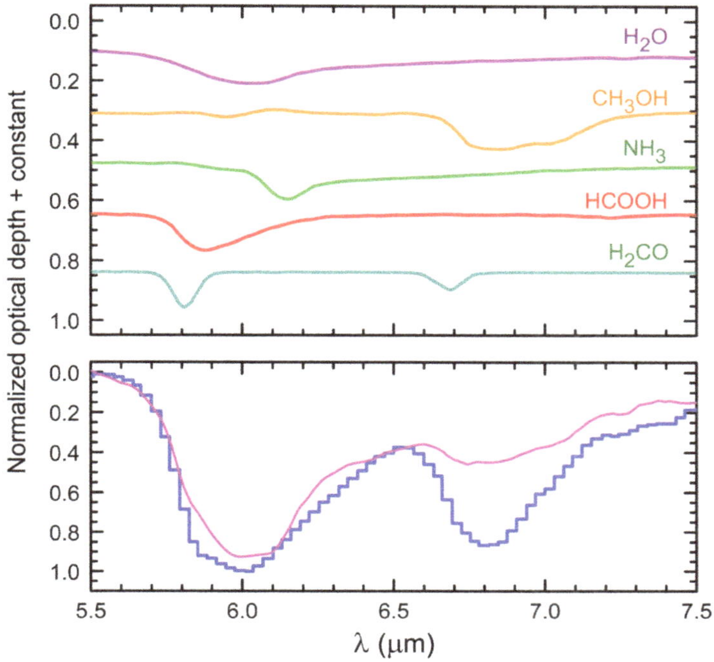

Figure 5.15. The 6.0 and 6.85 μm absorption features observed toward the low-mass YSO HH46 IRS (lower frame, blue), compared with laboratory spectra for some candidate ices (upper frame); data from Boogert et al. (2015) and references therein. The magenta curve (lower frame) represents a combination of laboratory spectra in proportions constrained by analyses of data across the entire infrared spectrum.

stretching modes at 3.53 and 9.75 μm (e.g., Gibb et al. 2004; Boogert et al. 2008). No realistic combination of the ices illustrated in Figure 5.15 provides an ideal fit to the observations, so other absorbers must be implicated. Weak 6.2 and 6.8 μm features observed in the diffuse ISM (Figure 5.5), attributed to vibrational modes in hydrocarbons, might also contribute to dense-ISM spectra; but non-detection of the 3.4 μm hydrocarbon dust signature in dense clouds (Section 5.3.6) suggests that this is unlikely to be significant. Furthermore, the amplitude of the 6.85 μm feature in dense clouds is well correlated with the column density of solid H_2O, consistent with an origin in the polar component of the ices.

The identity of the absorber(s) responsible for a major part of the 6.85 μm feature remains uncertain. A proposed candidate is the NH_4^+ ion, produced in the laboratory by photolysis of ice mixtures containing H_2O and NH_3 (Schutte & Khanna 2003). Production of NH_4^+ cations requires an equal number of anions to be produced. Candidates that might exist in interstellar ices include OCN^- and $HCOO^-$, both of which have some observational support, but neither seems sufficiently abundant to balance the $\sim$10% concentration of NH_4^+ needed to account for the 6.85 μm feature. The absorption profile of NH_4^+ in an H_2O-dominated matrix is broader than the observed feature (Gávez et al. 2010), and its central wavelength is sensitive to temperature. An active environment may be necessary to drive sufficient production of the ion, requiring a significant UV photon flux and/or thermal processing. Hence, detection of the 6.85 μm feature in cold, dark clouds (Boogert et al. 2008) seems inconsistent with this identification. In contrast, the 4.62 μm feature attributed to OCN^- ions is detected only in YSO environments (van Broekhuizen et al. 2005).

5.3.6 Refractory Dust in the Dense ISM

The cores of ice-mantled particles in dense clouds (Figure 5.14) are presumably members of the same dust populations detected in the diffuse ISM. If so, they are composed of refractory materials such as silicates and amorphous carbon, and their initial sizes should follow the same power-law distribution (Section 3.6.1), subject to subsequent modification by coagulation (Section 9.1.1), and mantle growth (Section 9.2.2). As such, they might be regarded simply as substrates, acquiring and dissipating ice mantles as they cycle in and out of molecular clouds. We conclude this chapter with an assessment of the validity of these assumptions.

The primary resource for assessing possible changes in the *silicate* population is the 9.7 μm absorption profile (e.g., Figure 5.6). In the diffuse ISM this feature exhibits a consistent profile shape, closely similar in different lines of sight, and is well matched by amorphous, anhydrous dust in the small-particle limit with mixed olivine-pyroxene stoichiometries (Section 5.2.1). Profiles observed in dense clouds tend to be systematically broader. Traditionally, two distinct spectral shapes, based on observations of silicate *emission* features toward μ Cephei (a dusty red supergiant) and the Trapezium (a cluster of luminous young stars), have been used to characterize this dichotomy, the latter typically yielding better fits to profiles observed in molecular clouds (e.g., Roche & Aitken 1984; Whittet et al. 1988; Bowey et al. 1998). Heated silicate grains in the Trapezium region are unlikely to have retained ice mantles, but might have preserved some imprint of their processing history in this environment (see Section 6.2.1 for further discussion and caveats).

Attempts to establish a representative 9.7 μm profile specific to *cold, quiescent* regions of the dense ISM face several obstacles. The breadth of the feature and its proximity to other absorptions cause uncertainties in continuum placement, and several ice features overlap the same spectral region. Examples include the 9.75 μm C–O stretching mode of CH_3OH (Table 5.2), and the broader H_2O libration mode centered near 12.5 μm (see Figure 5.4); these might appear to deepen and broaden, respectively, the apparent silicate profile if not fully accounted for. Moreover, the position and profile shape of the libration feature vary with the composition and crystallinity of the ice mixture (Hagen et al. 1983; Kitta & Krätschmer 1983), and might therefore differ systematically in the spectrum of a YSO compared with that of a background star. But the best-observed sources in dense clouds are, indeed YSOs. This leads to a further complication arising from the fact that, in many cases, YSO spectra include a blend of both circumstellar emission and interstellar absorption, and heated grains local to the source may be annealed, resulting in crystallization (e.g., Poteet et al. 2011). Spectra of background stars are easier to interpret but observationally far more challenging, because of their inherently declining continua at longer wavelengths (Figure 5.3).

Taking these factors into account, available data confirm that silicate profiles in cold molecular clouds exhibit real (albeit modest) departures from the mean diffuse-ISM profile (van Breemen et al. 2011). This implies real differences in the silicates themselves, arising in their physical form and/or chemical composition. Of these possibilities, physical differences seem more plausible: ad hoc shifts in stoichiometry may yield better fits but lack any logical explanation, whereas shifts in grain size are, indeed, predicted. For example, the presence of an ice mantle on a silicate core may enhance the sticking probability of grain–grain collisions, promoting further growth by coagulation (McClure 2009). However, the observed profiles indicate that the silicate-bearing grains do not grow much larger than $\sim 1\ \mu$m (van Breemen et al. 2011), i.e., they remain effectively in the small-particle limit at $\lambda \sim 10\ \mu$m. Grain growth may be accompanied by systematic changes in *shape*, and these also affect the profile. The analysis by van Breemen et al. suggests that the dichotomy may be explained if grains in molecular clouds are more rounded, i.e., more closely resembling spheres, compared with those in the diffuse ISM.

A related question concerns the behavior of the peak silicate optical depth, $\tau_{9.7}$, relative to extinction or reddening. The near-infrared color excess E_{J-K} is often used to measure reddening toward targets with high extinctions, lacking reliable data at shorter wavelengths (see Section 3.2.2). Results show that the silicate feature is systematically weaker per unit E_{J-K} in dense clouds compared with the diffuse ISM (Chiar et al. 2007, 2011; McClure 2009; van Breemen et al. 2011). This behavior is not fully understood; but it seems more likely to be caused by differences in extinction, as measured by E_{J-K}, than in $\tau_{9.7}$, given that the disparity in the silicate profile is relatively modest. Grain growth might reduce E_{J-K} relative to $\tau_{9.7}$, by making the near-infrared continuum extinction "grayer," without causing major changes in the silicate profile.

The primary infrared signature of refractory *carbonaceous* dust in the diffuse ISM is the C–H feature centered at 3.4 μm (Section 5.2.3). In dense clouds, the long-wavelength wing of the 3 μm H_2O-ice feature (Figure 5.12) produces a smooth gradient across this spectral range, and discrete absorptions superposed on it may be separated by fitting a local continuum. An example is shown in Figure 5.16 and compared with corresponding data for a representative diffuse-ISM spectrum. Absorption centered at 3.25 μm in the dense-ISM spectrum is attributed to aromatic hydrocarbons in the ices (Hardegree-Ullman et al. 2014; Chiar et al. 2021), and the diffuse-ISM profile also includes an aromatic component (Figure 5.10). Dramatic differences are evident in the 3.3–3.6 μm spectral range corresponding to C–H resonances in aliphatic hydrocarbons. The 3.4 μm diffuse-ISM feature appears to be either absent or severely modified in the dense-ISM spectrum, which instead exhibits a broad, asymmetric absorption peaking near 3.47 μm. The identity of the 3.47 μm feature is unclear. It is widely observed in molecular clouds, its amplitude correlates closely with that of the 3 μm H_2O-ice feature (Chiar et al. 1996; Brooke et al. 1999), and a corresponding polarization excess is also seen (Hough et al. 1996). These results imply that the 3.47 μm absorber(s) reside in the H_2O-dominated ice layers of aligned, core-mantle grains. Both carbonaceous molecules and ammonium hydrates have been discussed as possible candidates (Allamandola et al. 1992; Dartois et al. 2002; Mennella 2010).

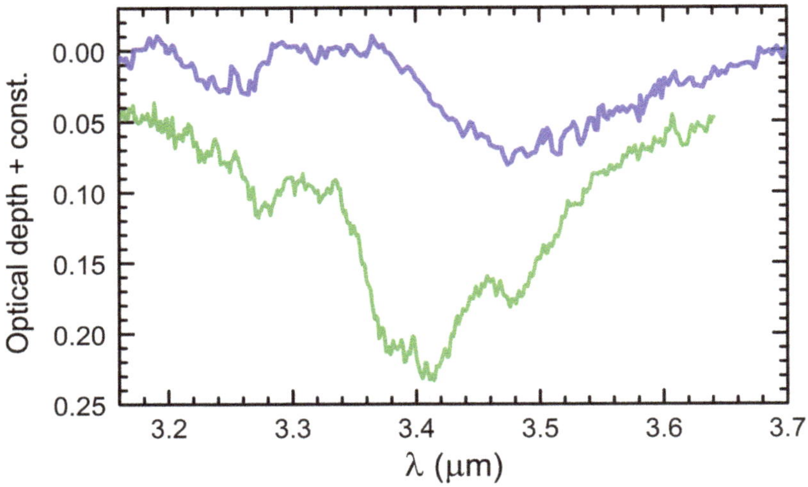

Figure 5.16. Comparison of C–H absorption profiles in dense and diffuse regions of the ISM. Residual optical depths are shown after removal of the 3 μm ice feature and its long-wavelength wing. Above (blue): S140, a young star in a molecular cloud (see also Figure 5.4). Below (green): the Quintuplet cluster toward the GC, representative of the diffuse ISM (see also Figure 5.10). The latter is displaced downward for display. Based on data from Hardegree-Ullman et al. (2014) and Chiar et al. (2013), respectively.

The systematic difference between diffuse and dense ISM profiles illustrated in Figure 5.16 may carry important information on the evolution of C-rich dust with environment. Two possibilities are suggested:

- The 3.4 μm and 3.47 μm absorbers are distinct and independent. The 3.4 μm feature, identified with aliphatic hydrocarbons on surfaces of bare C-rich grains, may be suppressed when the particles become coated with ices, because the chemical processes that lead to ice formation scavenge hydrogen (Grishko & Duley 2000). Simultaneously, 3.47 μm absorption appears as a signature of the ices.
- The 3.4 μm and 3.47 μm absorbers are both hydrocarbons, and there is an evolutionary link between them. The profile produced by a mix of aliphatic hydrocarbons varies with chain length (which determines the CH_2/CH_3 ratio), and with the degree to which each vibrational mode is suppressed or dampened by the proximity of other species such as ices. Mennella (2010) presents laboratory data in support of this scenario.

Neither of these possibilities is excluded by the existing evidence.

References

Adamson, A. J., Whittet, D. C. B., & Duley, W. W. 1990, MNRAS, 243, 400

Adamson, A. J., Whittet, D. C. B., Chrysostomou, A., et al. 1999, ApJ, 512, 224

Aitken, D. K. 1989, in 22nd ESLAB Symposium, Infrared Spectroscopy in Astronomy, ed. B. H. Kaldeich (ESA publication SP-290), 99

Allamandola, L. J., Sandford, S. A., Tielens, A. G. G. M., & Herbst, T. M. 1992, ApJ, 399, 134
Alonso-Herrero, A., Esquej, P., Roche, P. F., et al. 2016, MNRAS, 455, 563
Banwell, C. N., & McCash, E. M. 1994, Fundamentals of Molecular Spectroscopy (4th ed.; London: McGraw-Hill)
Bohren, C. F., & Huffman, D. R. 1983, Absorption and Scattering of Light by Small Particles (New York: Wiley), 141
Boogert, A. C. A., Ehrenfreund, P., Gerakines, P. A., et al. 2000, A&A, 353, 349
Boogert, A. C. A., Blake, G. A., & Tielens, A. G. G. M. 2002, ApJ, 577, 271
Boogert, A. C. A., Pontoppidan, K. M., Knez, C., et al. 2008, ApJ, 678, 985
Boogert, A. C. A., Chiar, J. E., Knez, C., et al. 2013, ApJ, 777, 73
Boogert, A. C. A., Gerakines, P. A., & Whittet, D. C. B. 2015, ARA&A, 53, 541
Bouilloud, M., Fray, N., Bénilan, Y., et al. 2015, MNRAS, 451, 2145
Bowey, J. E., Adamson, A. J., & Whittet, D. C. B. 1998, MNRAS, 298, 131
Brooke, T. Y., Sellgren, K., & Geballe, T. R. 1999, ApJ, 517, 883
Chen, T., Xiao, C. Y., Li, A., & Zhou, C. T. 2022, MNRAS, 509, 5231
Chiar, J. E., & Tielens, A. G. G. M. 2001, ApJ, 550, L207
Chiar, J. E., & Tielens, A. G. G. M. 2006, ApJ, 637, 774
Chiar, J. E., Adamson, A. J., Kerr, T. H., & Whittet, D. C. B. 1995, ApJ, 455, 234
Chiar, J. E., Adamson, A. J., & Whittet, D. C. B. 1996, ApJ, 472, 665
Chiar, J. E., Tielens, A. G. G. M., Whittet, D. C. B., et al. 2000, ApJ, 537, 749
Chiar, J. E., Adamson, A. J., Pendleton, Y. J., et al. 2002, ApJ, 507, 198
Chiar, J. E., Adamson, A. J., Whittet, D. C. B., et al. 2006, ApJ, 651, 268
Chiar, J. E., Ennico, K., Pendleton, Y. J., et al. 2007, ApJ, 666, L73
Chiar, J. E., Pendleton, Y. J., Allamandola, L. J., et al. 2011, ApJ, 731, 9
Chiar, J. E., Tielens, A. G. G. M., Adamson, A. J., & Ricca, A. 2013, ApJ, 770, 78
Chiar, J. E., de Barros, A. L. F., Mattioda, A. L., & Ricca, A. 2021, ApJ, 908, 239
Chu, L. E. U., Hodapp, K., & Boogert, A. 2020, ApJ, 904, 86
Coupeaud, A., Demyk, K., Meny, C., et al. 2011, A&A, 535, A124
Dartois, E., & d'Hendecourt, L. 2001, A&A, 365, 144
Dartois, E., d'Hendecourt, L., Thi, W., Pontoppidan, K. M., & van Dishoeck, E. F. 2002, A&A, 394, 1057
Day, K. L. 1981, ApJ, 246, 110
Demyk, K., Meny, C., Lu, X.-H., et al. 2017a, A&A, 600, A123
Demyk, K., Meny, C., Leroux, H., et al. 2017b, A&A, 606, A50
de Vries, M. S., Reihs, K., Wendt, H. R., et al. 1993, GeCoA, 57, 933
Dorschner, J., Friedemann, C., Gürtler, J., & Henning, T. 1988, A&A, 198, 223
Dorschner, J., Begemann, B., Henning, T., Jäger, C., & Mutschke, H. 1995, A&A, 300, 503
Draine, B. T., & Hensley, B. S. 2021, ApJ, 910, 47
Do-Duy, T., Wright, C. M., Fujiyoshi, T., et al. 2020, MNRAS, 493, 4463
Duley, W. W. 1974, Ap&SS, 26, 199
Duley, W. W., & Williams, D. A. 1981, MNRAS, 196, 269
Ehrenfreund, P., & van Dishoeck, E. F. 1998, AdSpR, 21, 15
Ehrenfreund, P., Boogert, A. C. A., Gerakines, P. A., et al. 1996, A&A, 315, L341
Ehrenfreund, P., Boogert, A. C. A., Gerakines, P. A., Tielens, A. G. G. M., & van Dishoeck, E. F. 1997, A&A, 328, 649
Fabian, D., Jäger, C., Henning, T., Dorschner, J., & Mutschke, H. 2000, A&A, 364, 282

Fogerty, S., Forrest, W., Watson, D. M., Sargent, B. A., & Koch, I. 2016, ApJ, 830, 71
Furton, D. G., Laiho, J. W., & Witt, A. N. 1999, ApJ, 526, 752
Gao, J., Jiang, B. W., & Li, A. 2010, EP&S, 62, 63
Gávez, O., Maté, B., Herrero, V. J., & Escribano, R. 2010, ApJ, 724, 539
Geballe, T. R., Pendleton, Y., Chiar, J., & Tielens, A. G. G. M. 2021, ApJ, 912, 47
Gerakines, P. A., & Hudson, R. L. 2020, ApJ, 901, 52
Gerakines, P. A., Whittet, D. C. B., Ehrenfreund, P., et al. 1999, ApJ, 522, 357
Gibb, E. L., Whittet, D. C. B., & Chiar, J. E. 2001, ApJ, 558, 702
Gibb, E. L., Whittet, D. C. B., Boogert, A. C. A., & Tielens, A. G. G. M. 2004, ApJS, 151, 35
Giuliano, B. M., Escribano, R. M., Martín-Doménech, R., Dartois, E., & Muñoz Caro, G. M. 2014, A&A, 565, A108
Greenberg, J. M., Li, A., Mendoza-Gomez, C. X., et al. 1995, ApJ, 455, L177
Godard, M., Geballe, T. R., Dartois, E., & Muñoz Caro, G. M. 2012, A&A, 537, 27
Gordon, K. D., Misselt, K. A., Bouwman, J., et al. 2021, ApJ, 916, 33
Grishko, V. I., & Duley, W. W. 2000, ApJ, 543, L85
Hagen, W., Tielens, A. G. G. M., & Greenberg, J. M. 1983, A&AS, 51, 389
Hardegree-Ullman, E. E., Gudipati, M. S., Boogert, A. C. A., et al. 2014, ApJ, 784, 172
Henning, Th. 2010, ARA&A, 48, 21
Hensley, B. S., & Draine, B. T. 2020, ApJ, 895, 38
Hensley, B. S., & Draine, B. T. 2021, ApJ, 906, 73
Hough, J. H., Sato, S., Tamura, M., et al. 1988, MNRAS, 230, 107
Hough, J. H., Chrysostomou, A., Messinger, D. W., et al. 1996, ApJ, 461, 902
Kebukawa, Y., & Alexander, C. M. 2011, GeCoA, 75, 3530
Kemper, F., Vriend, W. J., & Tielens, A. G. G. M. 2004, ApJ, 609, 826
Kemper, F., Vriend, W. J., & Tielens, A. G. G. M. 2005, ApJ, 633, 534
Kitta, K., & Krätschmer, W. 1983, A&A, 122, 105
Knacke, R. F., & Krätschmer, W. 1980, A&A, 92, 281
Lacy, J. H., Baas, F., Allamandola, L. J., et al. 1984, ApJ, 276, 533
Léger, A. 1983, A&A, 123, 271
Li, A., & Draine, B. T. 2001, ApJ, 554, 778
Li, A., & Greenberg, J. M. 2002, ApJ, 577, 789
Li, Q., Liang, S. L., & Li, A. 2014, MNRAS, 440, L56
McClure, M. 2009, ApJ, 693, L81
Mennella, V. 2010, ApJ, 718, 867
Mennella, V., Brucato, J. R., Colangeli, L., & Palumbo, P. 1999, ApJ, 524, L71
Mennella, V., Brucato, J. R., Colangeli, L., & Palumbo, P. 2002, ApJ, 569, 531
Min, M., Waters, L. B. F. M., de Koter, A., et al. 2007, A&A, 462, 667
Mutschke, H., Andersen, A. C., Clément, D., Henning, Th., & Peiter, G. 1999, A&A, 345, 187
Öberg, K. I., Boogert, A. C. A., Pontoppidan, K. M., et al. 2011, ApJ, 740, 109
Pendleton, Y. J., & Allamandola, L. J. 2002, ApJS, 138, 75
Pendleton, Y. J., Tielens, A. G. G. M., & Werner, M. W. 1990, ApJ, 349, 107
Pendleton, Y. J., Sandford, S. A., Allamandola, L. J., Tielens, A. G. G. M., & Sellgren, K. 1994, ApJ, 437, 683
Poteet, C. A., Megeath, S. T., Watson, D. M., et al. 2011, ApJ, 733, L32
Pontoppidan, K. M., Fraser, H. J., Dartois, E., et al. 2003, A&A, 408, 981
Pontoppidan, K. M., Boogert, A. C. A., Fraser, H. J., et al. 2008, ApJ, 678, 1005

Rawlings, J. M. C., & Williams, D. A. 2021, MNRAS, 500, 5117
Rieke, G. H., & Lebofsky, M. J. 1985, ApJ, 288, 618
Roche, P. F., & Aitken, D. K. 1984, MNRAS, 208, 481
Roche, P. F., Alonso-Herrero, A., & Gonzalez-Martin, O. 2015, MNRAS, 449, 2598
Rogantini, D., Costantini, E., Zeegers, S. T., et al. 2020, A&A, 641, A149
Sandford, S. A., Allamandola, L. J., Tielens, A. G. G. M., & Valero, G. J. 1988, ApJ, 329, 498
Sandford, S. A., Allamandola, L. J., Tielens, A. G. G. M., et al. 1991, ApJ, 371, 607
Sandford, S. A., Pendleton, Y. J., & Allamandola, L. J. 1995, ApJ, 440, 697
Schutte, W. A. 1999, in Formation and Evolution of Solids in Space, ed. J. M. Greenberg, & A. Li (Dordrecht: Reidel), 177
Schutte, W. A., & Khanna, R. K. 2003, A&A, 398, 1049
Shenoy, S. S., Whittet, D. C. B., Chiar, J. E., et al. 2003, ApJ, 591, 962
Smith, C. H., Wright, C. M., Aitken, D. K., Roche, P. F., & Hough, J. H. 2000, MNRAS, 312, 327
Smith, R. G., Sellgren, K., & Tokunaga, A. T. 1989, ApJ, 344, 413
Smith, R. G., Sellgren, K., & Tokunaga, A. T. 1993, MNRAS, 263, 749
Speck, A. K., Whittington, A. G., & Hofmeister, A. M. 2011, ApJ, 740, 93
van Breemen, J. M., Min, M., Chiar, J. E., et al. 2011, A&A, 526, 152
van Broekhuizen, F. A., Pontoppidan, K. M., Fraser, H. J., & van Dishoeck, E. F. 2005, A&A, 441, 249
Whittet, D. C. B. 2015, ApJ, 811, 110
Whittet, D. C. B., Bode, M. F., Longmore, A. J., Baines, D. W. T., & Evans, A. 1983, Natur, 303, 218
Whittet, D. C. B., Bode, M. F., Longmore, A. J., et al. 1988, MNRAS, 233, 321
Whittet, D. C. B., Duley, W. W., & Martin, P. G. 1990, MNRAS, 244, 427
Whittet, D. C. B., Smith, R. G., Adamson, A. J., et al. 1996, ApJ, 458, 363
Whittet, D. C. B., Boogert, A. C. A., Gerakines, P. A., et al. 1997, ApJ, 490, 729
Whittet, D. C. B., Gerakines, P. A., Hough, J. H., & Shenoy, S. S. 2001a, ApJ, 547, 872
Whittet, D. C. B., Pendleton, Y. J., Gibb, E. L., et al. 2001b, ApJ, 550, 793
Whittet, D. C. B., Shenoy, S. S., Bergin, E. A., et al. 2007, ApJ, 655, 332
Whittet, D. C. B., Cook, A. M., Chiar, J. E., et al. 2009, ApJ, 695, 94
Whittet, D. C. B., Cook, A. M., Herbst, E., Chiar, J. E., & Shenoy, S. S. 2011, ApJ, 742, 28
Whittet, D. C. B., Poteet, C. A., Chiar, J. E., et al. 2013, ApJ, 774, 102
Williams, D. A., Hartquist, T. W., & Whittet, D. C. B. 1992, MNRAS, 258, 599
Wright, C. M., Aitken, D. K., Smith, C. H., Roche, P. F., & Laureijs, R. J. 2002, in The Origins of Stars and Planets: The VLT View, ed. J. Alves, & M. McCaughrean (New York: Springer), 85
Wright, C. M., Do Duy, T., & Lawson, W. 2016, MNRAS, 457, 1593

D C B Whittet

Chapter 6

The Observed Properties of Dust—IV. Continuum and Line Emission

Emission from dust contributes a significant fraction of the total radiative output of the Milky Way and other galaxies, and plays an important role as a thermal regulator in many astrophysical situations and on many size scales. In a typical interstellar environment, a dust particle gains energy mainly by absorption of ultraviolet photons from the ambient radiation field, the absorbed energy re-emerging as infrared emission from the dust. Equilibrium is established if a grain emits power equal to that absorbed at some temperature that depends on its size and composition. Temperatures of order 15 K are predicted for grains in the size range needed to account for visible extinction in the diffuse interstellar medium, corresponding to continuum emission peaking in the far-infrared. Observations across the infrared spectrum led not only to confirmation of this prediction but also to the discovery of emission at much shorter infrared wavelengths, implying that a subset of the dust population is warmed to much higher temperatures. Models that account for this must include contributions from particles small enough to undergo substantial transient heating when individual UV photons are absorbed. The emitters include not only very small grains (nanoparticles) but also polycyclic aromatic hydrocarbon molecules (PAHs) identified by their characteristic spectral features.

This chapter reviews the observations and their significance, focusing primarily on particles energized by the general interstellar radiation field (ISRF). Additional discussions of emission observed in circumstellar environments appear in Sections 8.2.1–8.2.2 (evolved stars) and 9.3.5–9.3.6 (young stars). See Section 2.3 for a review of the physical processes and concepts pertaining to dust emission. The application of polarized emission from aligned grains to studies of galactic magnetic fields is discussed in Section 4.1.

doi:10.1088/2514-3433/ac7204ch6

6.1 Galactic Infrared Continuum Emission

6.1.1 Morphology

Diffuse emission from cold dust at far-infrared to submillimeter wavelengths traces the distribution of mass in the interstellar medium (Section 2.3.3). The global distribution of this emission over the entire sky is illustrated in Figure 6.1. The metric used for display is the optical depth at 353 GHz, deduced by fitting a modified blackbody curve to the observed flux in each pixel over the spectral range $100 < \lambda < 850$ μm, yielding an angular resolution of $\sim 5'$ (Planck Collaboration 2014). The emission in this spectral range is almost entirely from dust in thermal equilibrium with the ISRF. The map is dominated by the equatorial disk of the Milky Way, in which multiple clouds along the line of sight contribute to the flux in any given direction. Regions away from the disk reveal a complex pattern of individual clouds and striations, known as "infrared cirrus," mostly associated with local material within a few hundred parsecs of the Sun. Cirrus clouds are generally diffuse, but some have significant molecular content (Boulanger & Pérault 1988; Gillmon & Shull 2006).

Figure 6.2 shows an example of infrared imaging at shorter wavelengths, used to study the distribution of matter at much higher temperatures. The Tarantula Nebula is an H II region in the Large Magellanic Cloud (LMC), heated by embedded luminous stars (Section 1.3.3). The bright area to the left and below center contains the R136 starburst region, where massive stars are forming at a significantly higher rate than elsewhere in the LMC galaxy. The image is a color composite of observations by the Spitzer Space Telescope obtained in three infrared passbands, centered at 3.6, 4.5, and 8.0 μm and color-coded blue, green, and red, respectively. Whitish areas are bright in all three passbands. The 4.5 μm passband mostly measures continuum emission from hot, ionized gas, whereas the 3.6 and 8.0 μm

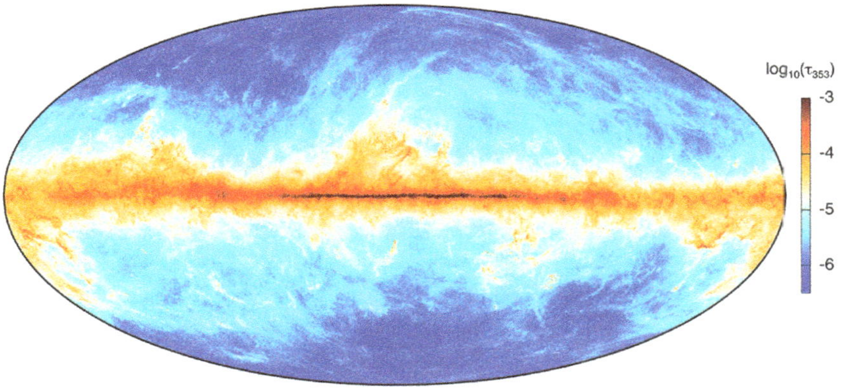

Figure 6.1. All-sky map in Mollweide projection, illustrating the distribution of dust optical depth at 353 GHz (850 μm) inferred from far-infrared observations of thermal emission. The map is orientated relative to the disk of the Milky Way (horizontal band) with the nucleus toward the center. Color coding is on a logarithmic scale, as indicted in the side bar. Image credit: reproduced with permission from Astronomy & Astrophysics (Planck Collaboration 2014).

Figure 6.2. Three-color composite image of the Tarantula Nebula, combining Spitzer images in passbands centered at 3.6, 4.5, and 8.0 μm (color-coded blue, green, and red, respectively). Magenta coloration (bluish red) traces the distribution of continuum and line emission from particles transiently heated by the ambient radiation field. Image credit: NASA/JPL-Caltech.

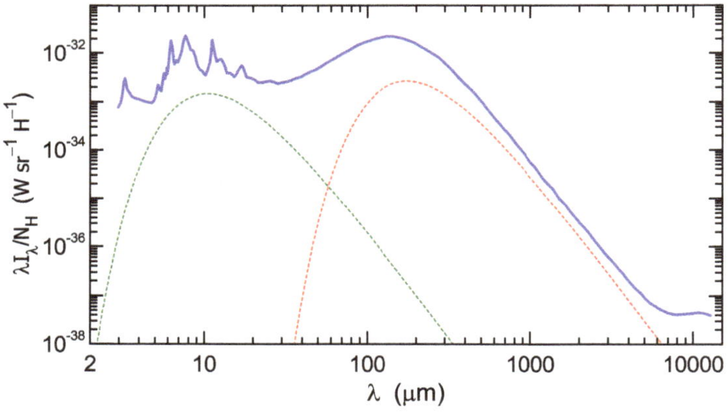

Figure 6.3. The mean spectral energy distribution of diffuse emission from dust in our Galaxy (blue curve), normalized relative to hydrogen column density, based on a review of previously published data by Hensley & Draine (2021). Also shown for comparison are modified blackbody curves of temperatures 15 K (red) and 250 K (green), each with constant emissivity index $\beta = 1.5$ and arbitrary vertical scaling.

passbands overlap strong spectral features of PAHs (Section 6.2.2). PAHs and other small particles are heated by individual energetic photons permeating the nebula to temperatures reaching 1000 K (Section 2.3.2); they cool by spectral line and/or continuum emission, both of which are typically most intense in the 8.0 μm passband (see also Figure 6.3). Thus, the magenta coloration (bluish red) traces the distribution of the smallest particles in the nebula.

6.1.2 Spectral Energy Distribution

The observed mean spectral energy distribution (SED) of diffuse emission from dust in the Milky Way is illustrated in Figure 6.3 (blue curve). Some four decades of wavelength are covered, from the near-infrared to the microwave region. Also shown for comparison are modified blackbody curves for two representative temperatures: cold dust at 15 K (red), and warmer dust at 250 K (green), with maxima near 150 μm and 10 μm, respectively. Dust temperatures sampled by the observations might extend anywhere from ~15 K for grains in thermal equilibrium to instantaneous values as high as ~1000 K for very small grains undergoing thermal transients (Section 2.3). The red and green curves serve to represent these distinct situations. Cold dust in thermal equilibrium with the ISRF is largely responsible for emission in the far-infrared (FIR) to millimeter range, but its contribution is negligible in the mid-infrared (MIR), orders of magnitude less than the observed intensity. At these shorter wavelengths, the emission is attributed to the cumulative effect of thermal transients arising in a size distribution of small grains, the amplitude of each event determined by the size and composition of the particle and the energy of the incident photon. A broad continuum peaking near 10 μm is produced, on which PAH emission features are superposed. Models for interstellar extinction (Section 3.6) are simultaneously constrained by these observations: the mid-infrared component is attributed primarily to small carbon grains and PAHs, the thermal emission at longer wavelengths to larger particles of both carbonaceous and silicate composition (Sodroski et al. 1997; Li & Draine 2001a; Zubko et al. 2004; Jones et al. 2013).

The SED in Figure 6.3 departs from the monotonic decline predicted by the Rayleigh-Jean approximation at the longest wavelengths, indicating the presence of an excess known as the anomalous microwave emission (AME). The source of this emission is not well understood. Synchrotron and Bremsstrahlung radiation from ionized gas may contribute to the microwave continuum, but these sources cannot account for the entire excess. The amplitude of the AME correlates with other measures of dust emission, suggesting an origin in some component of the dust. A proposed mechanism is electric dipole emission from spinning ultra-small grains. See Hensley & Draine (2021) for further discussion and references.

The observed emission from dust in the Milky Way provides a benchmark for comparison with other galaxies. In many cases, their SEDs are qualitatively similar, both normal and active galaxies exhibiting continuum peaks and/or line emission consistent with the presence of distinct cold dust and warmer small-grain populations (Rodriguez-Espinosa et al. 1996; Dale et al. 2001; Galliano et al. 2018). Such observations suggest a degree of similarity in the nature of the particles and the manner in which they reprocess stellar radiation in different systems. A major influence on the SED of a galaxy is its star formation history. The most prolific sources of MIR flux in a typical "normal" galaxy are regions of rapid, massive star formation activity (e.g., Figure 6.2; see also Section 1.4.6). Thus, the SED of a galaxy undergoing a phase of starburst activity will differ markedly from that of an otherwise identical galaxy in a quiescent phase of its evolution, characterized by differences in the contributions of the warmer versus cooler components. Modeling

SEDs has emerged as an important diagnostic tool, revealing signatures that distinguish between quiescence, starbursts, and nuclear activity across a wide range of galaxy types (e.g., Genzel & Cesarsky 2000; Galliano et al. 2018, 2021; Hunt et al. 2019; Dale et al. 2020).

6.1.3 Dust and Gas

The "big" grains responsible for the cold component of the SED (Figure 6.3) contain a high proportion of the total dust mass. This conclusion is inferred from the form and range of the size distribution function that enables good fits to observed extinction curves (Section 3.6.1; see also Kim et al. 1994). The distribution of dust measured by thermal emission from these grains may be compared with data for tracers of interstellar gas, such as radio-wavelength continuum and spectral line emissions in our own and other galaxies. The correlation with H I column density deduced from 21 cm line emission is generally very good in diffuse regions of the ISM (Boulanger et al. 1996; Planck Collaboration 2014), but inevitably breaks down in denser clouds where a significant fraction of the hydrogen is molecular. The integrated CO emission line intensity may be used as a proxy measure of H_2, using an appropriate value for the conversion factor (Bolatto et al. 2013), provided that CO remains mostly in the gas (see Section 9.2.4). Such observations enable the large-scale emission from dust associated with molecular, atomic and ionized phases of the ISM to be distinguished and mapped. Cold dust in the diffuse H I phase has been shown to contribute the dominant fraction ($\sim 70\%$) of the total FIR luminosity of the Milky Way, with dust in cold molecular and warm ionized gas contributing $\sim 20\%$ and $\sim 10\%$, respectively (Sodroski et al. 1989, 1994, 1997).

Comparisons between the amplitude of FIR emission and the total gas column in a given line of sight allow the dust-to-gas ratio to be estimated. The flux density at wavelength λ received in a beam of solid angle Ω may be expressed as

$$F_\lambda = \Omega\, \tau_\lambda B_\lambda(T_\mathrm{d}) \tag{6.1}$$

assuming that the dust is optically thin ($\tau_\lambda \ll 1$) in the FIR. The optical depth may thus be determined from the observed flux for dust at an assumed temperature (e.g., $T_\mathrm{d} \simeq 15$ K, as indicated by the peak of the spectral energy distribution). This is related to the mean dust density ρ_d along a pathlength L by

$$\tau_\lambda = \rho_\mathrm{d}\kappa_\lambda L \tag{6.2}$$

where κ_λ is the mass absorption coefficient of the dust (Section 2.3.3). The dust-to-gas ratio is then

$$Z_\mathrm{d} = 0.71\frac{\rho_\mathrm{d}}{\rho_\mathrm{H}} = 4.3 \times 10^{26}\frac{\tau_\lambda}{N_\mathrm{H}\kappa_\lambda} \tag{6.3}$$

where $N_\mathrm{H} = \rho_\mathrm{H}L/m_\mathrm{H}$ is the total hydrogen column density in the line of sight, and the factor 0.71 allows for the contributions of other gases (principally He) to the total mass of gas. The atomic component of N_H is measured directly, and the molecular component deduced from CO emission. Dust associated with ionized gas contributes negligibly to the optical depth in most diffuse lines of sight. As an example, consider data for dust and gas in the solar neighborhood of the galactic disk, which yields a mean ratio $\tau_{240}/N_H \approx 1.3 \times 10^{-29}$ m^2/nucleon for observations of dust emission at $\lambda = 240$ μm (Sodroski et al. 1994, 1997). Combining this value in Equation (6.3) with an estimate of $\kappa_{240} \approx 0.72$ m^2 kg^{-1} for an appropriate grain model, the mean dust-to-gas ratio is $Z_d \approx 0.008$.

The good agreement between this result and an independent estimate based on extinction ($Z_d \approx 0.007$; Section 3.2.6) is encouraging but perhaps fortuitous. Potential errors affecting estimates of Z_d based on Equation (6.3) are numerous. They include uncertainties in κ_λ, in the temperature-dependent evaluation of τ_λ from F_λ (Equation (6.1)), and in the determination of N_H, including possible errors in the CO → H_2 conversion and differences in sampling of gas relative to dust. Such factors may introduce systematic as well as random errors in Z_d that are hard to quantify. The uncertainties may be minimized in comparative studies, i.e., investigations of spatial variations in Z_d from one region to another, using a consistent set of methods and assumptions. In contrast, estimates based on extinction (Section 3.2.6) are less susceptible to systematic errors, because interstellar reddening and hydrogen line absorptions are observed in identical lines of sight, but also restricted in scope to relatively nearby regions of our Galaxy. Observations of extinction may be used to optimize the calibration of corresponding results based on FIR data (e.g., Planck Collaboration 2014).

Such investigations have found robust evidence for a notable trend: the dust-to-gas ratio declines systematically with increasing galactocentric radius in the disks of the Milky Way and other spiral galaxies (Sodroski et al. 1997; Pohlen et al. 2010; Casasola et al. 2020). This behavior is consistent with a corresponding trend in metallicity, as expected if the abundance of dust correlates with the availability of the dust-forming chemical elements (Section 7.1).

6.2 Infrared Spectral Emission Features

A heated grain may lose energy by spectral line emission if vibrational modes in the solid are activated (see Section 5.1.1 for an introduction to vibrational spectroscopy). This may occur when dust is sufficiently heated by absorption of radiation, either in proximity to an individual source or more generally by exposure to the ambient ISRF in diffuse regions of the ISM. However, only a small subset of the solid-state features detected in absorption (Chapter 5) have direct counterparts in emission. A primary reason for this is that the ices, pervasive in dark clouds (Section 5.3), are vaporized in situations that might otherwise give rise to emission features. Contrarily, populations of very small grains that are optically thin in the infrared may nevertheless exhibit prominent line emission if the relevant energy levels are efficiently "pumped" by absorption of photons at shorter wavelengths. The principal emission signatures attributed to cosmic dust arise in silicates and aromatic hydrocarbons.

6.2.1 Silicates

Silicates are sufficiently robust to survive under a wide range of physical conditions. Heating may induce structural changes, but temperatures typically well in excess of

1000 K are needed for significant melting and vaporization. Emission features corresponding to the principal silicate vibrational modes at 9.7 and 18.5 μm (Table 5.1) are seen when grains are heated to temperatures of a few hundred Kelvin, as commonly occurs in the envelopes of luminous young stars embedded in their parent clouds, in the extended atmospheres of cool, evolved stars with oxygen-rich circumstellar shells, and in the circumnuclear regions of dusty active galaxies. However, only the smallest (nanometer-sized) silicate grains are expected to reach such temperatures in the general ISM, remote from individual luminous sources.

By comparing the emission profiles in such lines of sight with each other and with the absorption profiles observed in cooler regions of the ISM, one may hope to gain insight into the properties of the silicates and their evolution in response to environment. It should be noted, however, that circumstellar emission by warm dust and foreground absorption by cold dust may often be superposed in the same line of sight. A general model for the spectral energy distribution of such a source therefore includes both emissive and absorptive terms (Gillett et al. 1975), such that the observed flux is

$$F_\lambda \propto \left[Q_\lambda B_\lambda(T_\mathrm{d})\right] e^{-\tau_\lambda} \tag{6.4}$$

where Q_λ and τ_λ are the emissivity and optical depth of warm and cold dust, respectively.

Emissivity profiles that have been used as standards for the 9.7 μm silicate feature are compared in Figure 6.4, determined from observations of the Trapezium cluster in the Orion nebula and the red supergiant μ Cephei. The emitting grains in the Trapezium H II region are presumed to be ambient interstellar grains heated by luminous stars in the cluster; those toward μ Cephei are presumed to have condensed in the atmosphere of the star itself. Of these, the μ Cephei profile appears to be the

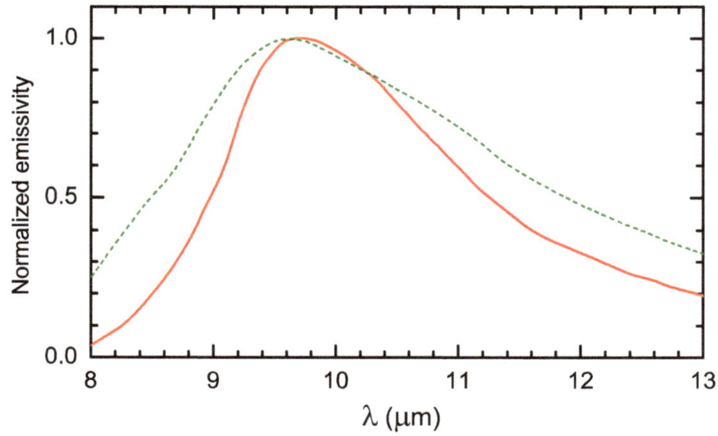

Figure 6.4. Silicate emissivity profiles obtained from observations of the red supergiant μ Cephei (solid red curve; Merrill & Stein 1976) and the Trapezium cluster (dashed green curve; Gillett et al. 1975). In each case, the emissivity is calculated from the observed flux (Equation (6.4)) and normalized to unity at the peak, assuming a temperature $T_\mathrm{d} = 250$ K for the blackbody continuum and negligible foreground extinction.

more broadly representative, providing a good match to observations of silicate absorption in diffuse regions of our Galaxy (Roche & Aitken 1984, 1985; Bowey et al. 1998, 2004; Chiar & Tielens 2006) and toward the nucleus of the luminous infrared galaxy NGC 4418 (Roche et al. 2015). This concurrence supports the conclusion that silicates in the diffuse ISM are chemically and structurally similar to at least some of those forming in the circumstellar outflows of cool, evolved stars with oxygen-rich atmospheres (see Sections 8.1.2 and 8.2.2 for further discussion).

The Trapezium profile is notably broader than that of μ Cephei (Figure 6.4), and as such it tends to yield better fits to observations of silicate absorption in dense molecular clouds. This outcome may be fortuitous, however. Problems that hinder efforts to measure absorption profiles specific to the silicate component of dust in dense clouds are discussed in Section 5.3.6. Fogerty et al. (2016) conclude that distinct populations of warm and cool silicate grains are needed to best represent the Trapezium profile, its width arising from temperature-dependent differences in peak emission, a situation that does not arise in a quiescent dense cloud. Variations in emissivity profiles from one line of sight to another may thus depend on conditions local to each source, so no single one can be considered representative.

Silicate emission observed in H II regions and circumstellar shells can be explained in terms of "classical" submicrometer-sized grains: small compared with the wavelengths of the features, but not necessarily small enough to undergo significant transient heating (Section 2.3.2). Do silicate *nanoparticles* also contribute? If they are abundant their emission features should be ubiquitous, not limited to regions in close proximity to stars. We should expect, for example, to see silicate emission superposed on the diffuse continuum and PAH line emission from the galactic disk (Figure 6.5), but if present it lies below detection limits. Calculations based on the

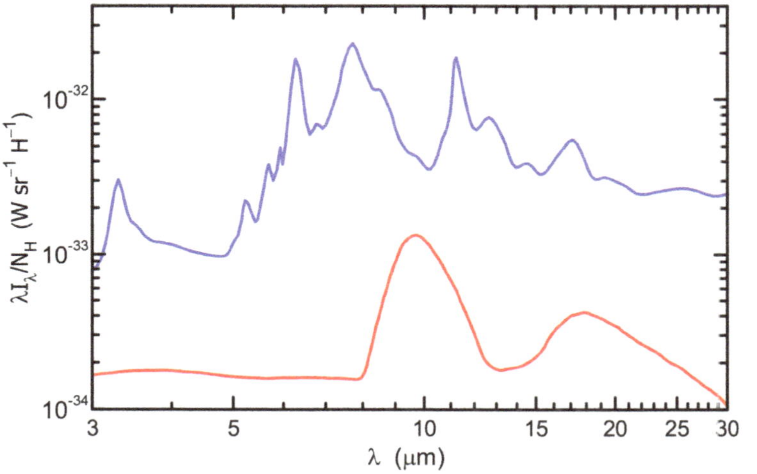

Figure 6.5. The observed mid-infrared spectral energy distribution of dust in the diffuse ISM (blue curve, expanded from Figure 6.3) compared with a model for emission from a population of transiently-heated silicate nanoparticles that might account for the anomalous microwave emission (red curve; Hensley & Draine 2017). The nanoparticles are assumed to contain 6% of the elemental Si depleted into dust in the diffuse ISM.

graphite-silicate model for interstellar dust allow an upper limit to be estimated on the abundance of silicate grains small enough to undergo sufficient transient heating in the general ISRF ($a \lesssim 2$ nm; Li & Draine 2001b). Up to about 10% of the interstellar Si could be in such particles without violating known observational constraints, which include their expected contributions to FUV extinction as well as to emission.

Renewed interest in silicate nanoparticles arose from the discovery of anomalous microwave emission and a proposal that it might be explained by electric dipole emission from ultra-small spinning grains (Section 6.1.2). A focus on silicates is motivated by a finding that the amplitude of the AME does not correlate with the observed emissions from PAHs, which might otherwise be considered likely candidates. Hensley & Draine (2017) conclude that silicates could be responsible for the AME if they are sufficiently abundant to contain at least 4% of the available Si, i.e., within the upper limit set by other observations. An example of the predicted emission spectrum from such a population is shown in Figure 6.5, compared with the observed SED over the same spectral range and calibrated on the same intensity scale. The peaks in the silicate emission fall significantly below the observed SED.

6.2.2 Polycyclic Aromatic Hydrocarbons

Discrete emission features in the mid-infrared spectra of bright nebulae were first detected during the pioneering years of infrared spectroscopy (Gillett et al. 1973; Soifer et al. 1976; Russell et al. 1977). The features where remarkable for being much broader than normal gas-phase nebular emission lines; lacking any obvious identifications, they became known as the "unidentified infrared" (UIR) features, a label that still sometimes persists today. However, these features have since become firmly associated with vibrational modes in polycyclic aromatic hydrocarbons (PAHs). PAHs are composed of C atoms arranged in linked hexagonal rings with H atoms (or other radicals) attached to peripheral sites; some simple examples are illustrated in Figure 6.6. A key factor in the emergence of the PAH hypothesis (Léger & Puget 1984; Allamandola et al. 1985) was the realization that such small particles may undergo substantial transient heating upon absorption of

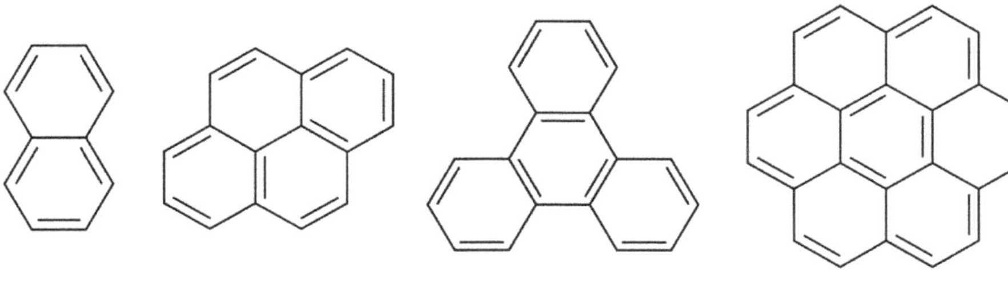

Figure 6.6. Schematic illustrations of some simple PAH structures. A carbon atom is located at each apex of the interlocking hexagons, with a peripheral hydrogen atom at each available valence site. From left to right: naphthalene (two rings, $C_{10}H_8$), pyrene (four rings, $C_{16}H_{10}$), triphenylene (also four rings, $C_{18}H_{12}$), and coronene (seven rings, $C_{24}H_{12}$).

Table 6.1. The Principal PAH Emission Features.

λ (μm)	Mode
3.3	C–H stretch
6.2	C–C stretch
7.7	C–C stretch + C–H in-plane bend
8.6	C–H in-plane bend
11.3	C–H out-of-plane mono bend
12.7	C–H out-of-plane trio bend

individual energetic photons (Section 2.3.2), thereby explaining the occurrence of emission features in environments where larger particles are too cold to produce them. PAHs have since became a major focus of research in all areas of astrophysics—laboratory investigations, theory and modeling as well as observations of sources ranging from nearby nebulae to distant galaxies. See Tielens (2008) and Peeters (2011) for reviews, and Joblin and Tielens (2011) for the proceedings of a conference devoted entirely to this topic.

The wavelengths of the principal features and their assignments are listed in Table 6.1, and some representative spectra are shown in Figure 6.7. PAHs are inferred to be present as a generic *class*, not as individual species, as the spectra do not yield unique identifications.[1] Many sizes and structures may be present, but compact, stable forms are likely to be favored under interstellar conditions. The infrared spectra place constraints on their size, ionization state and chemical purity. As a population, PAHs may be considered representative of the smallest particles in a broad size distribution of carbonaceous grains that contribute to different components of the infrared emission (see Table 6.2 and discussion below).

The observations show that PAHs are ubiquitous throughout the cosmos. In our Galaxy they are prominent if dense interstellar or (C-rich) circumstellar matter is subject to intense blue-UV irradiation, as is the case in photodissociation regions and planetary nebulae (see Figure 6.7 for examples), thus signaling their presence in environments surrounding both young and old stars. They are also detected in reflection nebulae, and in diffuse regions of the galactic disk, where they appear as structure superposed on the mid-infrared continuum from heated small grains in the SED (Section 6.1.2; Figures 6.3 and 6.5). They are readily mapped in the interstellar media of nearby external galaxies such as the Magellanic Clouds, and they provide a means of tracing physical conditions and star formation activity in all types of galaxy (Galliano et al. 2018; Chastenet et al. 2019; Cortzen et al. 2019). The features seen the starburst galaxy NGC 695, for example, are qualitatively very similar to those seen in the Orion Bar, a region of vigorous recent star formation in our own Galaxy (Figure 6.7).

[1] The first specific identification of a PAH was made at radio wavelengths; McGuire et al. (2021).

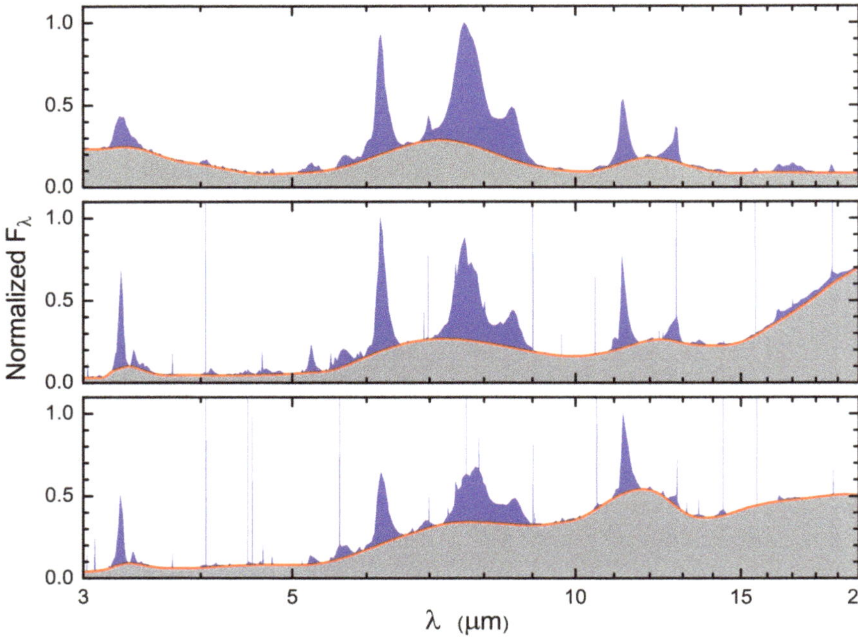

Figure 6.7. Spectra of diverse sources displaying strong PAH emission features. Two are located in our Galaxy: NGC 7027 (bottom), a planetary nebula, and the Orion Bar (center), a photodissociation region (Peeters et al. 2002; Peeters 2011). NGC 695 (top) is a spiral galaxy undergoing starburst activity, its spectrum adjusted to rest-frame (Brown et al. 2014). Spectral features (blue) include narrow gas-phase lines in addition to broader PAH features. Red curves are spline functions fitted to data apparently free of emission features. Underlying areas (gray) illustrate the occurrence of broad plateaus superposed on general continua.

Table 6.2. Overview of Infrared Emissions from C-rich Particles.

Carrier	Component	N_C	Size (nm)	f_C (ppm)
PAHs	MIR features	20–100	0.4–1	14
PAH clusters	MIR plateaus	100–1000	1–2	8
Small grains	MIR continuum	10^3–10^6	2–10	25
Larger grains	FIR continuum	$>10^6$	>10	$\gtrsim 35$

Notes: N_C is the number of C atoms per particle; f_C is the required abundance of C (parts per million) relative to total H.

Although the main features dominate the spectra, numerous weaker bands are also evident (a few of them attributed to fullerenes; see Section 6.2.3). PAH features have typical widths in the range 0.03–0.5 μm and are thus easily distinguished from the much narrower gas-phase atomic and ionic lines seen in emission nebulae (Figure 6.7). Much broader emissions are also evident: known as plateaus, they underlie the main groups, covering the approximate spectral ranges 3.2–3.6 μm, 6–9 μm and 11–13 μm.

The various PAH features and plateaus all belong to a common generic spectrum, in the sense that when one is present so generally are the others, although spatial variations occur in their relative strengths.

Variations also occur in profile shapes. The profiles of the various features differ from one to another in the same spectrum, and differences are also seen in the same feature from source to source. These variations are important as diagnostics of the nature of the particles and their response to environment, but also challenging to interpret. Some features (e.g., 3.3 and 8.6 μm) are fairly symmetric, others exhibit appreciable asymmetry, most often in the form of a steep blue rise and a more gradual red decline (e.g., 6.2 and 11.2 μm; the 12.7 μm feature is an exception). The broad, strong 7.7 μm "feature" is actually a blend of two or more components, and variations in their relative amplitudes alter the profile shape (Shannon & Boersma 2019). Constructing a consistent mathematical model to represent the emission profiles is therefore exacting. In isolation, an ideal harmonic oscillator has a Lorentzian emission profile, but in practice the vibrating atoms may be perturbed by interactions with near neighbors, thus modifying the profile. Accretion of PAHs into clusters may shift and broaden the profiles, and their positions and widths also depend on temperature. For an individual PAH experiencing a thermal transient, its instantaneous temperature is determined by the energy of the absorbed photon, the number of degrees of freedom in the particle, and the time elapsed since the excitation event (cf Equation (2.40) and Figure 2.6). An observed profile in a given environment might thus be considered as the superposition of many discrete features arising in PAHs at different temperatures and with different sizes and structures. See Pech et al. (2002), Tielens (2008), and Roser & Ricca (2015) for further discussion.

The relative strengths of the features may vary not only from source to source but also spatially within an extended source. Statistical analyses reveal the presence of two distinct groups: the 3.3 and 11.2 μm features tend to correlate more closely with each other than with other features; similarly, those at 6.2 and 8.6 μm appear to be an intercorrelated pair. Laboratory and theoretical studies provide a simple explanation of this behavior: neutral PAHs emit strongly at 3.3 and 11.2 μm, whereas PAH cations show dominant emission in the 5–10 μm region. This interpretation is consistent with observations of nebulae in proximity to stellar sources of ionizing irradiation, such as reflection nebulae, H II regions, photodissociation regions and planetary nebulae (e.g., Galliano et al. 2008; Boersma et al. 2018). Peak intensity ratios such as $I_{6.2}/I_{11.2}$ may be calibrated with respect to the ionization state of the PAHs (see Figure 6.8, left panel). These results provide a widely applicable tool for exploring physical conditions in other regions and in other galaxies (e.g., Rigopoulou et al. 2021). Ionized PAHs also differ from neutrals in displaying significant optical activity at visible wavelengths, leading to a proposal that they might account for the diffuse interstellar bands (Section 3.5.1), but no unambiguous identifications have yet been made.

A PAH is, of course, an aromatic hydrocarbon by definition, but the possibility exists for substitutions. As examples, an N atom might replace ≡C–H in the hexagonal structure, or a radical such as $-CH_3$, –CN, –CHO or –OH might replace a peripheral H. Such changes modify the spectra and may therefore be detectable.

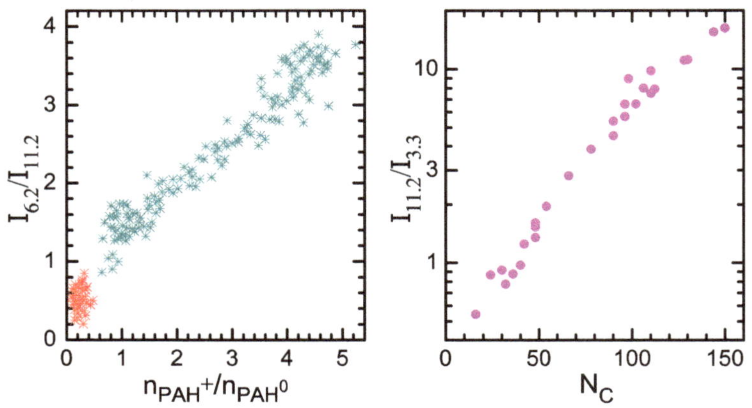

Figure 6.8. Examples of PAH intensity ratios displaying systematic variations with ionization state and size. Left: observed values of the ratio $I_{6.2}/I_{11.2}$ in the reflection nebula NGC 7023 (green) and the planetary nebula NGC 40 (red), calibrated relative to the ratio of PAH cations to neutrals (Boersma et al. 2018). Right: the dependence of $I_{11.2}/I_{3.3}$ on the number of C atoms in PAHs with different sizes and structures (Croiset et al. 2016). Based on analyses using the NASA Ames PAH IR Spectroscopic Database (Bauschlicher et al. 2018).

The aliphatic stretching mode of attached CH_3 is a probable source of the subfeature at 3.4 μm (Figure 6.7; see Joblin et al. 1996), its relative strength indicating that carbon in the particles has $\sim$2% aliphatic bonding (Tielens 2008). There is no conclusive evidence in the infrared spectra for other substitutions, but a detection of cyanonaphthalene, i.e., naphthalene (Figure 6.6) with CN in place of a peripheral H, has been reported from observations of rotational spectra at radio wavelengths (McGuire et al. 2021). Overall, the weight of evidence is that the carriers of the infrared emission features are predominantly pure aromatic hydrocarbons with few substitutions.

Variations in relative strength also yield information on particle size. The 3.3 and 11.2 μm features both arise primarily in neutral PAHs, as discussed above, thus the intensity ratio $I_{11.2}/I_{3.3}$ is insensitive to ionization state. It is, however, sensitive to size, increasing systematically with the number of C atoms in the molecule (Croiset et al. 2016 and references therein), as illustrated in Figure 6.8 (right panel). This result further enhances the potential of spatially-resolved spectrophotometry of PAHs as a means to probe physical conditions in extended objects such as photodissociation regions and reflection nebulae, using different line ratios to map the distributions of PAHs with respect to both size and ionization state. Table 6.2 suggests a hierarchy of sizes for PAHs and other carbonaceous particles that contribute to infrared line and continuum emission across the spectrum (adapted from Tielens 2008). The large polarizability of PAH molecules results in a bonding force between their aromatic planes, and this may lead to the formation of clusters. A PAH cation or anion bonds strongly to neutral PAHs by charge-induced dipole bonding, thus ionization may enhance clustering. The clusters are likely to form initially as stacks of two-dimensional PAHs (Rapacioli et al. 2006), rather than as soot-like nanoparticles in which aromatic units are connected randomly by aliphatic

Figure 6.9. Composite infrared image of NGC 7023. Passbands centered at 8.0 and 24 μm are coded green and red, mapping the distributions of ionized PAHs and small-grain continuum emission, respectively. Transitions from green to red indicate the boundaries of a photodissociation region around the central star. Blue colors correspond to shorter infrared wavelengths in which the photospheres of normal field stars are most prominent. Image credit: NASA/JPL-Caltech, based on data from the Spitzer and WISE missions.

links. Stacking is thus consistent with the finding noted above that the carriers of the emission bands have low concentrations of aliphatic C. Clusters have been proposed as carriers of the broad plateaus (Figure 6.7), and their size distribution presumably extends continuously into somewhat larger particles that contribute to the continuum emission. The nature of the particles in the various size regimes depend on growth, modification and destruction processes discussed in more detail in Chapters 8 and 9.

Figure 6.9 illustrates the distributions of PAH cations and very small grains in a well-studied nebula. NGC 7023 (the Iris Nebula) contains a photodissociation region (PDR) illuminated by a young star near the center of the image. Physical conditions change markedly across the boundaries of the PDR. PAHs are effectively ionized by incident UV photons inside the PDR, their distribution mapped by the intensity of emission features within a 6–10 μm passband, colored green in the image. Irradiation also alters the size distribution of the particles inside the PDR, causing disintegration of clusters and photo-evaporation of individual PAHs. With increasing distance from the central star, a transition is reached as the ionizing photon flux dips rapidly and the PAHs become mostly neutral. Intensity in the image is then dominated by continuum emission from transiently-heated small grains, mapped in red. See Rapacioli et al. (2005) and Croiset et al. (2016) for related discussion.

6.2.3 Fullerenes

Fullerenes are three-dimensional carbon allotropes, of which C_{60} is the best-known example. This quasi-spherical cage molecule has a polyhedral shell composed of interlocking hexagonal and pentagonal units. First detected by laboratory synthesis, fullerenes are now known to occur naturally in very small quantities on Earth, e.g., in certain C-rich minerals, and in space. They are highly stable and thus durable under a wide range of physical conditions.

Like PAHs, fullerenes such as C_{60} are subject to stochastic heating by UV photon absorption in the ISM. Neutral C_{60} has vibrational modes that predict the appearance of infrared features at 7.0, 8.5, 17.4, and 18.9 μm. Partial blending with much stronger PAH features adds to the challenge of detection, but their presence has been confirmed in astronomical spectra: they are observed in nebulae subject to intense UV radiation fields (e.g., Sellgren et al. 2010; Cami et al. 2010), and more tentatively in the diffuse ISM (Berné et al. 2017). Only a small fraction of all available interstellar carbon ($\lesssim$0.1%) needs to be in C_{60} to explain the strengths of the features. Berné & Tielens (2012) review proposed mechanisms for their production in space.

6.3 Extended Red Emission

So far in this chapter we have focused on the emissions at infrared and submillimeter wavelengths arising from *heating* of irradiated particles, i.e., resulting from excitation of vibrational energy levels. To emit thermally at shorter wavelengths ($\lambda \lesssim 1$ μm), the dust would need to be heated to temperatures $T_{\rm d} \gtrsim 2000$ K, exceeding sublimation temperatures for all but the most refractory solids. However, emission may also occur as the result of *photoluminescence*. This process is triggered by absorption of an energetic visible or UV photon, inducing an electronic transition from the ground state to a bound high-energy state in a molecule or the conduction band in a semiconductor. The particle then relaxes through a series of vibrational/rotational transitions until a state is reached whereby a return to the ground state is possible by means of an electronic transition upon emission of a lower-energy photon, typically at red or near-infrared wavelengths. Interstellar radiation attributed to this process is termed the extended red emission (ERE).

The ERE is difficult to observe because it is intrinsically faint and blended with other sources of diffuse radiation, such as scattering from dust and both line and continuum emission from hot nebular gas. These can be separated by careful analysis of high-quality data. Its spectrum exhibits a broad profile, extending from ~0.55 μm to the near-infrared, with a peak that varies in wavelength from source to source. A typical example is shown in Figure 6.10. The profile appears to be continuous and lacking in structure at the highest available spectral resolution (fluctuations in Figure 6.10 illustrate the noise level). The ERE has been detected in a wide range of circumstellar and interstellar environments in the Milky Way, and in other galaxies (see Witt & Vijh 2004 and Witt & Lai 2020 for reviews and extensive guides to previous literature). It is pervasive in environments where dust is exposed to significant UV irradiation, most conspicuously in emission or reflection nebulae around hot stars, but is also detectable at much lower intensity levels in the diffuse ISM (Gordon et al. 1998). Thus, the emitter appears to be a common constituent of the interstellar medium.

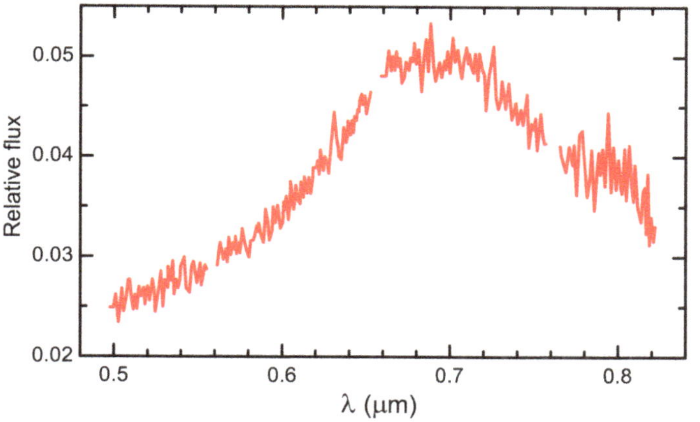

Figure 6.10. The profile of extended red emission observed in the reflection nebula NGC 7023, corrected for scattering and displayed as the ratio of nebular to stellar flux (adapted from Witt & Vijh 2004). Gaps at 0.56, 0.656, and 0.76 μm correspond to spectral regions contaminated by strong night-sky and nebular emission lines.

The identity of this constituent is unknown, but the observations provide important clues. The case for a *carbonaceous* carrier is strongly supported by statistics on the occurrence of ERE in planetary nebulae. The abundance of elemental carbon in a planetary nebula depends on the evolutionary history of the stellar progenitor, specifically the degree to which its atmosphere has been enriched in this product of internal nucleosynthesis (Section 7.1). A C-rich atmosphere promotes C-rich molecule formation and dust nucleation (Section 8.1). The ERE is seen preferentially in those that are so enriched.

Another significant clue arises from the nature of the radiation field. In principle, photoluminescence can be triggered by photons ranging in wavelength from the visible to the extreme ultraviolet. However, all circumstellar sources of strong ERE involve excitation by stars with surface temperatures $\gtrsim$10,000 K, whereas ERE is absent in nebulae illuminated by significantly cooler stars ($\lesssim$7000 K). This finding places rather tight limits of $10.5 < \varepsilon_{\rm ph} < 13.6$ eV on the energy range of photons needed to initiate the ERE process (Darbon et al. 1999; Lai et al. 2017). If the carrier is an ion, such high photon energies may be necessary to produce it by photo-ionization; subsequent photoluminescence may then be triggered by photons over a wider energy range (Witt & Lai 2020).

Evident parallels may be drawn from these results between the carriers of the ERE and those of the infrared emission features discussed in Section 6.2.2. Both appear to be carbon-rich and to require broadly similar levels of excitation to induce emission, often occurring in the same sources. However, they cannot be one and the same as their amplitudes are not well correlated. Detailed mapping of reflection nebulae reveals a spatial hierarchy with respect to increasing distance from the source of radiation, in the sequence ERE → ionized PAHs → neutral PAHs (Lai et al. 2017). The ERE is most intense in regions where molecular hydrogen is almost fully dissociated, so the gas is transparent to photons with energies in the optimal range for its production. These results tend to favor an origin for the ERE in isolated, ionized, dehydrogenated carbon molecules, containing perhaps 20–40 atoms (Witt & Lai 2020). A correspondence has been found between the strengths of the ERE and the diffuse interstellar bands, suggesting the possibility of a common origin (Lai et al. 2020). Both require a complexity of energy states over a similar spectral range.

References

Allamandola, L. J., Tielens, A. G. G. M., & Barker, J. R. 1985, ApJ, 290, L25
Bauschlicher, C. W., Ricca, A., Boersma, C., & Allamandola, L. J. 2018, ApJS, 234, 32
Berné, O., & Tielens, A. G. G. M. 2012, PNAS, 109, 401
Berné, O., Cox, N. L. J., Mulas, G., & Joblin, C. 2017, A&A, 605, L1
Boersma, C., Bregman, J., & Allamandola, L. J. 2018, ApJ, 858, 67
Bolatto, A. D., Wolfire, M., & Leroy, A. K. 2013, ARA&A, 51, 207
Boulanger, F., & Pérault, M. 1988, ApJ, 330, 964
Boulanger, F., Abergel, A., Bernard, J.-P., et al. 1996, A&A, 312, 256
Bowey, J. E., Adamson, A. J., & Whittet, D. C. B. 1998, MNRAS, 298, 131
Bowey, J. E., Rawlings, M. G., & Adamson, A. J. 2004, MNRAS, 348, L13
Brown, M. J. I., Moustakas, J., Smith, J.-D. T., et al. 2014, ApJS, 212, 18
Cami, J., Bernard-Salas, J., Peeters, E., & Malek, S. E. 2010, Sci., 329, 1180
Casasola, V., Bianchi, S., De Vis, P., et al. 2020, A&A, 633, A100
Chastenet, J., Sandstorm, K., Chiang, I.-D., et al. 2019, ApJ, 876, 62
Chiar, J. E., & Tielens, A. G. G. M. 2006, ApJ, 637, 774
Cortzen, I., Garrett, J., Magdis, G., et al. 2019, MNRAS, 482, 1618
Croiset, B. A., Candian, A., Berné, O., & Tielens, A. G. G. M. 2016, A&A, 590, A26
Dale, D. A., Helou, G., Contursi, A., Silbermann, N. A., & Kolhatkar, S. 2001, ApJ, 549, 215
Dale, D. A., Anderson, K. R., Bran, L. M., et al. 2020, AJ, 159, 195
Darbon, S., Perrin, J.-M., & Sivan, J.-P. 1999, A&A, 348, 990
Fogerty, S., Forrest, W., Watson, D. M., Sargent, B. A., & Koch, I. 2016, ApJ, 830, 71
Galliano, F., Madden, S. C., Tielens, A. G. G. M., Peeters, E., & Jones, A. P. 2008, ApJ, 679, 310
Galliano, F., Galametz, M., & Jones, A. P. 2018, ARA&A, 56, 673
Galliano, F., Nersesian, A., Bianchi, S., et al. 2021, A&A, 649, A18
Genzel, R., & Cesarsky, C. J. 2000, ARA&A, 38, 761
Gillett, F. C., Forrest, W. J., & Merrill, K. M. 1973, ApJ, 183, 87
Gillett, F. C., Forrest, W. J., Merrill, K. M., Capps, R. W., & Soifer, B. T. 1975, ApJ, 200, 609
Gillmon, K., & Shull, J. M. 2006, ApJ, 636, 908
Gordon, K. D., Witt, A. N., & Friedmann, B. C. 1998, ApJ, 498, 522
Hensley, B. S., & Draine, B. T. 2017, ApJ, 836, 179
Hensley, B. S., & Draine, B. T. 2021, ApJ, 906, 73
Hunt, L. K., et al. 2019, A&A, 621, A51
Joblin, C., & Tielens, A. G. G. M. 2011, PAHs and the Universe: A Symposium to Celebrate the 25th Anniversary of the PAH Hypothesis, EAS Publications Series, Vol. 46
Joblin, C., Tielens, A. G. G. M., Allamandola, L. J., & Geballe, T. R. 1996, ApJ, 458, 610
Jones, A. P., Fanciullo, L., Köhler, M., et al. 2013, A&A, 558, A62
Kim, S. H., Martin, P. G., & Hendry, P. D. 1994, ApJ, 422, 164

Lai, T. S.-Y., Witt, A. N., & Crawford, K. 2017, MNRAS, 469, 4933
Lai, T. S.-Y., Witt, A. N., Alvarez, C., & Cami, J. 2020, MNRAS, 492, 5853
Léger, A., & Puget, J. L. 1984, A&A, 137, L5
Li, A., & Draine, B. T. 2001a, ApJ, 554, 778
Li, A., & Draine, B. T. 2001b, ApJ, 550, L213
McGuire, B. A., Loomis, R. A., Burkhardt, A. M., et al. 2021, Sci., 371, 1265
Merrill, K. M., & Stein, W. A. 1976, PASP, 88, 285
Pech, C., Joblin, C., & Boissel, P. 2002, A&A, 388, 639
Peeters, E. 2011, in IAU Symp. 280, The Molecular Universe, ed. J. Cernicharo, & R. Bachiller (Cambridge: Cambridge Univ. Press), 149
Peeters, E., Hony, S., Van Kerckhoven, C., et al. 2002, A&A, 390, 1089
Planck Collaboration, 2014, A&A, 571, A11
Pohlen, M., Cortese, L., Smith, M. W. L., et al. 2010, A&A, 518, L72
Rapacioli, M., Joblin, C., & Boissel, P. 2005, A&A, 429, 193
Rapacioli, M., Calvo, F., Joblin, C., et al. 2006, A&A, 460, 519
Rigopoulou, D., Barale, M., Clary, D. C., et al. 2021, MNRAS, 504, 5287
Roche, P. F., & Aitken, D. K. 1984, MNRAS, 208, 481
Roche, P. F., & Aitken, D. K. 1985, MNRAS, 215, 425
Roche, P. F., Alonso-Herrero, A., & Gonzalez-Martin, O. 2015, MNRAS, 449, 2598
Rodriguez-Espinosa, J. M., Pérez Garcia, A. M., Lemke, D., & Meisenheimer, K. 1996, A&A, 315, L129
Roser, J. E., & Ricca, A. 2015, ApJ, 801, 108
Russell, R. W., Soifer, B. T., & Merrill, K. M. 1977, ApJ, 213, 66
Sellgren, K., Werner, M. W., Ingalls, J. G., et al. 2010, ApJ, 722, L54
Shannon, M. J., & Boersma, C. 2019, ApJ, 871, 124
Sodroski, T. J., Dwek, E., Hauser, M. G., & Kerr, F. J. 1989, ApJ, 336, 762
Sodroski, T. J., Bennett, C., Boggess, N., et al. 1994, ApJ, 428, 638
Sodroski, T. J., Odegard, N., Arendt, R. G., et al. 1997, ApJ, 480, 173
Soifer, B. T., Russell, R. W., & Merrill, K. M. 1976, ApJ, 210, 334
Tielens, A. G. G. M. 2008, ARA&A, 46, 289
Witt, A. N., & Lai, T. S.-Y. 2020, Ap&SS, 365, 58
Witt, A. N., & Vijh, U. P. 2004, in ASP Conf. Ser. 309, Astrophysics of Dust, ed. A. N. Witt, G. C. Clayton, & B. T. Draine (San Francisco, CA: ASP), 115
Zubko, V., Dwek, E., & Arendt, R. G. 2004, ApJS, 152, 211

AAS | IOP Astronomy

Dust in the Galactic Environment (Third Edition)

D C B Whittet

Chapter 7

The Observed Properties of Dust—V. Element Depletions

It is well-known that hydrogen and helium together comprise some 98% by mass of all observed (baryonic) matter in the universe. As inhabitants of a rocky planet, we are accustomed to an atypical situation in which the other 2% predominates, and interstellar dust is likewise composed primarily of important planet-building elements such as C, O, Mg, Si, and Fe. Spectroscopy of interstellar gas provides an indirect but powerful means of quantifying the contributions of these and other "heavy" elements to the dust (see Section 1.2.6 for an introductory overview). The presence of a given element in solid form is inferred from its observed gas-phase abundance, which may be far less than predictions based on standard abundances that include all possible forms and phases. The observed depletions place fundamental constraints on grain models that are highly complementary to compositional indicators based on optical properties, discussed in previous chapters. The sensitivity of depletion to environment also helps to elucidate the nature of the mechanisms by which gas and dust exchange material in the interstellar medium. An essential starting point is to assess the origin and distribution of the condensible elements, i.e., those likely to be incorporated into solids, and thereby to establish an appropriate set of standard abundances (Section 7.1). The observed depletions and their implications are reviewed in subsequent sections.

7.1 The Condensible Elements

The set of chemical elements we now observe in the cosmos is the outcome of processes that created, modified and distributed them as the universe evolved. Simultaneous advances in cosmology, nuclear physics, and stellar astrophysics during the 20th century enabled robust models to be developed that describe these processes. According to the standard cosmological model, a short-lived period of nucleogenesis in the expanding fireball created a universe initially composed almost

doi:10.1088/2514-3433/ac7204ch7

exclusively of H and He; conditions needed for significant production of heavier elements were reached subsequently in stellar cores and cataclysms. This section outlines the processes that led to the availability of condensible elements in the ISM, and assesses how their abundances evolve in space and time. See Trimble (1991, 1997) and Wallerstein et al. (1997) for comprehensive reviews of nucleosynthesis processes.

7.1.1 Origins

The nuclei of heavier elements and isotopes are synthesized in stellar cores by fusing lighter ones together. Core temperatures $\gtrsim 10^7$ K are needed to ignite such reactions, because they must overcome the coulomb force that inhibits fusion of positively-charged atomic nuclei, and progressively heavier elements thus require progressively higher temperatures. Once ignited, these reactions are generally exothermic, i.e., there is a net release of nuclear binding energy, thereby producing the source not only of a star's luminosity but also of outward thermodynamic pressure that maintains its stability by balancing the inward force of gravity.

Depending on its mass, a star may undergo a series of "burning" phases during its lifetime, in which the product of one phase is the fuel of the next. Normal main-sequence stars like the Sun subsist by hydrogen burning, in which networks of reaction pathways (the p-p chain and CNO cycle) convert H into He. This phase is the most enduring because H is not only abundant but also has the highest "octane" of all nuclear fuels in terms of energy release per nucleon. Eventually, H becomes depleted in the core, which then shrinks until internal temperatures and densities reach levels needed to ignite He $\rightarrow$ C fusion (the triple-α process). Simultaneously, the outer layers of the star expand and it evolves to become a red giant. O may also be formed during this phase by addition of another α particle to C. When He becomes depleted in the core, single low-mass stars like the Sun are unable to generate core conditions necessary to ignite further burning cycles, typically ending their lives as white dwarfs after shedding their outer layers to form planetary nebulae.

Stars significantly more massive than the Sun may undergo further burning cycles in which elements such as C, N, and O are fused to produce still heavier elements such as Mg, Si, S and, ultimately, Fe. No further exothermic reactions are then possible because Fe has the greatest binding energy per nucleon of all elements. The outcome of Fe accumulation in the core is a supernova explosion driven by catastrophic core collapse. The resultant shock wave propagates outward, transforming layers surrounding the core into a soup of highly energized particles; many of the compound nuclei formed in earlier fusion cycles are dissociated, generating copious supplies of protons, neutrons, and α particles. Observations and models are in reasonable agreement on the abundances of the elements emerging from this melee (e.g., Kobayashi et al. 2006). Destructive and constructive reactions operate in parallel, tending to favor the more stable products. The availability of free neutrons enables production of elements heavier than Fe, which occurs in a step-wise fashion: neutron capture (resulting in a unit increase in atomic mass) is followed by β-decay (emission of an electron, leading to a unit increase in atomic number). All products merge into the ISM as the supernova remnant expands. High-energy cosmic rays emanating from supernovae may also drive fission reactions in the ISM, a primary source of the Li, Be, B group of elements.

These processes occur in individual stars and on different timescales, each according to its mass. Additional mechanisms of element synthesis arise in close binary systems, of which the most significant as a source of new elements is the type Ia supernova. Such an event is triggered by accretion from a companion star onto a white dwarf until it reaches a critical mass for thermonuclear runaway, leading to copious production of elements such as Si and Fe. See the review by Johnson (2019) for further discussion of all relevant processes and their likely contributions to the origins of each chemical element in the periodic table.

7.1.2 The Astration Cycle

The widespread availability of nucleosynthesis products in the interstellar medium depends on a feedback cycle of stellar mass loss (the astration cycle). The first generations of stars formed from gas composed of primordial H and He. As stars evolve, they return a portion of their mass to the ISM, infusing it with new elements that then become available for inclusion in subsequent generations of stars and their planetary systems. A star loses mass continuously throughout its life in the form of a stellar wind emanating from its atmosphere, as exemplified by the Sun's solar wind. During the main-sequence phase, however, the wind dissipates only a minuscule fraction of a star's mass and contains no new elements, fusion products remaining locked in its core. But during later phases of stellar evolution mass-loss rates tend to be orders of magnitude higher (Section 8.4.1), and winds may contain nucleosynthesis products transported to the surface by deep convection currents.

Subsequently, depending on its mass, the core collapses to become a white dwarf, a neutron star, or a black hole, and the outer layers of the star are ejected as a planetary nebula or a supernova remnant. Elements falling into the collapsed core are effectively lost from the feedback cycle, unless exhumed by an external event such as accretion leading to thermonuclear runaway, but the expanding envelope carries products of nucleosynthesis into the surrounding ISM. The infusion rate depends on the density of the stellar population, its initial mass function, and the star formation rate, resulting in both spatial and temporal variations within normal spiral galaxies such as the Milky Way. Metallicity tends to increase systematically with declining galactocentric distance (e.g., Sharda et al. 2021), implying that such galaxies tend to evolve progressively outward from the center.

A schematic illustration of the temporal trend in heavy-element abundances over the lifespan of the Milky Way is shown in Figure 7.1. Colored rectangles represent typical distributions in metallicity and age for different stellar populations. Metallicity is defined as

$$\left[\frac{\mathrm{Fe}}{\mathrm{H}}\right] = \log\left\{\frac{N_{\mathrm{Fe}}}{N_{\mathrm{H}}}\right\} - \log\left\{\frac{N_{\mathrm{Fe}}}{N_{\mathrm{H}}}\right\}_{\odot} \tag{7.1}$$

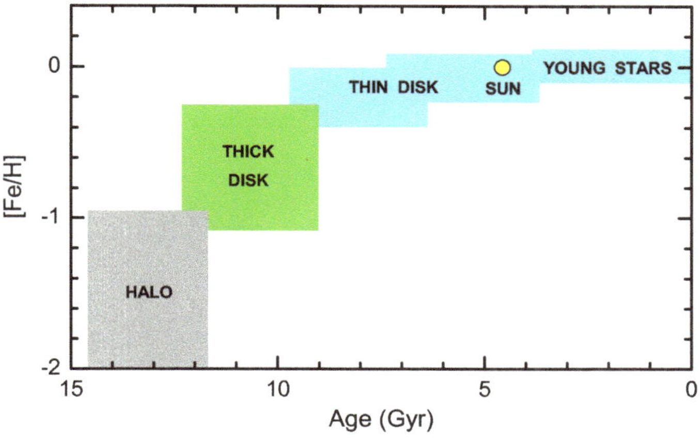

Figure 7.1. Schematic illustration of the rise in metallicity over the lifetime of our Galaxy (adapted from Trimble 1997). Rectangles represent typical distributions for stars in different populations and age groups. The locus of the Sun is also shown.

i.e., as an iron abundance, although in practice observations of several elements may be included in the analysis. Results are calibrated with respect to the solar abundance (see below), thus negative values of [Fe/H] indicate subsolar metallicity. The populations in Figure 7.1 range from ancient, metal-poor globular clusters in the outer spheroidal halo of the Galaxy to newly-formed stars in the thin component of the disk. The general trend suggests a period of relatively rapid early enrichment, driven by short-lived massive stars that evolved to become supernovae, followed by a gradual leveling off, such that the metallicity of the youngest stars is not markedly different from that of the Sun. See Section 7.1.4 for further discussion.

7.1.3 The Solar Abundances

The relative abundances of the individual elements constitute a fundamental data set of major significance. It is both a test-bed for models of nucleosynthesis and a window into evolutionary history. Figure 7.2 presents a snapshot of the abundance pattern in the sample most readily accessible to us: results for our solar system provide a data set that is both comprehensive and precise, widely used as the standard for comparison with other regions of the cosmos (see Asplund et al. 2009 for a review). Uniquely, it is based on laboratory analyses of meteoritic remnants from the protoplanetary disk as well as on stellar spectroscopy. Abundances may thus be determined more accurately for more elements than is the case for any other sample. As a normal, single, main-sequence star, the Sun is assumed to have preserved a reliable imprint of abundances in the nebula from which it formed in the disk of our Galaxy, some 4.6 Gyr ago. All but the lightest elements were forged in previous generations of stars and recycled into the ISM prior to the Sun's birth.

Abundances in Figure 7.2 are displayed on a logarithmic scale relative to $N_H = 10^{12}$, i.e., the height of the column for element X is $\log \epsilon_X = \log(N_X/N_H) + 12$. This format

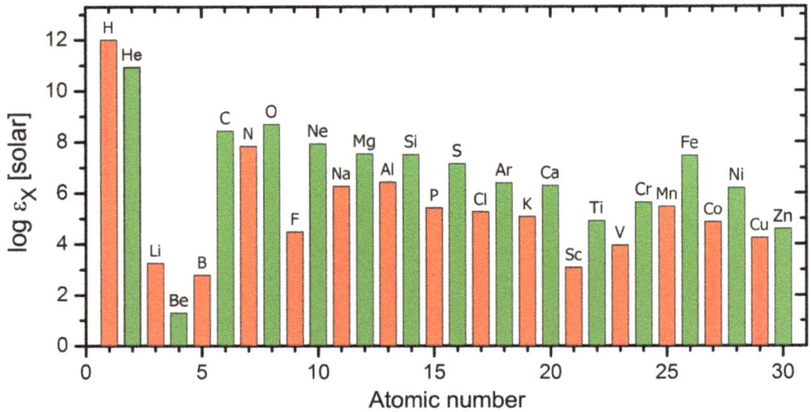

Figure 7.2. Histogram of mean elemental abundances in the solar system, displayed in order of increasing atomic number from $Z = 1$ to 30 (H to Zn). Abundances are logarithmic by number of atoms relative to $N_H = 10^{12}$. Columns for odd and even-numbered elements are colored red and green, respectively. Based on data from Asplund et al. (2009).

is common in the astronomical literature and has the minor advantage that values are positive for all but the rarest elements. Only the first 30 are shown in Figure 7.2 as this set includes all those likely to make any appreciable contribution to the mass of cosmic dust. The pattern corroborates the standard model of nucleosynthesis, outlined in Section 7.1.1. From left to right, the abundant primordial elements H and He are followed by a trough at $Z = 3–5$ attributed to the intrinsic fragility of the Li, Be, B group, which are typically consumed in stars. A notable feature of the pattern for higher elements ($Z \geqslant 6$) is the "odd–even" effect: those with even-numbered Z values (green columns) are systematically more abundant than their neighbors. Nuclei with even Z values are composed of integral numbers of α-particles. Thus, their abundances are consistent with a formation sequence that builds elements by α-addition and thereby creates nuclei of greatest stability. This effect extends to the prominent iron peak at $Z = 26$, corresponding to the end-point of exothermic nucleosynthesis, beyond which abundances tail off rapidly; all elements with $Z > 30$ (not shown) have abundances $N_X/N_H < 10^{-8}$.

As previously mentioned, the solar abundances are based on laboratory studies of certain primitive meteorites as well as on spectroscopic observations of the Sun's atmosphere.[1] It is informative to compare these two distinct sets of data, shown graphically in Figure 7.3. Meteorites are collisional fragments of asteroids. Those used for this purpose are type CI carbonaceous chondrites, the most volatile-rich of the meteorite classes; thus, they are considered to be the least altered of all such remnants from the Sun's protoplanetary disk, assembled from the same mix of elements that went into the Sun. All elements in the solar atmosphere are now of

[1] The silicon abundance is typically used to tie these two sets of data together (Asplund et al. 2009). Abundances in the planetary science literature are generally normalized to $N_{Si} = 10^6$ instead of $N_H = 10^{12}$.

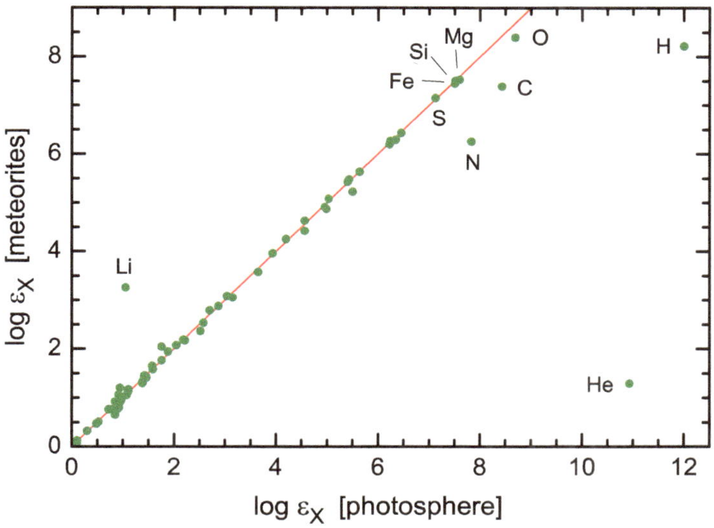

Figure 7.3. Plot of element abundances in the solar photosphere versus those in meteorites (type CI carbonaceous chondrites), based on data from Asplund et al. (2009). Uncertainties are typically smaller than the plotting symbols. The red diagonal line represents exact agreement. Several elements discussed in the text are labeled.

course fully vaporized, regardless of their previous state; those in meteorites are largely condensates from the protoplanetary disk, albeit potentially modified by subsequent loss of volatiles. Figure 7.3 thus indicates the extent to which each element in the disk was condensed and retained within solids.

A large majority of the elements, including major mineral-forming metals such as Mg and Fe, plot on or very close to the line of exact agreement in Figure 7.3. As expected, these elements maintain a robust presence in solids under interplanetary conditions. In contrast, noble gases are retained in minuscule quantities: only He appears in Figure 7.3 because the others are out of range ($N_X/N_H < 10^{-12}$) on the meteoritic axis! Another notable exception is lithium, and in this case the meteoritic value is considered the more reliable as Li is easily destroyed in stars. Of greatest interest in the present context is the behavior of the "CHON" group, which show differing degrees of retention in the meteorites.

If the Sun's atmosphere correctly represents the endowment of CHON elements in the birth cloud, their offsets toward lower abundances in meteorites (Figure 7.3) reflect differing volatilities of the solids that accommodate them. The meteorites likely formed inside or close to the "snow line" of the Sun's protoplanetary disk, i.e., in an environment where ices were largely sublimated (Lecar et al. 2006) or subjected to photochemical processing (Section 5.2.3). Under such conditions, H is retained only as a minor constituent of non-volatile solids such as hydrated minerals and hydrocarbons. In contrast, oxygen is retained more efficiently, owing to the pervasive presence of O-bearing minerals such as silicates and oxides in the meteorites. Solid carbon appears in many guises, including carbonates, carbides, graphite and diamond, but the bulk of the eponymous carbonaceous fraction of the chondrites is composed of organic refractory matter and amorphous carbon. Nitrogen is accommodated mostly within organics and notably less prevalent in meteorites than C or O. We might surmise from these results that cosmic dust in the ISM at the time and location of the Sun's birth was likewise composed principally of the elements C, O, Mg, Si, S, and Fe, i.e., those with $N_X/N_H > 10^{-5}$ in the chondrites.

Do the solar abundances provide an appropriate standard for the present-day ISM? To what extent have interstellar abundances in the solar neighborhood of the Milky Way been enhanced and modified since the birth of the Sun? These are crucial questions underpinning studies of element depletions in the ISM, which we shall next attempt to address.

7.1.4 Abundances in Young Stars

Galaxies become progressively enriched in heavy elements over time as the astration process cycles through successive generations of stars (Section 7.1.2). The Sun formed when the Galaxy was approximately two thirds of its present age. Thus, we expect to see clear differences when we compare solar abundances with those in much older or much younger stars, which should be metal-poor and metal-rich, respectively, relative to the Sun. This prediction is readily confirmed for the oldest stars (see Figure 7.1), but evidence that it holds true for the youngest ones is equivocal. Observations that allow us to address this question focus on two stellar samples: the early-type stars (spectral types O and B), and solar analogs (single F and G type main-sequence stars).

Early-type stars formed very recently in astrophysical terms, within the past ~ 100 Myr, and their atmospheres should thus provide a reliable guide to modern abundance levels in the ISM. Initial results yielded a surprise, however: rather than exhibiting the expected excess over solar values, abundances in early-type stars were found to be *subsolar*, typically by ~ 0.2 dex, equivalent to $\sim 63\%$ by number of atoms (Savage & Sembach 1996; Snow & Witt 1996; Sofia & Meyer 2001). More recent work has mitigated this anomaly somewhat, owing to enhanced analytical techniques (Nieva & Przybilla 2012) and a downward revision of the recommended solar abundances for C, N, and O (Lodders 2003; Asplund et al. 2009, 2021). Another concern is that C, N, and O abundances observed in early-type stars might be underestimated because models for stellar atmospheres do not fully account for their ionization states (Lyubimkov 2013). All things considered, there is no compelling evidence that abundances in early-type stars differ markedly from solar values.

F and G stars are far more numerous and have far greater longevity, thus they are widely used as tracers of galactic enrichment trends (e.g., Bensby et al. 2014). Their spectra enable detailed line-by-line matching with the solar spectrum (Bedell et al. 2018), yielding optimal precision in the comparisons. Samples limited to G type stars in the solar neighborhood show that the Sun's metallicity lies in the top 25% of the distribution (Adams 2010). Samples limited to the youngest F and G type stars (with estimated ages $\lesssim$2 Gyr) yield abundances that do not deviate markedly from solar values (see Hensley & Draine 2021 for a review).

In summary, these results suggest that the Sun might be a reasonable standard for the local ISM by pure serendipity: it appears to have been born with an enhancement of the same order as the degree of local galactic enrichment over its 4.6 Gyr lifespan. Proposed explanations include the possibility that the Sun has migrated from a birth site $\sim$ 3 kpc closer to the galactic center than its present location, or that the solar nebula was enriched by ejecta from a nearby supernova explosion. The latter possibility gains empirical support from certain isotopic abundance anomalies in meteorites, most readily explained by nuclear processing in SNe (Adams 2010). Stellar metallicity appears to be closely linked to the presence and architecture of planetary systems (Adibekyan et al. 2012; Bedell et al. 2018), leading to conjecture that the Sun's enhancement might have anthropic significance.

7.1.5 Abundances from X-Ray Observations

X-ray absorption spectroscopy is a powerful and direct technique for measuring element abundances in the ISM. Both gas and dust are semi-transparent in this region of the spectrum, and photon energies are much higher than chemical binding energies, so the spectral response depends primarily on the elemental composition of the material. X-ray absorption spectroscopy is thus uniquely sensitive to all reservoirs of an element in the intervening medium, both dust and gas, both ions and neutrals. The dust produces fine-structure lines near photoelectric absorption edges that carry information on composition and structure: see Lee et al. (2009) and Costantini et al. (2019) for detailed reviews.

The observations require high sensitivity and spectral resolution, thus sampling is limited to lines of sight toward reasonably bright X-ray sources. Suitable targets in our home Galaxy are luminous X-ray binary systems (e.g., Gatuzz et al. 2014, 2016; Psaradaki et al. 2020; Rogantini et al. 2020), but it should be noted that they are not randomly distributed, the majority lying toward the Galactic Center. Extragalactic X-ray sources have also been observed as background continua for absorptions in the Milky Way (Baumgartner & Mushotzky 2006) and in the host galaxy (Winter et al. 2007). Spectral coverage of the energy range $0.2 < \varepsilon_{\rm ph} < 10$ keV is needed to detect all of the important condensible elements: those available for study with current facilities include O, Mg, Si, and Fe, but regrettably not C (Costantini et al. 2019). In general, results available to date for our Galaxy appear to corroborate those based on optical stellar spectroscopy. Interstellar O, Mg and Si abundances are close to solar, and of these, Mg and Si reside almost entirely ($\gtrsim$95%) in solids, consistent with a silicate composition (Rogantini et al. 2020). The interstellar Fe abundance is less well constrained and might be somewhat subsolar (Gatuzz et al. 2016). Fe also resides almost entirely in solids, but only partly in silicates (Westphal et al. 2019), leaving the remainder unexplained. Spectroscopy of the Fe K-edge absorption may be the key to future unlocking of this uncertainty (Rogantini et al. 2018).

7.1.6 Converging on a Standard

Elemental depletions are most widely studied in the solar neighborhood of the galactic disk, within a few kiloparsecs of the Sun (Section 7.2). The reliability of the results depends implicitly on assumptions regarding the average abundances of the condensible elements, both relative and absolute, in this galactic region. Caveats accompany any choice for such a standard, for reasons discussed above. The options are summarized here and enumerated in Table 7.1.

- **The Sun.** Our solar system yields a data set thought to be precise and internally consistent for most elements (Section 7.1.3), although uncertainties regarding the crucial C, N, O group persist (Asplund et al. 2009, 2021). A modest correction is needed to allow for diffusion and gravitational settling of heavier elements over the Sun's lifespan (Lodders 2003). Applying data for a 4.6 Gyr-old star to the present-day ISM may seem counterintuitive, but evidence suggests that it is a reasonable choice.
- **Early-type stars.** Because these stars are young they should be reliable indicators of recent abundance levels in the ISM, and they have not lived long enough to migrate far from their birth sites. As a sample, their significance is somewhat compromised by a relatively small number of suitable targets (in comparison to F and G stars) and concerns regarding the reliability of assumed ionization states.
- **F and G stars.** This sample is much larger, but also far more diverse in terms of age and spatial distribution of potential birth sites. In general, the ages of individual stars are poorly constrained, so subsamples restricted to members below a certain age may not be entirely reliable.

Table 7.1. Abundances of Major Condensible Elements.

	Solar System			Young Stars	
Element	Solar	Protosolar	PS + GCE	O, B Type	F, G Type
C	245	288	324	214	290
N	67.6	79.4	—	62	—
O	490	575	682	575	517
Na	2.0	2.3	—	—	—
Mg	35.5	41.7	53.9	36.3	43.0
Al	2.9	3.4	3.5	—	3.5
Si	34.7	40.7	44.6	31.6	40.2
S	15.5	18.2	17.2	—	19.5
Ca	2.2	2.6	3.2	—	3.0
Cr	0.45	0.52	—	—	—
Fe	29.5	34.7	43.7	33.1	31.7
Ni	1.7	1.9	2.1	—	1.8

Notes: Values are in parts per million (ppm) by number relative to H. All stable isotopes are included. See text for further details.

Abundance values for each of these possibilities are listed in Table 7.1. The twelve most abundant condensible elements ($N_X/N_H \gtrsim 0.5$ ppm) are included. Solar values (column 2) are from Lodders (2003). They are in close agreement for most elements with more recently recommended values by Asplund et al. (2021), and are preferred here for consistency with the extensive studies of observed depletions by Jenkins (2009) and Ritchey et al. (2018). Protosolar values (column 3) include an adjustment to correct for diffusion and gravitational settling, with the goal of recovering values representative of the birth cloud; again, these are taken from Lodders (2003). Note that this correction is applicable to meteoritic as well as photospheric data because the former are normalized relative to the latter using the Si abundance. Also shown in Table 7.1 (column 4) are protosolar values with an additional adjustment based on a model for galactic chemical enrichment since the birth of the solar system (PS + GCE), as recommended by Hensley & Draine (2021). Finally, values for O and B stars are from Nieva & Przybilla (2012), and those for young F and G stars are weighted averages of results cited by Hensley & Draine (2021). It is striking that the PS + GCE abundances are systematically higher than those observed in either sample of young stars for almost all elements. Application of this adjustment thus tends to yield overestimates of abundances in the current ISM, for reasons discussed in Section 7.1.4. Protosolar abundances (Table 7.1, column 3) are assumed in the following section.

7.2 The Observed Depletions

7.2.1 Methods and Metrics

The term depletion refers to the under-abundance of a gas-phase element with respect to its standard reference abundance, resulting from its assumed presence in dust. It may be specified as a depletion index, defined for some element X as

$$D(\mathrm{X}) = \log\left\{\frac{N_\mathrm{X}}{N_\mathrm{H}}\right\}_\mathrm{gas} - \log\left\{\frac{N_\mathrm{X}}{N_\mathrm{H}}\right\}_\mathrm{ISM} \tag{7.2}$$

where the ISM term represents the assumed reference abundance of X (Section 7.1.6). We thus expect $D(\mathrm{X}) \approx 0$ (to within estimated uncertainties) for undepleted elements, becoming more negative for those with greater depletions. It is useful also to define the fractional depletion:

$$\delta(\mathrm{X}) = 1 - 10^{D(\mathrm{X})} \tag{7.3}$$

which is bound by the limits $\delta(\mathrm{X}) = 0$ (all atoms in the gas) and $\delta(\mathrm{X}) = 1$ (all atoms in the dust). The abundance of element X in the dust is then

$$\mathcal{A}_\mathrm{d}(\mathrm{X}) = \left\{\frac{N_\mathrm{X}}{N_\mathrm{H}}\right\}_\mathrm{dust} = \delta(\mathrm{X})\left\{\frac{N_\mathrm{X}}{N_\mathrm{H}}\right\}_\mathrm{ISM}. \tag{7.4}$$

The gas-phase column densities N_X and N_H (Equation (7.2)) are evaluated from analysis of interstellar absorption lines observed in stellar spectra. Coverage should include all applicable ionization states. Most of the relevant features occur in the far to mid ultraviolet region of the spectrum ($91.2 < \lambda < 320$ nm), and are thus accessible to observation only from space. Early-type stars are the most suitable targets because their spectral energy distributions peak in this wavelength range. High-resolution spectrometers on board the Copernicus satellite, the Hubble Space Telescope, and the Far Ultraviolet Spectroscopic Explorer have provided a wealth of observational data (see Savage & Sembach 1996 and Jenkins 2014 for reviews).

The column density of an absorber producing an unsaturated absorption line is related to the observed line strength by

$$N_X = \left(\frac{4\epsilon_0 m_e c}{e^2}\right)\frac{W_\nu}{f} \qquad (7.5)$$

where W_ν is the equivalent width of the line in frequency units, f is the intrinsic oscillator strength of the transition, and the various physical constants have their usual meanings. Equation (7.5) describes the linear region of the curve of growth (see, e.g., Spitzer 1978, p. 52–55): each absorbing atom along the line of sight sees essentially the full continuum level, such that an increase in the number of absorbers causes a proportionate increase in W_ν. However, if the absorption is too strong (optical depth $\tau_\nu \gtrsim 1$ at the line center), it is said to be saturated and Equation (7.5) underestimates the true column density. The curve of growth allows estimates of N_X for saturated lines, but results are inherently far less accurate. Thus, the ideal transitions have intrinsic strengths that are neither too small nor too large, such that absorption lines are readily detectable but not saturated. The accuracy to which f can be measured is a further source of uncertainty; best estimates are sometimes revised, requiring updates to previously published results to generate a homogeneous data set (Jenkins 2009).

Because of these issues, the depletions of some elements are known to considerably greater precision than others. The most problematic of those expected to contribute a major fraction of the dust mass is carbon. Correct evaluation of N_C requires summation of values for two ionization states. The first ionization potential of C (11.26 eV) is less than that of H (13.60 eV), so most of the available gas-phase carbon is in C II in diffuse clouds and in C I in denser clouds (gas-phase CO becomes important at higher densities than those considered here). The abundance of C I is relatively easy to measure, that of C II much more difficult because the strong permitted resonance lines at 103.6 and 133.4 nm are invariably saturated (Hobbs et al. 1982; Sofia et al. 2011). Data on interstellar C II abundances are mostly derived from observations of a very weak, unsaturated semi-forbidden C II line at 232.5 nm (Hobbs et al. 1982; Cardelli et al. 1996; Sofia et al. 1997, 2004). In contrast to the situation for carbon, oxygen (13.62 eV), and nitrogen (14.53 eV) have higher first ionization potentials than hydrogen, and only the neutral species need be considered. Most metals have first ionization potentials in the range 5–8 eV and second ionization potentials >13.6 eV, so the singly-ionized forms are predominant in interstellar clouds (Ca is a notable exception, with a second ionization potential of 11.9 eV). Both Na and Ca are problematic because their most readily accessible spectral signatures (the well-known sodium D and calcium H and K lines) do not represent their primary ionic forms.

Interstellar extinction places practical constraints on the lines of sight in which depletions can be investigated. Suitable spectra are most readily obtained for lightly and moderately reddened stars ($E_{B-V} \lesssim 0.5$); the environments sampled thus tend to be diffuse regions, in which H is predominantly atomic, although data exist for some lines of sight containing significant proportions of molecular gas. A useful metric for characterizing the dominant environment in a line of sight close to the galactic disk is the mean number density of hydrogen nuclei (Spitzer 1985), summed over atomic and molecular phases:

$$\langle n_{\rm H} \rangle = \frac{N({\rm H\,I}) + 2N({\rm H_2})}{L} \qquad (7.6)$$

where the pathlength L is specified by the distance to the star. A column of low mean density, $\langle n_{\rm H} \rangle < 0.2 \times 10^6$ m^{-3}, is unlikely to intercept a cloud containing a cool phase; but the presence of cool atomic or molecular gas along a line of sight (Section 1.3.2) elevates $\langle n_{\rm H} \rangle$ to values typically $\gtrsim 10^6$ m^{-3}. Assuming that depletion occurs when atoms encounter and stick to dust grains, we expect levels to increase systematically with $\langle n_{\rm H} \rangle$. Further insight is provided by the Doppler motions of gas-phase interstellar lines. To take a well-studied example, spectra of the star ζ Oph exhibit two components, at heliocentric radial velocities of −27 km s^{-1} and −15 km s^{-1}, that contain warm neutral and cold molecular gas, respectively. The depletions measured in these components show clear differences, being systematic higher in the colder, denser cloud.

An important development in the quest to tie depletion measures to environment is the study of *differential* depletions, introduced by Jenkins (2009). This approach emphasizes relative changes in depletion, rather than absolute values, as a means of investigating how each element behaves in response to ambient conditions. It thereby circumvents systematic errors associated with the choice of standard reference abundances. When the observed depletion index of one element is plotted against that of another, combining data for many lines of sight, a linear correlation is typically seen, with slopes and intercepts that differ depending on which elements are paired. This behavior implies that the overall amplitude of the depletion for each sight-line can be characterized by a single parameter, designated F_* (the depletion factor). The subscript indicates that F_* applies to the line of sight (toward a star) rather than to a specific element. It represents a weighted mean depletion, evaluated after removal of offsets arising from the fact that depletions are typically nonzero and vary from element to element under conditions in which they are minimized (see Jenkins 2009 for in-depth discussion). F_* is tied to ambient conditions by normalization, such that it is close to zero for lines of sight with lowest depletions and unity for a prototypical high-depletion region (chosen to be the −15 km s^{-1} cloud toward ζ Oph). The depletion index observed across many lines of sight is then linearly correlated with F_*:

$$D({\rm X}) = B_{\rm X} + A_{\rm X}(F_* - z_{\rm X}) \qquad (7.7)$$

where values of the coefficients A_X, B_X and z_X are unique to each element X. Jenkins (2009) tabulates values of these coefficients for 16 elements, deduced from analysis of a large, homogenized data set containing 243 stars. Additional evaluations for some rarer elements are available from Ritchey et al. (2018). As expected, the depletion factor increases with the mean number density of hydrogen (Equation (7.6)), yielding an empirical relation:

$$F_* = 0.461 \log\langle n_H \rangle - 1.99. \tag{7.8}$$

Observations of relatively abundant heavy elements such C, O, Mg, Si, and Fe (Table 7.1) are naturally of greatest interest as they dominate the mass (and hence presumably the chemical composition) of the solid particles that sequester them. Nevertheless, observations of much rarer elements are also important to complete the picture. They include metals such as Cd and Ti that span the range from near-zero depletion to among the highest known.

7.2.2 Average Depletions in the Milky Way

Representative results are presented in Table 7.2, based on observations of early-type stars within a few kiloparsecs of the Sun. The depletion index D(X) and the corresponding dust-phase abundance $\mathcal{A}_d$(X) are defined in Equations (7.2) and (7.4) with $\mathcal{A}_d$(X) denoted here in parts per million (ppm) relative to H. Three distinct environments are considered, corresponding to values of the depletion factor $F_* = 0.0$, 0.5 and 1.0. As noted in Section 7.2.1 above, values $F_* \simeq 0$ correspond to lines of sight with the lowest observed depletions, containing warm diffuse and hot

Table 7.2. Summary of Depletion Data

Element	D(X)	$\mathcal{A}_d$(X)	D(X)	$\mathcal{A}_d$(X)	D(X)	$\mathcal{A}_d$(X)
	$F_* = 0.0$		$F_* = 0.5$		$F_* = 1.0$	
C	−0.11	65	−0.16	90	−0.21	112
N	−0.11	(18)	−0.11	(18)	−0.11	(18)
O	−0.01	13	−0.12	142	−0.24	241
Na	−0.15	0.7	—	—	−0.9	2.0
Mg	−0.27	19	−0.77	35	−1.27	39
Al	−0.7	2.7	−1.4	3.3	−2.5	3.4
Si	−0.22	16	−0.79	34	−1.36	39
S	+0.16	(0)	−0.28	9	−0.72	15
Ca	−1.4	2.5	−2.5	2.6	−3.7	2.6
Cr	−0.83	0.4	−1.55	0.5	−2.27	0.5
Fe	−0.95	31	−1.59	34	−2.24	35
Ni	−0.94	1.7	−1.68	1.9	−2.43	1.9
Z_d	0.0028		0.0054		0.0071	

Notes: Depletion indices, dust-phase abundances (ppm) and resultant dust-to-gas mass ratios are listed for environments with low, intermediate and high values of F_*. See text for details.

intercloud gas (Section 1.3) in which all but the most refractory elements remain undepleted. Values $F_* \simeq 0.5$ are typical of the diffuse ISM, in which hydrogen is mostly atomic, and $F_* \simeq 1.0$ corresponds to a "diffuse molecular" cloud ($A_V \sim 1$), dense enough for significant $H \rightarrow H_2$ conversion but still below the threshold for detectable ice absorption (Section 5.3.2). Values of $D(X)$ in Table 7.2 are calculated from Equation (7.7) using coefficients determined by Jenkins (2009) for most elements; the exceptions are Na, Al, and Ca (not included in Jenkins' study), for which other sources of data were assessed (e.g., Crinklaw et al. 1994; Savage & Sembach 1996; Welty et al. 1999).

Also listed in Table 7.2 (final row) are depletion-based estimates of the dust to gas ratio in each environment, given by the summation

$$Z_d = 0.71 \times 10^{-6} \sum \mathcal{A}_d(X)m(X) \tag{7.9}$$

where $m(X)$ is the atomic mass number, $\mathcal{A}_d(X)$ is the dust-phase abundance in parts per million, and the factor 0.71 allows for the presence of He in the gas. Nitrogen is not included in the summation because its small and apparently uniform depletion seems more likely to arise from an offset caused by the choice of reference abundance than from depletion into solids (see Section 7.2.3 for further discussion). An independent estimate of Z_d based on extinction in the diffuse ISM ($Z_d \approx 0.007$; Section 3.2.6) is similar to the depletion-based estimates for higher F_* values, as expected if most of the extinction occurs in clouds.

Estimated relative contributions of the major condensible elements to the total dust mass are illustrated in Figure 7.4 for each environment. Two caveats should be noted: the data are most susceptible to systematic errors arising from the choice of standard abundances when depletions are small ($F_* \simeq 0$); and values for carbon are subject to the largest observational uncertainties for reasons explained in Section 7.2.1. Nevertheless, some significant trends are suggested. Fe, the most refractory of the "big five" elements, dominates at $F_* = 0$ and becomes "diluted" by enhanced depletions of other elements in environments with higher F_* values. The contributions of Mg and Si are more stable, but that of O shows dramatic shifts with environment. Systematic error might account for the remarkably small O content at $F_* = 0$, but cannot explain its general increase with F_*. This trend is hard to understand if minerals such as silicates and oxides are the primary repositories, a result that has important implications for grain models, further discussed in Section 7.3.

7.2.3 Correlation with Condensation Temperature

It is helpful to express the propensity of an element to form solids in terms of its condensation temperature (T_{cond}). This is defined for each element in a gas of solar composition as the temperature at which 50% of the atoms condense into solids in some form under chemical and thermodynamic equilibrium (Lodders 2003 and references therein). A pressure of 10^{-4} bar is assumed, appropriate to the solar nebula at 1 AU, and of the same order as pressures in the grain-forming atmospheres of red giant stars (Section 8.1). A wide range of physical conditions and non-equilibrium processes are implicated in the formation and destruction of interstellar

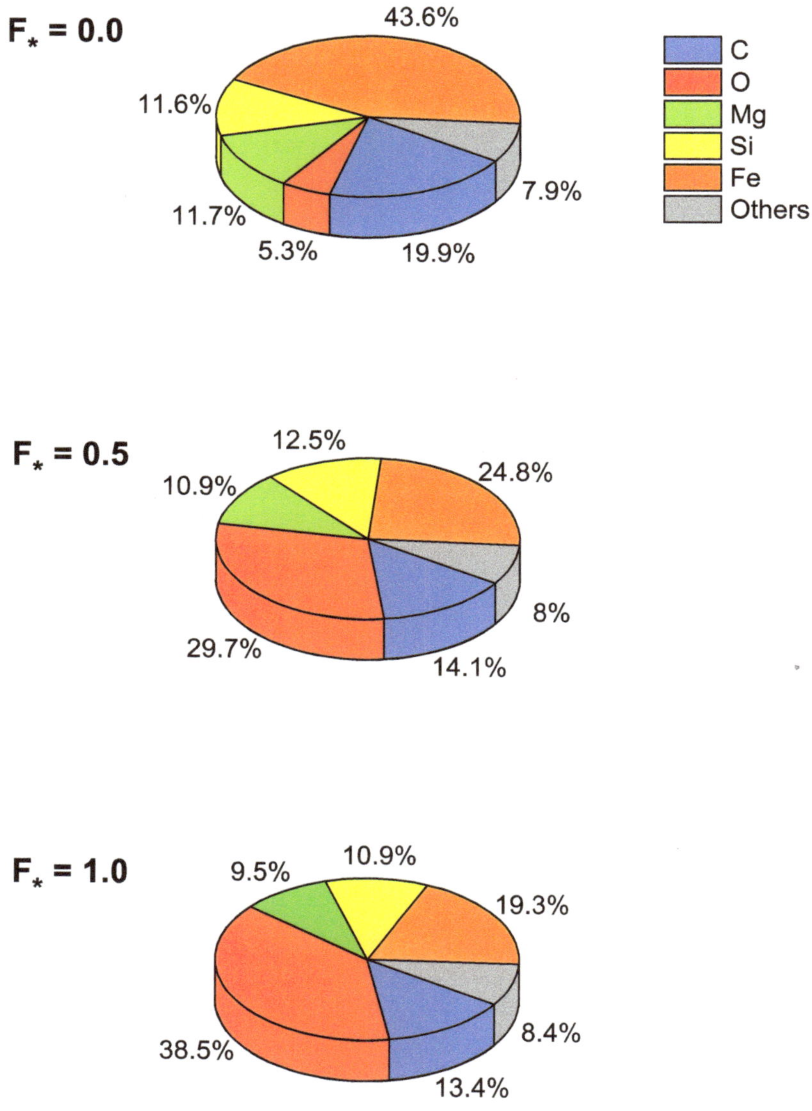

Figure 7.4. Pie charts of estimated contributions to the total dust mass of major condensible elements in low, moderate and high-depletion environments, based on results in Table 7.2.

particles, but T_{cond} is nevertheless a useful surrogate for the relative affinity of an element to form solids and be resistant to vaporization. As usual, C is an exception: refractory phases such as graphite do not condense in an atmosphere of solar composition.

The occurrence of a correlation between T_{cond} and the depletion index $D(\text{X})$ was first demonstrated by Field (1974), using data for the well-studied diffuse molecular cloud toward ζ Oph (see Savage & Sembach 1996 for an updated version).

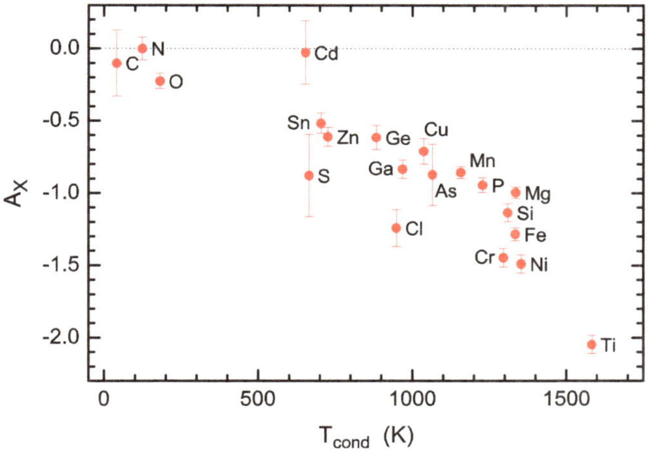

Figure 7.5. The correlation of depletion slope with condensation temperature. $A(X)$ values are from Jenkins (2009) and Ritchey et al. (2018); T_{cond} values are from Lodders (2003).

Refractory metals such as Al, Ca, Ti, Fe, and Ni ($T_{cond} > 1300$ K) were found to have fractional depletions >99%, far greater than those of more volatile elements with lower condensation temperatures. To investigate the sensitivity of such trends to environment it is important to compare data for many lines of sight, but only a few have been studied to levels of precision and spectral coverage comparable with the observations of ζ Oph. A suitable method for exploring trends across multiple sight-lines, enabled by the work of Jenkins (2009), is to consider the *rate of change* in depletion for each element with respect to condensation temperature. This is illustrated in Figure 7.5, which plots T_{cond} versus the coefficient A_X, i.e., the slope of the linear relation between $D(X)$ and F_* (Equation (7.7)).

Three elements in Figure 7.5 (C, N, and Cd) have null values of the depletion slope to within estimated uncertainties. This occurrence might imply one of two scenarios: either the element is essentially undepleted across the board, or it is sequestered in refractory solids that are resistant to a net change of state in the environments being sampled. The inference may be different in each case, as we next discuss.

The primary carbon-bearing condensates in an atmosphere of solar composition are ices such as CH_4, CO_2, and CO. Under interstellar conditions, these volatiles are viable in small particles (as distinct from the interiors of bulk objects such as comets) only in the protected environments of cold molecular clouds. However, there is ample evidence, discussed in the preceding chapters, for the presence of refractory C-bearing solids such as graphite and hydrocarbons in the diffuse ISM: indeed, it would be impossible to account for the observed properties of the dust if C were undepleted. Likely sources of refractory carbon include condensation in the winds of stars with C-enriched atmospheres, and energetic processes that reduce ices and organic matter to amorphous carbon. C is sufficiently abundant that it does not need to be heavily depleted to make a significant contribution to the dust mass.

In the case of nitrogen, a different explanation is indicated. Both observations and condensation models imply that it remains essentially undepleted in the ISM outside of dense molecular clouds (Savage & Sembach 1996; Lodders 2003; Jenkins 2009). Only at temperatures cold enough for ices such as N_2, NH_3, and $NH_3{\cdot}OH_2$ (ammonium hydrate) to form does it gain an appreciable presence in solids. Organic refractory matter might be considered as a potential exception; but the null slope is then hard to explain because this material is unlikely to remain stable across the range of environments in which depletions are measured, and other observations suggest that it is no more than a minor constituent of interstellar dust (Section 5.2.3). N is therefore omitted from estimates of Z_d in Section 7.2.2.

Available data for cadmium, one of the rarest elements to be detected in the ISM, provide little or no evidence of depletion at any F_* value. This behavior seems anomalous as it differs from that of other metals with similar condensation temperatures. See Ritchey et al. (2018) for further discussion and a summary of previous work.

Another element requiring special mention in this context is sulfur. Results have typically been interpreted as evidence that S is undepleted in the ISM, i.e., resident almost exclusively in the gas phase, even in lines of sight where other elements are highly depleted (e.g., Savage & Sembach 1996; Welty et al. 1999). This would be surprising, however, given that it is expected to condense readily into solid compounds such as FeS at $T_{cond} \sim 650$ K in a gas of solar composition. Indeed, S appears to have been sequestered efficiently into such solids in the solar nebula, as indicated by the concurrence of its meteoritic and photospheric abundances (see Figure 7.3). Problems that hinder accurate measurement of its abundance in the ISM, such as line saturation and uncertainties in the distribution between ionization states, are reviewed in detail by Jenkins (2009), whose recommended values are used here. They suggest that S is at least mildly depleted in diffuse and semi-dense clouds. Its locus in Figure 7.5 is consistent with the general trend to within considerable uncertainty.

Overall, the results demonstrate a clear link between the observed depletion of an element and its inherent affinity for the solid phase. Condensation models yield predictions for the chemical content of solids forming in a gas of solar composition: most of the Mg and Si goes into silicates and most of the Fe goes into metal alloys. Silicates and other mineral compounds account for some but not all of the O.

7.2.4 Depletions in Other Galaxies

The metal-content of a galaxy depends on its age, mass, structure, and star formation history, and is therefore a vital indicator of evolutionary status. The metallicities of nearby galaxies such as the Magellanic Clouds are reasonably well-known, and element depletions may be evaluated in lines of sight toward individual stars, enabling insightful comparisons with results for the Milky Way. Observations of local galaxies may, in turn, inform our interpretation of results for more distant galaxies (Jenkins & Wallerstein 2017; Galliano et al. 2021). Vital targets for studying the evolution of metallicity over cosmic time are gas-rich galaxies with redshifts typically in the range $2 \lesssim z \lesssim 4$, detected by the absorption lines they produce in the spectra of more distant quasars (damped Lyα systems; see Wolfe et al. 2005 for a review). The goal here is to evaluate heavy-element abundances across all phases, but the observations give access only to the gaseous component: the extent to which they are modified by depletion into dust is unconstrained. This problem can be mitigated if a universal pattern of relative depletions can be established for the condensible elements. For example, the abundance ratio of a modestly-depleted element such as Zn to a more heavily-depleted element such as Fe might provide a reliable proxy for general depletion (De Cia et al. 2016).

The Small Magellanic Cloud (SMC) is of particular interest in this context as the most accessible example of a low-metallicity system (~20% solar). Reference abundances for the gas-phase ISM in the SMC are derived from observations of young stars: to within the uncertainties, these appear to be generally consistent with "dilute solar" values for the metallic elements, i.e., they are present in solar proportions relative to each other but systematically diminished relative to H (Jenkins & Wallerstein 2017). Depletions are then deduced using the method described in Section 7.2.1. Results show that most elements, including Si, S, Cr, Fe, Ni, and Zn, exhibit depletion behavior consistent with that seen in the Milky Way. Significant differences are found for a few elements: Ti has a shallower depletion slope in the SMC than in the Milky Way, Mn has a steeper one. The cause of these disparities is unknown. One possibility is that they might reflect different chemical affinities with respect to the substrates available for condensation. The SMC is known to be deficient in elemental carbon and to have a much lower mass-fraction of dust in very small carbonaceous particles compared with the Milky Way (Chastenet et al. 2019). The smallest particles in the size distribution dominate the available surface area, and a systematic bias in composition may thus affect the depletion pattern.

The average metallicity of the Large Magellanic Cloud (~50% solar) is intermediate between those of the SMC and the Milky Way. Depletions in the LMC have been studied by the same method for a similar suite of elements, yielding a pattern broadly consistent with that seen in our Galaxy (Roman-Duval et al. 2021): depletions increase systematically with column density and correlate closely with each other for most elements, suggesting a common response to changes in environment. A comparison of results for all three systems (LMC, SMC, and Milky Way) shows that the relation between depletions of Fe and other elements is almost invariant (Roman-Duval et al. 2022), establishing a consistent pattern of behavior across substantial ranges of metallicity, gas density, and dust-to-gas ratio. Gas-phase column densities are relatively easy to measure for Fe because of its numerous UV transitions with a wide range of oscillator strengths. Thus, Fe provides a useful proxy for other elements that are more difficult to observe in more distant galaxies.

7.3 Implications for Grain Composition

The abundances and depletions of the condensible elements impose crucial constraints on models developed to explain the optical properties of dust. The missing atoms are presumed to be sequestered almost entirely into solids,[2] and the observations thus establish limits on their availability and inform our assumptions regarding the chemical composition of the particles. Trends in the data suggest that the dust includes (i) a transient component that changes state according to ambient conditions, and (ii) a robust component that remains mostly solid. The ISM is in essence a crucible into which stars inject material with some inherent chemical composition and dust-to-gas ratio, each according to its evolutionary state. Stardust particles are then subjected to constructive and destructive processes as they cycle through different environments, acquiring atoms from the gas inside clouds and releasing them in regions exposed to harder radiation fields and shocks. As the highest densities covered by the observations do not extend into the domain where ice absorption features are readily detectable, it is natural to assume that the transient component consist of compounds with volatilities intermediate between those of ices and highly refractory solids such as metallic iron. In this section we focus on the behavior of the five elements (C, O, Mg, Si, Fe) expected to contribute most of the mass of dust in the solar neighborhood of the Milky Way.

7.3.1 Carbon

Carbon appears to be moderately depleted across the range of environments sampled by the observations. Results in Table 7.2 suggest a dust-phase abundance of about 100 ppm in relatively high-depletion sight-lines ($F_* \gtrsim 0.5$), apparently insufficient to explain all of the phenomena attributed to carbonaceous grains. The overall requirements are for (i) some 60–80 ppm in very small particles (PAHs, PAH clusters, graphite) to account for the mid-infrared spectral line and continuum emissions (Section 6.2.2) and UV absorption bump (Section 3.4.5); and (ii) at least a further 100 ppm in the larger grains responsible for far-infrared continuum emission (Section 6.1) and general extinction (Section 3.6). There is, however, considerable latitude in the depletion data for this element, for reasons previously discussed. An upward revision of the reference abundance and/or a reduction in gas-phase column densities, within the margins of uncertainty, could allow for as much as $\sim$ 200 ppm of carbon in the dust, as proposed by Sofia et al. (2011), enough to satisfy the above requirements.

7.3.2 Magnesium and Silicon

These elements exhibit closely similar behavior, as expected if their preferred solid forms are the same, e.g., silicates such as $MgSiO_3$ and $MgFeSiO_4$. Silicates are subject to destruction by shocks in low-density regions, which may account for the observed reductions in $\mathcal{A}_d(\mathrm{Mg})$ and $\mathcal{A}_d(\mathrm{Si})$ as F_* becomes small (Table 7.2 and Figure 7.6). The behavior of Fe is significantly different, implying that most of it is sequestered in other, more robust forms. Thus, the depletion data support the conclusion that interstellar silicates tend to be magnesium-rich, broadly consistent with their spectral properties (Section 5.2.1).

---

[2] Carbon monoxide, by far the most abundant gas-phase interstellar molecule after H_2, accounts for no more than a few ppm of its constituent elements in diffuse molecular clouds (Snow & McCall 2006).

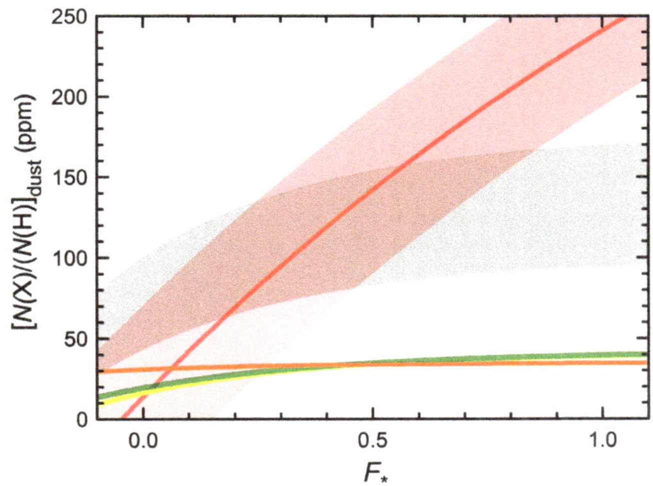

Figure 7.6. Comparison of dust-phase abundance trends with respect to depletion factor, F_*. The curves correspond to X = Mg (green), Si (yellow), Fe (orange), and O (red), based on data from Jenkins (2009). The limits of probable uncertainty for O are indicated by the pink band. The gray area represents the estimated range for oxygen accommodated into refractory solids such as silicates and metal oxides (see text for details).

The contribution of silicates to the total mass of dust may be estimated for an assumed mean composition. For simplicity, consider a mixture containing $MgSiO_3$ and $MgFeSiO_4$ units in equal numbers, so the constituent elements Mg, Si, Fe, O are present in proportions 1:1: 0.5:3.5, respectively. This mix seems plausible as it includes both pyroxene and olivine stoichiometries, incorporates some Fe while remaining Mg-rich, and requires numerically equal depletions of Mg and Si atoms, consistent with observations. When the depletion is high ($F_* = 1$; Table 7.2), we have $\mathcal{A}_d(\mathrm{Mg}) \simeq \mathcal{A}_d(\mathrm{Si}) \simeq 39$ ppm. Adding Fe and O in the above proportions and carrying out the summation (Equation (7.9)) over just these four elements, the estimated mass of silicates is roughly half ($\sim 53\%$) of the total depleted mass. An independent estimate based on the ratio of silicate optical depth to extinction ($\tau_{9.7}/A_V$; Section 5.2.1) yields a broadly similar result. The availability of Si thus seems adequate to account for the observed silicate features, but with little scope for inclusion in other forms.

7.3.3 The Oxygen Crisis

Perhaps the greatest challenge posed by the depletion data is to explain the behavior of oxygen, the most abundant of all the condensible elements (Jenkins 2009; Whittet 2010). The environments represented by the observations range from intercloud gas to clouds of moderate density, as previously discussed, thus we expect all solids to be largely non-volatile; but the depletion of O may exceed levels that can be accommodated in such compounds.

The problem is illustrated in Figure 7.6, which compares the variation of dust-phase abundance with depletion factor for O, Mg, Si, and Fe. Also shown are estimated lower and upper bounds for O contained in silicates and oxides (the area between these limits is shaded gray). The limits are set by considering the probable range of the oxidation ratio O/M, where M is the total number of abundant metallic elements (Mg, Si, and Fe; the contributions of others are negligible). A value of O/M = 1.5 is applicable to $MgSiO_3$ and Fe_2O_3, and O/M = 1.2 is a reasonable average for a mixture of monoxides (SiO, MgO, FeO) and silicates such as $MgFeSiO_4$ with olivine stoichiometry. The upper bound in Figure 7.6 assumes that all depleted Mg, Si and Fe atoms are fully oxidized to O/M = 1.5; the lower bound assumes an oxidation level of O/M = 1.2 and that only Mg and Si are oxidized.

As the curves in Figure 7.6 are based on general trends, they cannot resolve changes in state for specific compounds under specific conditions. Nevertheless, an important inference can be made: *the rate of change in oxygen depletion with F_* is too steep to be explained by its interaction with the metals.* The divergence is most evident in high-depletion environments ($F_* \sim 1$), where the dust contains at least 80 ppm more oxygen than can be accommodated in mineral compounds with high levels of oxidation; the discrepancy doubles to ~160 ppm if the levels are more moderate. Thus, some other reservoir for depleted oxygen must be present. Because this conclusion is based on the rate at which the depletion changes, not on its absolute value toward any given star, it cannot be explained by a systematic error in the standard reference abundance.

What form might this reservoir take? Once the metals are fully saturated with oxygen, the only other elements abundant enough to combine with O in sufficient numbers are H, C, and O itself. Volatiles such as CO, CO_2, and O_2 cannot condense or survive as solids in even the densest environments considered here, and pervasive gas-phase molecules such as CO account for no more than a few ppm of oxygen in such regions. OH and H_2O molecules in hydrated minerals can also be discounted because observational evidence indicates that interstellar grains are mostly anhydrous (Section 5.2.1), and unreasonably high levels of hydration would be needed to account for the shortfall. The only remaining candidates appear to be carbonaceous dust and bulk solid H_2O.

The possibility that carbonaceous particles might sequester sufficient oxygen to explain the anomaly is discussed by Whittet (2010) and Jones & Ysard (2019). The numerous small particles in the size distribution present the greatest surface area for attachment of impinging atoms and molecules. OH radicals, for example, might replace H atoms at peripheral sites of aromatic structures such as PAHs or graphite. However, surface attachment at a level sufficient to explain the excess depletion of O at $F_* = 1$ would require O to be at least as abundant as C in these grains, which is clearly unrealistic. Sufficient depletion seems possible only if the O is incorporated into larger particles that contain most of the mass and volume of the dust. Organic refractory matter is an attractive candidate because it has a plausible production mechanism as the result of star formation in dense clouds (Section 9.3.3), and may release its O gradually as it carbonizes on exposure to harsh interstellar radiation fields and cosmic rays in the diffuse ISM, consistent with the general trend. However, a limit on the oxygen content of organic solids is placed by spectroscopy, specifically the strengths of the infrared absorptions arising from O–H and C=O vibrational modes (Section 5.2.3); non-detection of such features implies that organic refractory matter in the diffuse ISM consists mostly of hydrocarbons (Pendleton & Allamandola 2002).

The remaining candidate is H_2O, by far the most abundant and least volatile of the ices known to be present in molecular clouds. The limiting factor in this case is non-detection in diffuse-ISM spectra of absorptions arising from the vibrational modes of H_2O, most notably the strong O–H resonance at 3 μm (see Figure 5.3). High-depletion lines of sight such as that toward ζ Oph have extinctions ($A_V \sim 1$) significantly below thresholds for ice detection in molecular clouds ($A_V^0 \gtrsim 2$; Section 5.3.2). This objection might be mitigated if the ices reside in or on very large grains (e.g., Poteet et al. 2015; Wang et al. 2015). Models show that the spectral features become progressively broader and shallower with increasing particle size, such that it may be possible to "hide" significant quantities of H_2O in the grains if they are sufficiently large. For example, Poteet et al. (2015) estimate that thick ice mantles on grains of radius $a \sim 3$ μm toward ζ Oph might contain $\sim$25 ppm of O without conflicting with available spectroscopic data, provided that the smaller grains remain ice-free.

As they disperse into the diffuse ISM from dense molecular clouds, grains may lose volatiles thermally by sublimation or non-thermally by photodesorption (Öberg et al. 2009), and these processes tend to occur at size-dependent rates. Grains small enough to undergo significant transient heating on absorption of individual energetic photons will rapidly lose any surface volatiles; however, large, thickly ice-mantled grains or composite particles containing pockets of internal ice will be more resilient, maintaining lower mean temperatures and evincing a measure of self-shielding from photodesorption. As an extreme example of such behavior, consider icy bodies of cometary size ($\sim$1 km). Comets retain volatiles over cosmic timescales under interstellar conditions remote from a star (e.g., Altwegg et al. 2019; Mumma & Charnley 2011). Comets of interstellar provenance have been identified (e.g., Bailer-Jones et al. 2020; Pfalzner et al. 2021) but their space density is unknown. An intriguing solution to the oxygen crisis might be the existence of giant grains of cometary composition ("dirty snowballs") that are slow to release their volatiles on exposure to interstellar radiation fields. Such grains would produce wavelength-independent extinction in the diffuse ISM if they are present in significant numbers: see Section 3.2.5 for related discussion.

7.3.4 Iron

Fe, the most abundant heavy metal, has a strong affinity for the solid phase. It maintains high levels of depletion in low-density environments, even as those of Si and Mg are declining (Figure 7.6), and must therefore reside primarily in a robust component of the dust. This behavior corroborates results from X-ray spectroscopy (Westphal et al. 2019), indicating that no more than about a third of the Fe is chemically bonded into silicates.[3] Other potential repositories for depleted iron include metallic Fe, oxides, carbides, and sulfides. Of these, sulfides seem least likely to be important because S is less abundant than Fe and also far less depleted in the ISM. An essential point to consider in assessing these candidates is the distribution of Fe between gas and solid phases in the source material, i.e., in supernova ejecta and in the outflows of asymptotic-giant-branch stars (Section 8.1). A major and perhaps dominant fraction of the Fe reaches the ISM in gaseous form, its transition into solids dependent on subsequent attachment to pre-existing grains at low temperature (Dwek 2016).

Unfortunately, the optical properties of interstellar dust do not place strong constraints on the nature of its iron content. Pure metallic Fe enhances the continuum opacity but is otherwise lacking in spectral signatures. Metallic Fe particles may condense in the atmospheres of dust-forming stars, and thereby influence their observed spectral energy distributions (Kemper et al. 2002; Gail et al. 2020); but pure metals are unlikely to remain pristine in the ISM, where they are subject to oxidation on timescales of order 10^6 years (Jones 1990). Some Fe-bearing oxides exhibit vibrational resonances in the wavelength range $15 < \lambda < 30$ μm, the resultant spectrum dependent on the precise composition (Henning et al. 1995). Pure wüstite (FeO), for example, has a broad feature centered near 20 μm, and corresponding structure is seen in the circumstellar spectrum of a dusty oxygen-rich star (Posch et al. 2002). However, no such features have been identified in the ISM, perhaps because attempts to detect shallow absorptions in this region of the spectrum are hampered by blending with the broad O–Si–O bending mode of silicates (e.g., Chiar & Tielens 2006; Poteet et al. 2015). Hence, the evidence for iron oxide dust in the ISM remains circumstantial, albeit supported by the ubiquity of Fe and its affinity for oxygen (Jones 1990), and by identification of presolar oxide grains in primitive meteorites (Nittler & Ciesla 2016).

The remaining possibility is that carbonaceous grains might sequester substantial quantities of iron. Carbides such as Fe_3C may condense in the atmospheres of carbon stars, but are considered to be no more than minor constituents of all stardust injected into the ISM (e.g., Gail et al. 2009). Major sources of elemental Fe (type Ia supernovae) drive outflows that appear to be dust-free (Dwek 2016). Subsequent depletion of Fe in the ISM is then dependent on encounters with pre-existing dust grains, and, as previously discussed, the most likely interactions are with the smallest ones that account for most of the available surface area. The electrostatic properties of these nanoparticles may influence the probability of adsorption. Whereas large grains tend to bear a net positive charge resulting from electron loss by the photoelectric effect (Section 2.4.2), small grains tend to be electrostatically neutral or weakly negative. Gas-phase Fe is predominantly singly-ionized in the diffuse ISM, and will thus attach preferentially onto negative or neutral substrates, i.e., onto small grains. Weingartner & Draine (1999) conclude

[3] Additional Fe may be accommodated in silicate grains containing Fe-rich *inclusions*, i.e., physically-embedded subunits that might survive disruption of the aggregate. Such particles may form in stellar outflows (Gail et al. 2020), and their magnetic properties may induce alignment in the ISM (Section 4.3.3).

that adsorption onto carbonaceous nanoparticles might account for some but probably not all of the depleted Fe.

To summarize, it seems probable that the Fe content of the dust has many guises that include both condensates formed in stellar outflows and adsorbates formed in the ISM. The answer to the question "Where is the interstellar iron?" (Zhukovska et al. 2018) may be "all of the above." Silicate stardust, independent Fe-rich grains in various stages of oxidation, and surface deposits on nanoparticles may each contribute to its observed depletion.

References

Adams, F. C. 2010, ARA&A, 48, 47
Adibekyan, V. Z., Santos, N. C., Sousa, S. G., et al. 2012, A&A, 543, A89
Altwegg, K., Balsiger, H., & Fuselier, S. A. 2019, ARA&A, 57, 113
Asplund, M., Grevesse, N., Sauval, A. J., & Scott, P. 2009, ARA&A, 47, 481
Asplund, M., Amarsi, A. M., & Grevesse, N. 2021, A&A, 653, A141
Bailer-Jones, C. A. L., Farnocchia, D., Ye, Q., Meech, K. J., & Micheli, M. 2020, A&A, 634, A14
Baumgartner, W. H., & Mushotzky, R. F. 2006, ApJ, 639, 929
Bedell, M., Bean, J. L., Melendez, J., et al. 2018, ApJ, 865, 68
Bensby, T., Feltzing, S., & Oey, M. S. 2014, A&A, 562, A71
Cardelli, J. A., Meyer, D. M., Jura, M., & Savage, B. D. 1996, ApJ, 467, 334
Chastenet, J., Sandstorm, K., Chiang, I.-D., et al. 2019, ApJ, 876, 62
Chiar, J. E., & Tielens, A. G. G. M. 2006, ApJ, 637, 774
Costantini, E., Zeegers, S. T., Rogantini, D., et al. 2019, A&A, 629, A78
Crinklaw, G., Federman, S. R., & Joseph, C. L. 1994, ApJ, 424, 748
De Cia, A., Ledoux, C., Mattsson, L., et al. 2016, A&A, 596, A97
Dwek, E. 2016, ApJ, 825, 136
Field, G. B. 1974, ApJ, 187, 453
Gail, H.-P., Zhukovska, S. V., Hoppe, P., & Trieloff, M. 2009, ApJ, 698, 1136
Gail, H.-P., Tamanai, A., Pucci, A., & Dohmen, R. 2020, A&A, 64, A139
Galliano, F., Nersesian, A., Bianchi, S., et al. 2021, A&A, 649, A18
Gatuzz, E., García, J., Mendoza, C., et al. 2014, ApJ, 790, 131
Gatuzz, E., García, J. A., Kallman, T. R., & Mendoza, C. 2016, A&A, 588, A111
Henning, T., Begemann, B., Mutschke, H., & Dorschner, J. 1995, A&AS, 112, 143
Hensley, B. S., & Draine, B. T. 2021, ApJ, 906, 73
Hobbs, L. M., York, D. G., & Oegerle, W. 1982, ApJ, 252, L21
Jenkins, E. B. 2009, ApJ, 700, 1299
Jenkins, E. B. 2014, The Life Cycle of Dust in the Universe: Observations, Theory, and Laboratory Experiments (LCDU2013), POS, Vol 207
Jenkins, E. B., & Wallerstein, G. 2017, ApJ, 838, 85
Johnson, J. A. 2019, Sci, 363, 474
Jones, A. P. 1990, MNRAS, 245, 331
Jones, A. P., & Ysard, N. 2019, A&A, 627, A38
Kemper, F., de Koter, A., Waters, L. B. F. M., Bouwman, J., & Tielens, A. G. G. M. 2002, A&A, 384, 585
Kobayashi, C., Umeda, H., Nomoto, K., Tominaga, N., & Ohkubo, T. 2006, ApJ, 653, 1145
Lecar, M., Podolak, M., Sasselov, D., & Chiang, E. 2006, ApJ, 640, 1115

Lee, J. C., Xiang, J., Ravel, B., Kortright, J., & Flanagan, K. 2009, ApJ, 702, 970
Lodders, K. 2003, ApJ, 591, 1220
Lyubimkov, L. S. 2013, Ap, 56, 472
Mumma, M. J., & Charnley, S. B. 2011, ARA&A, 49, 471
Nieva, M.-F., & Przybilla, N. 2012, A&A, 539, 143
Nittler, L. R., & Ciesla, F. 2016, ARA&A, 54, 53
Öberg, K. I., Linnartz, H., Visser, R., & van Dishoeck, E. F. 2009, ApJ, 693, 1209
Pendleton, Y. J., & Allamandola, L. J. 2002, ApJS, 138, 75
Pfalzner, S., Aizpuru Vargas, L. L., Bhandare, A., & Veras, D. 2021, A&A, 651, A38
Posch, T., Kerschbaum, F., Mutschke, H., Dorschner, J., & Jäger, C. 2002, A&A, 393, L7
Poteet, C. A., Whittet, D. C. B., & Draine, B. T. 2015, ApJ, 801, 110
Psaradaki, I., Costantini, E., Mehdipour, M., et al. 2020, A&A, 642, A208
Ritchey, A. M., Federman, S. R., & Lambert, D. L. 2018, ApJS, 236, 36
Rogantini, D., Costantini, E., Zeegers, S. T., et al. 2018, A&A, 609, A22
Rogantini, D., Costantini, E., Zeegers, S. T., et al. 2020, A&A, 641, A149
Roman-Duval, J., Jenkins, E. B., Tchernyshyov, K., et al. 2021, ApJ, 910, 95
Roman-Duval, J., Jenkins, E. B., Tchernyshyov, K., et al. 2022, ApJ, 928, 90
Savage, B. D., & Sembach, K. R. 1996, ARA&A, 34, 279
Sharda, P., Krumholz, M. R., Wisnioski, E., et al. 2021, MNRAS, 502, 5935
Snow, T. P., & McCall, B. J. 2006, ARA&A, 44, 367
Snow, T. P., & Witt, A. N. 1996, ApJ, 468, L65
Sofia, U. J., & Meyer, D. M. 2001, ApJ, 554, L221
Sofia, U. J., Cardelli, J. A., Guerin, K. P., & Meyer, D. M. 1997, ApJ, 482, L105
Sofia, U. J., Lauroesch, J. T., Meyer, D. M., & Cartledge, S. I. B. 2004, ApJ, 605, 272
Sofia, U. J., Parvathi, V. S., Babu, B. R. S., & Murthy, J. 2011, AJ, 141, 22
Spitzer, L. 1978, Physical Processes in the Interstellar Medium (New York: Wiley)
Spitzer, L. 1985, ApJ, 290, L21
Trimble, V. 1991, A&ARv, 3, 1
Trimble, V. 1997, OLEB, 27, 3
Wallerstein, G., Iben, I., Parker, P., et al. 1997, RvMP, 69, 995
Wang, S., Li, A., & Jiang, B. W. 2015, MNRAS, 454, 569
Weingartner, J. C., & Draine, B. T. 1999, ApJ, 517, 292
Welty, D. E., Hobbs, L. M., Lauroesch, J. T., et al. 1999, ApJS, 124, 465
Westphal, A. J., Butterworth, A. L., Tomsick, J. A., & Gainsforth, Z. 2019, ApJ, 872, 66
Whittet, D. C. B. 2010, ApJ, 710, 1009
Winter, L. M., Mushotzky, R. F., & Reynolds, C. S. 2007, ApJ, 655, 163
Wolfe, A. M., Gawiser, E., & Prochaska, J. X. 2005, ARA&A, 43, 861
Zhukovska, S., Henning, T., & Dobbs, C. 2018, ApJ, 857, 94

D C B Whittet

Chapter 8

The Life Cycle of Dust—I. Circumstellar Origins

The life cycles of stars and dust are closely linked (see Section 1.4.5 for an overview). In this chapter, we focus on old stars as sources of new grain material. The atmospheres and winds of luminous stars such as red giants and supergiants contain environments conducive to the nucleation and growth of refractory dust grains ("stardust"). Depending on the mass and evolutionary status of the star, the condensates may be carbon-rich or oxygen-rich, and grain formation may occur over extended periods lasting a million years or very rapidly in a cataclysmic event such as a nova or supernova outburst. Periods of grain formation are typically accompanied by episodes of copious mass-loss, generating dusty outflows that merge into the surrounding interstellar medium. A striking example is illustrated in Figure 8.1.

This chapter begins with a review of the physical conditions and chemical processes that control the formation of dust in circumstellar environments (Section 8.1). Stardust research is a domain in which astronomical observations and laboratory studies of extraterrestrial materials are complementary and highly synergistic: results from these areas of investigation are reviewed in Sections 8.2 and 8.3, respectively. A vital and persistent question concerns whether stellar outflows represent a dominant or relatively modest contribution to the total inventory of interstellar dust. The available evidence is reviewed in the final section.

8.1 Dust Formation in Stellar Outflows

8.1.1 Theoretical Basis

Dust formation in stellar atmospheres was first discussed in response to emerging evidence for its presence in the circumstellar environments of stars with low surface temperatures (see Bode 1988 for a review of early work). Luminous stars located in the upper right-hand region of the Hertzsprung–Russell diagram (Figure 8.2) were implicated. Subsequent research has identified asymptotic giant branch (AGB) stars and core-collapse supernovae as major sources, corresponding to the most prolific

doi:10.1088/2514-3433/ac7204ch8

Figure 8.1. Composite portrait of the planetary nebula NGC 6302 (the Butterfly Nebula), combining UV and visible-light images obtained by the Wide Field Camera 3 of the Hubble Space Telescope. Filters are used to isolate individual spectral lines diagnostic of physical conditions in the gas (Kastner et al. 2022). The nebula contains matter ejected from a central star at speeds of up to $\sim$ 600 km s^{-1} over a period of at least 2000 years, as the star evolved from the asymptotic giant branch to become a hot white dwarf. The star is hidden from view behind a dusty equatorial disk that constricts the outflow, resulting in the nebula's bipolar structure. White areas in the image trace gas energized by shocks, attributed to recent, fast-moving ejecta that overtakes and collides with slower-moving matter ejected during an earlier phase of the star's evolution. Image credit: NASA/ESA/Hubble Space Telescope.

phases of dust production for stars of intermediate (1–8 $M_{\odot}$) and high (>8 $M_{\odot}$) initial mass, respectively (Höfner & Olofsson 2018; Sarangi et al. 2018).

We begin by considering the nature of circumstellar matter around a dust-forming red giant (discussion specific to supernovae is differed to Section 8.1.6). A schematic illustration of an AGB star and its environment is shown in Figure 8.3. Elemental C and O are produced by internal nucleosynthesis and mixed into the convective envelope. Taking the center of the star as the origin, the stellar surface (photosphere) is located at radial distance $r = R_*$, and the surrounding atmosphere is subject to dynamical disturbances caused by stellar pulsations and shocks. The atmospheric layer immediately above the photosphere may contain a chromosphere, in which the gas is substantially hotter than in the photosphere (e.g., Gail & Sedlmayr 1987; Guerrero & Ortiz 2020), but at greater distances the temperature is

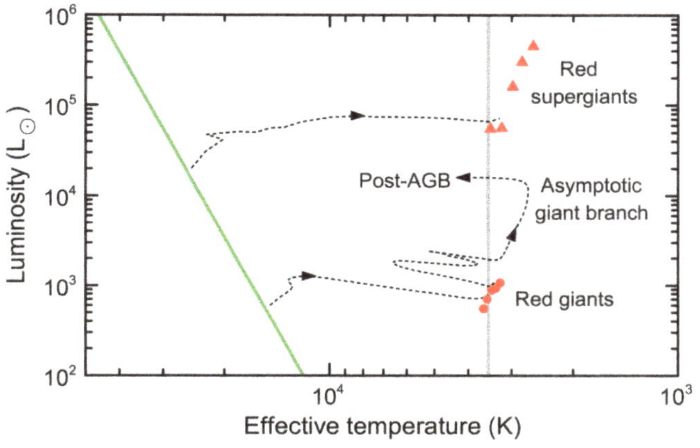

Figure 8.2. Hertzsprung–Russell diagram (effective temperature versus luminosity) showing the post-main-sequence evolution of typical stars that may form dust in their outflows. The main sequence is shown in green, and the dashed curves are evolutionary tracks for single stars of masses 5 $M_\odot$ (below) and 15 $M_\odot$ (above). Red symbols are mean loci for stars in the standard spectral-type sequence from M2 to M6 for giants (luminosity class III, circles) and supergiants (luminosity class I, triangles). The vertical gray line indicates an effective temperature of 3500 K, below which stars typically begin to make dust. A 5 $M_\odot$ star may cross this threshold briefly during its initial (first ascent) red giant phase, but the principal episode of dust production occurs later when it reaches the asymptotic giant branch.

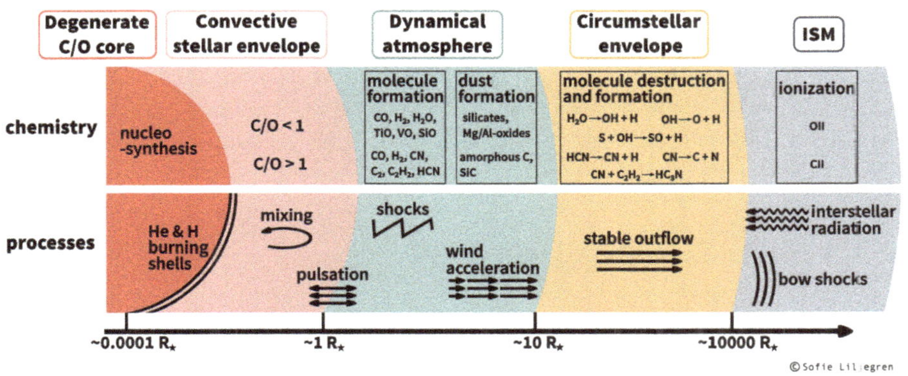

Figure 8.3. Schematic illustration of an AGB star and its circumstellar environment, including regions and processes discussed in the text. Boundaries between layers are indicated in units of stellar radius ($R_* \sim 1$ AU $\sim 200\ R_\odot$ for a typical AGB star). Image credit: created by and courtesy of Sofie Liljegren; reproduced with permission from Astronomy and Astrophysics Review (Höfner & Olofsson 2018).

likely to decline monotonically with r. Simple molecules such as H_2, H_2O, CO, and SiO become stable, and dust forms as gas temperatures drop below the condensation temperatures of major solids ($T \lesssim 1500$ K). The grains are driven outward from the star by radiation pressure and transfer momentum to the gas by collisions. The number density of gas in the stellar wind also declines with r due to dilution, such that $n \propto r^{-p}$, with $p = 2$ for uniform expansion at constant speed. The outer boundary of the circumstellar envelope may be considered as the radial distance $r \gtrsim 10^4 R_*$, at which temperatures and densities approach interstellar values. Dependent on the density of the local ISM and the star's relative motion, a significant bow shock may be produced (Cox et al. 2012).

Grain formation is expected to be most rapid near the transition zone from dynamical atmosphere to circumstellar envelope ($r \sim 10\ R_*$; Figure 8.3). Here, gas densities are typically $n \sim 10^{19}\ \mathrm{m}^{-3}$, orders of magnitude higher than those in even the densest interstellar clouds, and temperatures fall below the condensation temperatures of common refractory solids. This confluence of physical conditions ensures that the atmospheres of such stars are efficient producers of dust. The gaseous wind flowing into this zone contains a mix of atomic, ionic, and molecular gas in proportions that vary according to ambient conditions. Ionization inhibits nucleation, because positively charged particles repel each other, but the flux of ionizing radiation is expected to be minimal unless there is a significant contribution from the chromosphere or from a hot binary companion star. Pulsational instabilities inherent to these aging stars drive periodic variations in physical conditions and may induce cyclic phases of grain nucleation, growth and ejection (Gail et al. 2016; Bladh et al. 2019).

The formation of solid particles in a gas that undergoes cooling has been discussed in terms of classical nucleation theory (e.g., Salpeter 1974), originally formulated to describe the condensation of liquid droplets in the Earth's atmosphere. The theory assumes that the system maintains thermal equilibrium. Condensation of some species X occurs when its partial pressure in the gas exceeds the vapor pressure of X in the condensed phase, and is most efficient at temperatures appreciably below the nominal condensation temperature. An individual unit of X in the gas phase (a monomer) is a constituent atom or molecule. Random encounters between monomers lead to the formation of clusters; for equilibrium, the number density of clusters containing i monomers is

$$n_i = n_1 \exp(-\Delta E_i/kT) \tag{8.1}$$

where n_1 is the number density of individual monomers and ΔE_i is the thermodynamic free energy associated with the formation of the cluster. For small clusters, ΔE_i increases with i, and thus n_i decreases. Above some critical size, however, the addition of further monomers is energetically favorable; the clusters become stable and grow to a maximum size limited by the availability of monomers. Classical nucleation is thus a two-step process: (i) formation of clusters of critical size, and (ii) growth of clusters into macroscopic particles.

Classical nucleation theory is a useful starting point, but its applicability as a description of grain formation in stellar atmospheres must be addressed. Pressures in the nucleation zone of a red-giant wind are typically less than in the terrestrial atmosphere by at least a factor of 10^4. Thus, the timescale for nucleation to occur may be longer than the timescale in which the physical conditions undergo appreciable change, and condensation is not then an equilibrium process

(Donn & Nuth 1985). Furthermore, solid particles are typically held together by strong valence bonds rather than by the weak polarization forces that bind liquid droplets, and the formation of grains thus involves chemical reactions as well as physical clustering. Whereas classical theory describes the nucleation and growth of particles that are chemically identical to the original monomers, chemical reactions integral with the nucleation process may lead to the formation of solids (such as silicates) for which no equivalent monomer is available in the gas. Nevertheless, Paquette & Nuth (2011) conclude that the assumptions inherent in classical nucleation theory have only a second-order effect on the predicted size distribution and number density of grains forming in circumstellar outflows. See also Agúndez et al. (2020) for further discussion.

8.1.2 The C/O Ratio

The molecular chemistry of the gas in the nucleation zone of a stellar atmosphere is largely determined by a single parameter, the numerical abundance of C relative to O. The precise value of this ratio is crucial in determining the nature of the available monomers and hence the composition of the dust that grows from them. This condition arises from the fact that carbon monoxide, the most abundant of all molecules containing condensible elements, forms readily and remains stable in circumstellar gas at temperatures below about 3000 K. The binding energy of CO (11.1 eV) is sufficiently high that the bond is not readily broken in simple gas-phase reactions. Thus, despite its status as a radical, CO is chemically rather unreactive. Furthermore, CO cannot itself condense into solids in such environments. The abundance of CO in the gas is thus effectively limited, in most situations, only by the abundances of the constituent atoms: whichever element out of C and O is less abundant tends to remain sequestered in gaseous CO, and is therefore unavailable to form solids.

Stars that show observational evidence for circumstellar dust production (Section 8.2) may be divided into two broad categories, based on determinations of their atmospheric C/O ratios: those with C/O <1 (classified type M in the normal spectral sequence), and those with C/O >1 (carbon stars, type C). Carbon is produced by helium burning in stellar interiors during the course of their post-main-sequence evolution (Section 7.1.1), and enrichment of their atmospheres occurs when and if this product is dredged up to the surface by deep convection currents and pulsational instabilities (Figure 8.3). For stellar populations with metallicities similar to that of the Sun, the initial C/O ratio is about 0.5 (Table 7.1); in the atmospheres of carbon stars, it is enhanced typically to values in the range 1.0–1.5. The prediction of nucleation models, largely confirmed by observations, is that M-type stars produce only oxygen-rich dust and C-type stars produce only carbon-rich dust.

The distinction between these situations is not absolute, however. As a star evolves, it may undergo successive phases of mass loss that synchronize with periods of dredge-up, leading to time-dependent changes in the nature of its dust production. A transition occurs when C/O ~ 1 (examples are classified S-type) in which C and O effectively block each other from participation in grain formation. The efficiency of dredge-up for any given star depends on its initial mass, composition, and subsequent mass-loss rate, and not all will ultimately acquire carbon-rich atmospheres. In a population of approximately solar metallicity, carbon stars tend to occupy a rather narrow range in mass (~1.5–3 $M_\odot$; e.g., Abia et al. 2020). Stars of lower mass are less likely to dredge up sufficient carbon; those of higher mass tend to consume it by C → N conversion, caused by CNO-cycle reactions in an H-burning shell.

8.1.3 Oxygen-rich Stars

Figure 8.4 illustrates the temperature-pressure (T, P) phase diagram for an atmosphere of solar composition (C/O ≈ 0.5). Gas streaming outward from the star follows a trajectory from upper right to lower left. The most abundant monomers that might participate in the production of solids as the gas cools are Fe, Mg, SiO, and H_2O (e.g., Gail & Sedlmayr 1986; Ferrarotti & Gail 2006; Gail et al. 2013). However, the first phase of grain formation involves rarer elements such as Al, Ca, and Ti, which form oxides such as alumina (Al_2O_3) and perovskite ($CaTiO_3$) at $T \sim 1500$ K. Although contributing little in terms of mass, these high-temperature condensates might facilitate the deposition of more abundant solids by providing nucleation centers (Onaka et al. 1989).

The major condensation phase begins as the temperature falls below about 1200 K, with the nucleation and growth of amorphous SiO clusters. The clusters chemisorb

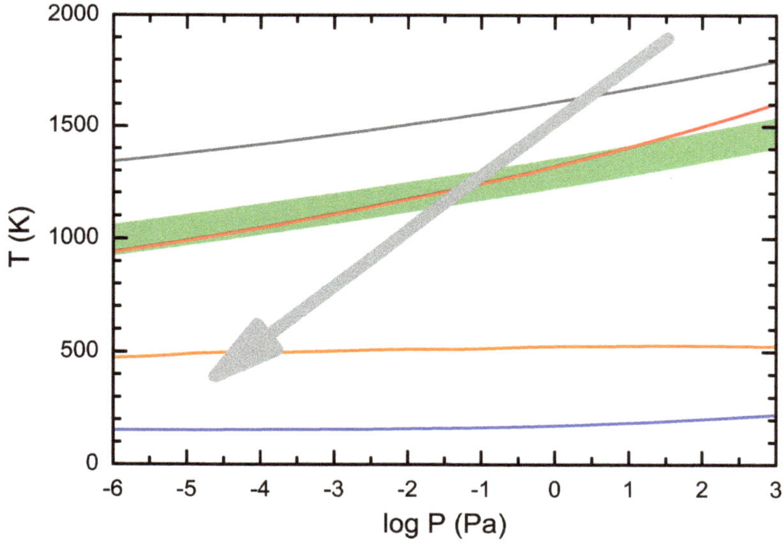

Figure 8.4. Temperature-pressure phase diagram illustrating stability limits of solids condensing in an atmosphere of solar composition. The gray arrow indicates the general trajectory of changing physical conditions in the outflow of a red giant. The green band corresponds to the zone in which magnesium silicates of mixed stoichiometries ($MgSiO_3$, Mg_2SiO_4) become stable. Other curves correspond to alumina (Al_2O_3, gray), metallic iron (red), wüstite (FeO, orange), and H_2O-ice (blue). Adapted from Salpeter (1977) with additional data from Barshay & Lewis (1976) and Gail et al. (2013).

other monomers, and abstraction of O from H_2O leads to further oxidization. Subsequent annealing may then result in the growth of linked SiO_3 chains with attached Mg cations (enstatite, $MgSiO_3$) or individual SiO_4 tetrahedra joined by cations (forsterite, Mg_2SiO_4). In an atmosphere of solar composition, one might expect silicates of mixed Mg/Fe content to dominate as thermodynamically favored end products, but a kinetically controlled formation process favors Mg silicates because they tend to form more rapidly, and there may be insufficient time to reach the most energetically favorable configuration (Nuth 1996; Rietmeijer et al. 1999). Iron might solidify as a pure metal, but laboratory simulations suggest that FeO monomers cluster more efficiently than Fe (Kimura et al. 2017). At $T \sim 700$ K, essentially all of the metallic elements are likely to have condensed into solids in some form or another. Finally, water-ice may condense in an optically-thick outflow at $T \lesssim 200$ K. A macroscopic grain emerging from such an atmosphere is thus likely to have a layered structure, dominated by Mg silicates, metallic iron and/or Fe oxides, perhaps deposited on a refractory "seed" nucleus, and coated with a thin surface layer of ice.

It is of interest to compare models for the condensation of solids in O-rich red giant winds with corresponding models for protoplanetary disks. Prevailing physical conditions may be similar, and in the case of the solar nebula the same mix of chemical elements is assumed, so the phase diagram in Figure 8.4 is relevant to both situations. It is therefore unsurprising that solids predicted to condense are broadly similar (e.g., Barshay & Lewis 1976; Lodders 2003). However, the dynamic evolution of the solar nebula was different, such that the condensation process may have been closer to thermodynamic equilibrium. Disparities in the composition and structure of the solids are therefore expected. Under equilibrium conditions, for example, FeO and H_2O may react with magnesium silicates at $\lesssim$400 K to form hydrated minerals such as serpentine, common constituents of primitive meteorites, whereas silicate dust in the ISM appears to be anhydrous (Section 5.2.1).

8.1.4 Carbon Stars

The equivalent phase diagram for a carbon-rich atmosphere is shown in Figure 8.5. A C/O ratio of 1.1 is assumed (the effect of varying C/O is also shown in the case of solid carbon). O is expected to be entirely sequestered into CO gas and therefore unavailable to form solids, as previously discussed. The first condensates to form in the cooling outflow are rare metallic carbides such as TiC (notable for their presence in presolar meteoritic stardust; Bernatowicz et al. 2005). However, unlike the situation in an O-rich atmosphere, chemical reactions involving metals do not regulate the primary phase of dust formation, which begins as temperatures drop below about 1500 K. Carbon itself is the dominant product (amorphous and graphitic forms have essentially the same stability limits). SiC particles may appear simultaneously or at somewhat lower temperatures, and metallic Fe solidifies at $T \sim 1200$ K. Iron carbide (Fe_3C) is also a potential product but is unlikely to be more than a trace constituent, its abundance limited by prior condensation of C.

The shape of the condensation curve for carbon is governed by the nature of the dominant monomer, which ranges from atomic and molecular carbon to

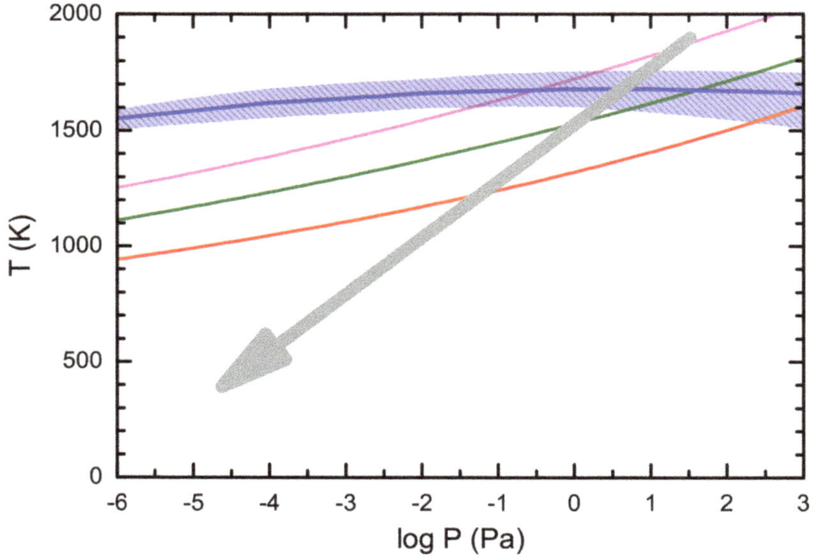

Figure 8.5. Similar to Figure 8.4 but for an atmosphere enriched in carbon, with solar abundances assumed for other elements. Colored curves correspond to graphitic carbon (blue), TiC (magenta), SiC (green), and metallic iron (red) for C/O =1.1. The blue shaded band illustrates the effect on carbon condensation of varying C/O from 1.05 (lower bound) to 1.2 (upper bound). Based on data from Lodders & Fegley (1995).

hydrogenated molecules such as C_2H_2 and CH_4, according to physical conditions and resultant chemical processes in the gas (Salpeter 1977). In the condensation zone of a red giant, acetylene (C_2H_2) is usually the dominant monomer. The kinetic processes that lead to the production of dust in this environment are analogous to soot production by combustion of hydrocarbons (Frenklach & Feigelson 1989; Tielens 2008), in which the products assemble into structures containing linked hexagonal rings. However, the molecular structure of the principal monomer is H–C≡C–H, the simplest example of a saturated *linear* molecule involving carbon bonding with alternate single and triple bonds. In order to construct aromatic hydrocarbons, the triple (sp) bond must be replaced by a double (sp^2) bond, a change that may result from a hydrogen abstraction reaction such as

$$C_2H_2 + H \rightarrow C_2H + H_2 \tag{8.2}$$

in which the product radical is structured C=CH, and is therefore effectively a ring segment (see, e.g., Figure 6.6). The ring may be closed by chemical reactions that attach two further monomers. Once formed, the ring is stable and can grow cyclically by abstraction of peripheral H atoms and attachment of further segments. This growth process may lead to the construction of planar PAH molecules containing multiple rings. However, competing reactions attach non-aromatic units to the rings. The likely outcome of the growth process is amorphous carbon, in which randomly-grouped ring clusters are connected by bridging units with linear (sp) or tetrahedral (sp^3) bonding (e.g., Santoro et al. 2020). This is, indeed, consistent with the known properties of soot particles. Amorphous carbon lacks the long-range order found in crystalline forms with sp^2 (graphite) or sp^3 (diamond) bonding.

The nucleation process outlined above assumes that the atmosphere of the carbon star remains hydrogen-rich, as is usually the case. Rare exceptions include the R Coronae Borealis (RCB) class of variable stars, which are anomalously deficient in hydrogen for reasons that are not yet fully understood (Montiel et al. 2018). Dust is known to form rapidly in episodic bursts in the atmospheres of RCB stars, evidenced by dramatic dimming of their light curves attributed to a sudden onset of circumstellar extinction. Both the physics and the chemistry of grain formation are likely to diverge from those of typical carbon stars, nucleation occurring in regions closer to the photosphere and at significantly higher temperature, where no molecules are present in the gas (Whitney et al. 1993). C itself is then the only abundant monomer. The atoms may assemble into aromatic structures with unsaturated peripheral sites, forming amorphous carbon grains with low hydrogen content.

8.1.5 Planetary Nebulae

The core of a star on the asymptotic giant branch ultimately collapses to become a hot white dwarf, simultaneously ejecting its outer layers to form a planetary nebula (PN). The morphologies of these nebulae differ dramatically, ranging from classic rounded forms that inspired their moniker to bipolar structures such as that illustrated in Figure 8.1 (see Kwitter & Henry 2022 for a comprehensive review). The origin of this diversity is not fully understood, but may be attributed in part to differences in mass (stars that produce bipolar nebulae are typically more massive) or the influence of a binary companion (Kastner et al. 2018). Each example provides a snapshot of a brief (10,000 yr) period in a star's evolution, in which the dusty circumstellar outflow created during the AGB phase is illuminated by its newly-unveiled core. Mass loss rates increase with luminosity as the star ascends the AGB, formation of the PN preceded by a "superwind" phase in which gas and dust stream outward at speeds $\sim$10 km s^{-1}. As the outer layers of the star dissipate, the compact core is exposed. The surface temperature of the core may be as high as 100,000 K, generating an intense flux of high-energy photons that ionizes the expanding envelope. As previously noted, grain nucleation tends to be inhibited in ionized gas, and little or no new dust production is expected to occur in the planetary nebula itself. The expanding gas may be accelerated to speeds $\sim$100 km s^{-1}, an order of magnitude higher than earlier wind speeds. Material from previous phases of mass loss may thus be swept up and reprocessed (see Figure 8.1), and some grain materials may be destroyed. However, Stasinska & Szczerba (1999) argue that the timescale for grain destruction in planetary nebulae is longer than their lifetimes. In common with their precursors, PNe may be grouped according to C/O ratio, which corresponds to the nature of their observable dust content.

8.1.6 Core-collapse Supernovae

The progenitors of core-collapse supernovae are stars of initial mass $\gtrsim$8 $M_\odot$. They evolve to become supergiants after leaving the main sequence (Figure 8.2), meanwhile maintaining hydrogen-rich photospheres, while the products of successive nuclear fusion cycles accumulate inside (Section 7.1.1). A supernova is triggered when internal accumulation of Fe reduces energy production to a point where the core undergoes gravitational collapse. The expanding remnants of these cataclysms are major sources of important dust-forming elements, most notably O, Mg, and Si (Johnson 2019). The extent to which these and other elements condense into solids prior to reaching the ISM has long been a topic of debate, but there is an accumulation of observational and experimental evidence, reviewed in the following sections, for significant dust formation in the ejecta. Progress toward the development of robust theoretical models for nucleation, growth and survival in this unique environment is discussed in a pair of review papers by Sarangi et al. (2018) and Micelotta et al. (2018), to which the reader is referred for detailed discussion. Some key points are summarized here.

As in other potential sources of circumstellar dust, a fundamental requirement is the presence of gas containing elements with an affinity for the solid phase under conditions of temperature and pressure that allow solids to nucleate and grow. In the wind of a red giant, such conditions may occur or recur on timescales of up to a million years; in a supernova they persist for less than a decade (Dwek et al. 2019). Dust is predicted to form in the cavity of hot gas that expands behind the initial blast wave caused by the explosion. Nucleation is inhibited by a flux of Compton electrons, energized by γ-ray photons released by decay of radioactive ^{56}Ni in the explosion, which ionizes and dissociates potential monomers. The cavity is also subject to disturbance by a reverse shock, arising from the impact of the blast wave on the outer layers of the progenitor and its circumstellar envelope (e.g., Bocchio et al. 2016). Depending on the composition and initial size of the condensates, processing by the reverse shock may lead to substantial ablation or even total destruction of the dust if it forms in a uniform outflow. However, the likely presence of denser clumps of gas in the outflow enhances rates of growth and provides a measure of protection. The shock wave decelerates as it propagates into a clump, thus greatly reducing rates of ablation.

Dust formation in supernovae is obviously not an equilibrium process. It may be simulated by application of chemical kinetics, considering evolving physical conditions in stratified layers of ejecta (Sarangi & Cherchneff 2013). The cycle begins with the formation of simple molecules such as CO and SiO. The ensuing chemistry is then dependent on the C/O ratio (Section 8.1.2). Core-collapse SNe are prolific sources of oxygen overall, but the degree of mixing with carbon varies from layer to layer, and dissociation of CO by Compton electrons may allow some C-rich dust production even in an O-rich gas (Clayton 2013; Liu et al. 2018). Potential monomers include many of those discussed in the context of red giant winds and, similarly, the condensates may range in composition from silicates and oxides to pure metals, sulfides, carbides, and carbon. Grains formed in this environment carry a memory of their origins, in the form of unique isotopic signatures, enabling identification of presolar examples in remnants of the Sun's protoplanetary disk (Section 8.3).

8.1.7 Exploding White Dwarfs

White dwarfs are the inert, degenerate cores of expired low and intermediate-mass stars. If left in isolation, they undergo no further thermonuclear activity that might exhume their heavy elements or kindle new production. In a close binary system, however, accretion of matter onto a white dwarf from a companion star may trigger such an event. If the white dwarf reaches a critical mass, it detonates as a type Ia supernova. Explosive nucleosynthesis then leads to copious production of Fe and other metals near the iron peak (Figure 7.2; see Johnson 2019). However, unlike core-collapse supernovae, type Ia events seem unlikely to be significant sources of dust: models suggest that grains either fail to condense or are subsequently destroyed by shocks (Dwek 2016; Kimura et al. 2017; Sarangi et al. 2018), consistent with a general lack of observational evidence for intrinsic dust in their expanding remnants (Section 8.2.1). The importance of type Ia SNe to dust production is the creation of relevant elements rather than dust itself. See also Section 7.3.4 for related discussion.

A distinct but related phenomenon is a nova outburst. This is caused by explosive thermonuclear ignition of matter accreting onto the outer layers of the white dwarf from its binary companion. In this case, however, the white dwarf maintains a subcritical mass and is not destroyed. The outburst ejects a rapidly expanding shell of ionized gas. Dust grains may begin to condense as the gas cools, evidenced by simultaneous occurrences of visual dimming and excess infrared emission some 30–100 days after the event (see Bode 2010 for a review). As before, grain formation is expected to commence when the expanding shell cools to temperatures $\lesssim$1200 K at which refractory solids composed of abundant elements become stable. Efficient dust production requires that the density is above some critical value at the condensation radius, and this depends on the total mass of ejected material. The availability of neutral atomic and molecular monomers is initially limited by UV radiation from the white dwarf, but the first condensates in a relatively dense flow will provide self-shielding if the dust becomes optically thick. Thus, some novae are copious sources of dust, while others produce little or none. Possible condensates include both oxygen-rich and carbon-rich compounds, ranging from silicates and SiC to hydrocarbons and solid carbon. However, as dust-producing novae are relatively rare events, they account for only a small proportion ($\sim$ 1%) of all stardust injected into the ISM (Gehrz et al. 1998).

8.2 Observations of Stardust

Observational techniques used to explore the nature and composition of dust grains in the circumstellar outflows of evolved stars are analogous to those used to study dust in the interstellar medium, described in previous chapters. The primary phenomena available for observation are infrared emission from heated dust, solid-state spectral emission and absorption features, and extinction of starlight. Models constrained by such observations are fairly successful in elucidating the physical properties of the particles and identifying at least some of the constituent materials. Important results are reviewed in this section. Complementary insights gained from analyses of presolar stardust in meteorites are reviewed in Section 8.3.

8.2.1 Infrared Continuum Emission

Infrared continuum emission greatly in excess of that expected from the Rayleigh–Jeans "tail" of a normal stellar photosphere is a defining characteristic of stars with circumstellar dust. Ultraviolet, visible, and near-infrared radiation from the photosphere is absorbed, and the heated grains emit at longer wavelengths. The spectral energy distribution (SED) of the emergent radiation depends on several factors, including the luminosity and spectrum of the star itself, the radial distribution of matter in the expanding envelope, and the composition and temperature of the dust. Radiative transfer models developed to simulate this emission are complex and dependent on numerous variables, but nonetheless important because they help to elucidate the dust formation process, the properties of the particles, and the mass-loss rates of the stars that produce them (e.g., Groenewegen 2012).

Information specific to the composition of the dust arises from both the general form of the SED and the presence of any distinct spectral features. In the idealized case of an isothermal dust shell of temperature T_d, the radiative flux from the dust varies as $F_\lambda \propto Q_\lambda B_\lambda(T_d)$, where Q_λ is the grain emissivity and $B_\lambda(T_d)$ is the Planck function (Section 2.3.3). In reality, of course, T_d varies with radial distance in the shell, from the condensation temperature of the primary solid phase at its inner boundary to interstellar values as $r \to \infty$ (Section 8.1.1), resulting in a composite spectrum. At sufficiently long wavelengths, however, B_λ is described by the Rayleigh–Jeans approximation and the spectral shape converges to

$$F_\lambda \propto \lambda^{-(\beta+4)} \tag{8.3}$$

independent of T_d. The spectral index of the emissivity, β, varies with the composition and crystallinity of the grains, as discussed in Section 2.3.4.

A wealth of observational data is available on which to test models for dust emission from post-main-sequence stars, both in the Milky Way and in local-group galaxies (e.g., Groenewegen 2012; Groenewegen & Sloan 2018 and references therein). Occasionally, emission may originate from ambient interstellar dust that happens to lie near the star, as appears to be the case for certain first-ascent red giants (Jura 1999); but dust shells around more evolved objects with $T_{eff} \lesssim 3500$ K (Figure 8.2) are ubiquitous and evidently of circumstellar origin. The observed spectral energy distributions of stars with both O-rich and C-rich atmospheres can generally be explained by models that assume expanding shells in which dust of the appropriate composition condenses at temperatures $\lesssim$1500 K. Examples are shown in Figures 8.6 and 8.7. In each case, starlight is strongly attenuated by an optically-thick envelope resulting from extended periods of dust formation during the AGB phase of the star's evolution. Grains in the outflows of such stars appear to have predominantly (but not exclusively) amorphous structure, characterized in O-rich stars by the silicate spectral features, and in C-rich stars by emissivity indices $\beta \sim 1$, consistent with the predictions of kinetic chemical models (Donn & Nuth 1985; Paquette & Nuth 2011). Some oxygen-rich stars show spectral evidence for partial crystallization of silicates in the outflows, a topic further discussed in Section 8.2.2.

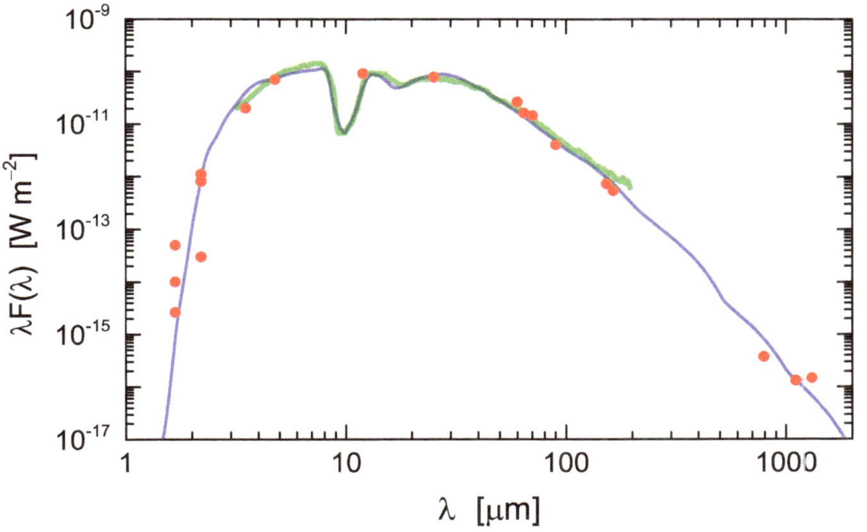

Figure 8.6. Spectral energy distribution of the oxygen-rich AGB star OH 26.5 + 0.6. Observational data include spectra from the short and long-wavelength spectrometers of the Infrared Space Observatory (green) and photometry from various sources compiled by Groenewegen (2012, red). Silicate features centered near 10 and 19 μm appear in absorption. The blue curve is a model based on calculations for olivine ($MgFeSiO_4$) particles of radius $a = 0.2$ μm (see Groenewegen 2012 for further details).

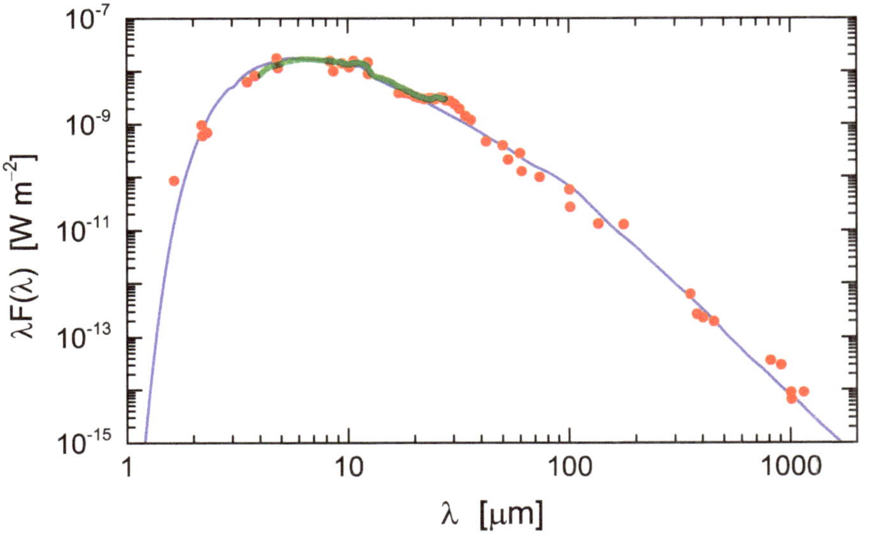

Figure 8.7. Similar to Figure 8.6 but for the carbon star IRC +10 216. The data include spectra from the Infrared Space Observatory (green curve; Thompson et al. 2006) and photometry (red circles), the latter corresponding approximately to the maximum phase of the star's variability. The model (blue curve) assumes amorphous carbon grains of radius $a = 0.16$ μm with a 3% admixture of SiC (Groenewegen 1997). See also Guélin et al. (2018) for further studies of IRC +10 216 and its environment.

The dust responsible for infrared emission from AGB stars such as those illustrated in Figures 8.6 and 8.7 is almost always unambiguously circumstellar in origin, but the situation for supernovae is more complex. The precursors of core-collapse SNe, the most important probable sources of supernova stardust (Section 8.1.6), evolve very rapidly and often detonate in relatively dense regions of the ISM, having had insufficient time to migrate far from their birth sites. Care must therefore be taken to distinguish between radiation emanating from dust created in the explosion and from pre-existing interstellar or circumstellar dust close to the source. Serendipitously, SN 1987A, the most closely studied of all core-collapse supernova to date, detonated in a low-density region of the Large Magellanic Cloud, in a line of sight clear of significant local or foreground extinction.

Evidence for dust formation in the ejecta of SN 1987A first appeared some 500 days after detonation. A systematic increase in mid-infrared flux was observed, consistent with the presence of grains of temperature $T_d \sim 600$ K (Wooden et al. 1993 and references therein). This was accompanied by an apparently simultaneous onset of a decline in visible light, steeper than the predicted rate of decay for the SN type, attributed to extinction. Subsequent monitoring of the infrared spectrum has shown a gradual cooling of the dust, evidenced by a shift of the peak toward longer wavelengths; by 2010 the dust temperature was approaching interstellar values ($T_d \sim 20$ K; Matsuura et al. 2011; Cigan et al. 2019; see Figure 8.8).

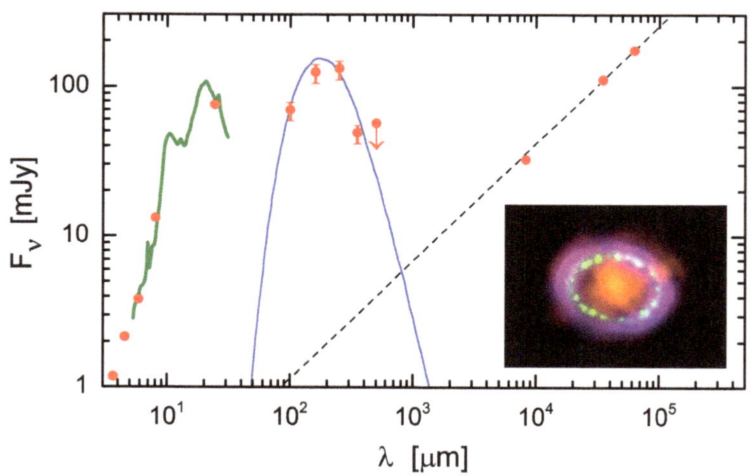

Figure 8.8. Spectral energy distribution of Supernova 1987A, based on infrared, submillimeter and radio data collected between 6000 and 9000 days after detonation (Matsuura et al. 2011 and references therein). The green curve is a spectrum from the Spitzer Space Telescope; red circles are photometric data from the Spitzer and Herschel Space Telescopes and ground-based observations. The blue curve is a dust model ($T_d \approx 20$ K) fitted to the far-infrared data. The dashed line represents synchrotron emission from hot ionized gas. A contemporaneous image is also shown, superposing data at submillimeter, visible and X-ray wavelengths (color-coded red, green, and blue, respectively). Image credits: ALMA (ESO/NAOJ/NRAO/A. Angelich); Hubble Space Telescope (NASA/ESA/R. Kirshner and P. Challis/Harvard-Smithsonian Center for Astrophysics/Gordon and Betty Moore Foundation); Chandra X-ray Observatory (NASA/CXC/Penn State/K. Frank).

Meanwhile, as dust internal to the expanding shock gradually cooled, evidence emerged for a spatially distinct reservoir in its path. This equatorial ring of circumstellar material around SN 1987A (Figure 8.8, inset image) is presumed to be a product of mass-loss during an earlier phase of the progenitor's evolution. First illuminated by the ultraviolet flash from the explosion, it was subsequently energized by the arrival of the shock, causing it to brighten across the spectrum. Shocked gas in the ring is a source of both X-ray and radio synchrotron radiation, its bead-like structure attributed to clumping. Mid-infrared radiation peaking near 20 μm in the SED (green curve in Figure 8.8) emanates from circumstellar dust in the ring, its spectral profile consistent with emission from silicate particles of temperature $T_d \sim 180$ K (Bouchet et al. 2006; Arendt et al. 2016). Dust in this hostile environment is gradually being decimated by sputtering; its estimated total mass ($\sim 3 \times 10^{-6}$ $M_\odot$) is minuscule compared with that of new dust formed behind the shock (red central area of the inset image), which has a probable range of 0.2–0.7 $M_\odot$ (Cigan et al. 2019).

Studies of recent historical supernovae in our Galaxy paint a somewhat consistent picture. Of these, SN 1572 (Tycho) and SN 1604 (Kepler) are classified type Ia (see Section 8.1.7), expected on theoretical grounds to produce little or no intrinsic dust, and observations of dust emission in their remnants can be explained entirely by ambient interstellar or circumstellar matter (Douvion et al. 2001; Williams et al. 2012). SN 1054 (progenitor of the Crab Nebula) is considered to be a core-collapse supernova, albeit with a relatively low estimated mass ($\sim$ 8–10 $M_\odot$), roughly half that of the SN 1987A progenitor. The estimated mass of dust attributed to condensation in SN 1054 is also relatively small, no more than about 0.05 $M_\odot$ (De Looze et al. 2019). Cassiopeia A is likewise considered to be the remnant of a core-collapse SN (its eruption date unknown, estimated circa 1680). It is located at a distance of $\sim$ 3.4 kpc in a direction close to the galactic disk, thus the line of sight contains copious quantities of interstellar matter, and this adds to the challenge of identifying dust intrinsic to the remnant. Detailed analyses of spatially-resolved infrared and submillimeter data reveal concentrations of cold dust in the unshocked ejecta of Cassiopeia A with an estimated total mass in the range 0.3–0.6 $M_\odot$ (De Looze et al. 2017), comparable with that of SN 1987A.

8.2.2 Infrared Spectral Features

Infrared spectroscopy provides a potentially powerful and direct means of identifying grain constituents in stellar ejecta, as previously discussed in the context of interstellar particles (Chapter 5 and Section 6.2). The principal features attributed to circumstellar dust and PAHs are listed in Table 8.1. Various subsets of the listed features are commonly observed in red supergiant (RSG) and asymptotic giant branch (AGB) stars, and in planetary nebulae (PNe). In contrast, there is little evidence to date for spectral features that can be attributed to intrinsic dust in supernovae. Silicate emission was detected from the shocked circumstellar ring of SN 1987A (the spectrum in Figure 8.8 exhibits 9.7 and 19 μm peaks on a steeply-rising continuum) but not in the central condensation zone when it was warm

Table 8.1. Infrared Circumstellar Dust Features of Evolved Stars and Planetary Nebulae

Wavelengths (μm)	Carrier	a/e?	ISM?	Object Types
Oxygen-rich:				
9.7, 19	a-silicates	a/e	Yes	RSG, AGB, PNe
10.0, 11.2, 16.3, 19.5, 23.5, 27.5, 33.5	c-forsterite	e	No?	RSG, AGB
10–17	a-Al_2O_3?	a/e	No	RSG, AGB
13	c-Al_2O_3?	e	No	RSG, AGB
20	FeO? MgO?	e	No	RSG, AGB
3.1, 6.0, 11.5	H_2O ice	a	Yes	OH/IR stars
44, 62	H_2O ice	e	No	OH/IR stars
Carbon-rich:				
3.3, 6.2, 7.7, 8.6, 11.3, 12.7	PAHs	e	Yes	PNe
11.2	SiC	a/e	No	AGB, PNe
21	?	e	No	PNe
30	MgS?	e	No	AGB, PNe

Notes: Features are grouped by the C/O ratio of the source. "PNe" here includes both planetary nebulae and their immediate precursors (post-AGB stars, proto-PNe). Only the principal features of PAHs and a typical crystalline silicate (c-forsterite) are listed. Information is included on whether each feature is generally seen in absorption or emission and whether it has a known interstellar counterpart. Note that a small subset of C-rich objects also display silicate features.

enough to be observable in the mid-infrared (Wooden et al. 1993). This situation might be interpreted as evidence that supernova dust is generally featureless (e.g., pure carbon or metallic iron rather than silicates), but it might also arise from masking of emission features by self-absorption in optically-thick ejecta (Dwek & Arendt 2015). Moreover, at least one core-collapse supernova has been observed to develop a silicate signature at the expected epoch of dust condensation (Kotak et al. 2009). Large quantities of elemental Mg and Si are generated in these events, and observations indicate that they become increasingly depleted from the gas, strengthening the case for silicates as primary condensates in the ejecta (Dwek et al. 2019).

The sources most widely observed to show circumstellar dust signatures are AGB stars and the planetary nebulae that they engender. Features may appear in net emission or net absorption, dependent on the radiative environment, the temperature and density profiles of the envelope, and the optical properties of the dust, factors that vary with the evolutionary state of the source. AGB stars exhibit distinctive infrared spectra governed by their atmospheric C/O ratios (Section 8.1.2), as illustrated in Figures 8.6 and 8.7. OH 26.5 + 0.6 belongs to the OH/IR subgroup of O-rich AGB stars undergoing rapid mass-loss (Section A.3), generating an optically-thick circumstellar shell rich in silicates (Justtanont et al. 1996); IRC +10 216 is a carbon star in an otherwise analogous evolutionary phase. Their SEDs differ most clearly in the mid-infrared, with silicates detected only in the former.

Extracting precise information from the spectra on the composition of the dust is challenging for several reasons. The observed SED may be modeled as a combination of distinct stellar and circumstellar components, after correcting for foreground extinction. The stellar component typically varies systematically with time, so ideally observations from different instruments and at different wavelengths should be matched to the same phase of the light curve. Broad gas-phase molecular bands that resemble solid-state features may be present in the spectrum; however, these are generally most prominent at wavelengths below 8 μm, whereas the dust features lie mostly between 8 and 30 μm. The circumstellar component may include a superposition of absorption and emission features arising in different regions along the line of sight. The opacity of the envelope increases as dust accumulates, and if it becomes optically thick the stellar photosphere is hidden from view. Simultaneously, the dust from which we receive most radiation becomes cooler, and absorption in the spectral features increases relative to emission. With all these factors taken into account, it is possible to deduce realistic spectral models that elucidate the composition of the dust, although some features still lack secure identifications. See Messenger et al. (2013), Golriz et al. (2014) and Shepard & Speck (2021) for related discussion.

Silicate features are invariably prominent in the spectra of dusty oxygen-rich stars. In some cases, the profiles of the 9.7 μm and 19 μm features are smooth and distinct, resembling those seen in the ISM, as expected if amorphous, Mg-bearing silicates are the primary constituents of the dust. Indeed, the circumstellar 9.7 μm feature observed in emission toward the red supergiant μ Cephei (Section 6.2.1) has been used as a standard reference profile in models fitted to its interstellar counterpart in absorption. In other cases, the spectra are more complex, signaling differences in silicate mineralogy and/or additional contributions from other compounds. Observed source-to-source variations in the spectra provide a basis for classification schemes that characterize the evolutionary state of the envelope and constrain the chemical and mineralogical properties of the dust (Begemann et al. 1997; Sloan & Price 1998). Representative examples from a study by Golriz et al. (2014) of AGB stars with silicate features in emission are shown in Figure 8.9 (upper frame). Spectra based on laboratory data of some candidate materials are also shown.

The silicate-dominated spectra present several interpretative challenges. One notable issue is the emissivity (or opacity) of the dust in the spectral range between the principal Si–O stretching and bending modes ($12 < \lambda < 16$ μm), which typically exceeds levels consistent with silicates alone, either amorphous or crystalline. A plausible explanation is the expected presence of aluminum in the particles. This highly condensible element (Section 7.2.3) is among the first to solidify in a cooling atmosphere. Its predicted initial form in an O-rich outflow (Section 8.1.3) is the refractory oxide alumina (Al_2O_3); some might also be incorporated into aluminosilicates (Mutschke et al. 1998) or spinel ($MgAl_2O_4$; DePew et al. 2006). The spectral profile of amorphous Al_2O_3 (Figure 8.9) peaks near 12 μm, where it blends with the silicate stretching mode and adds opacity between the two principal features. Furthermore, in its *crystalline* mineral form (corundum), Al_2O_3 produces a sharper, narrower feature centered near 13 μm, providing a tentative identification of a dust feature at this wavelength seen in many O-rich AGB stars, including one of those shown in Figure 8.9 (blue curve); see Begemann et al. (1997), Sloan et al. (2003) and Golriz et al. (2014) for further discussion and examples. Other metal oxides that may

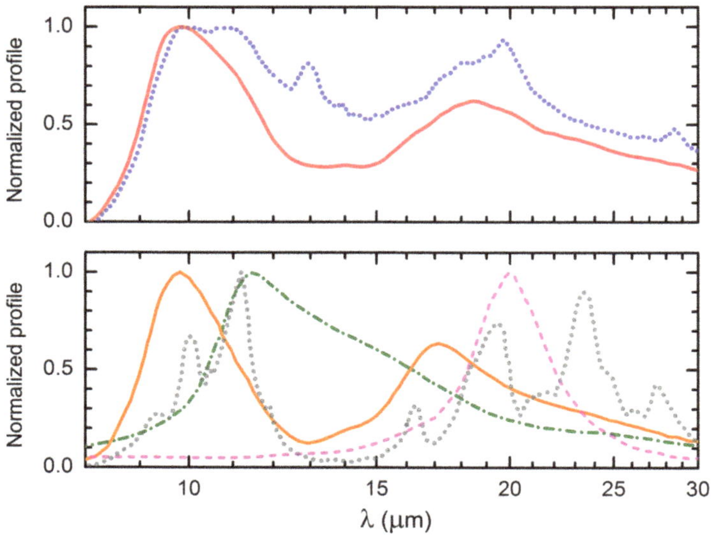

Figure 8.9. Contrasting dust emission features in two oxygen-rich AGB stars, compared with data for some candidate materials. Upper frame: spectral profiles of c32-16 (solid red) and NGC 6522-5 (dotted blue), from the analysis by Golriz et al. (2014) of observations obtained by the Spitzer Space Telescope. Lower frame: synthetic spectra based on the optical properties of amorphous olivine ($MgFeSiO_4$, solid orange, Dorschner et al. 1995), crystalline forsterite (Mg_2SiO_4, dotted gray, Jäger et al. 1998), amorphous alumina (Al_2O_3, dotted–dashed green, Begemann et al. 1997), and wüstite (FeO, dashed magenta, Henning et al. 1995). Each curve is normalized to unity at peak.

contribute to the observed profiles include FeO and MgO (Henning et al. 1995; Posch et al. 2002), both of which have features peaking near 20 μm; thus, if present, they overlap the silicate bending mode.

Spectral signatures of crystallization provide evidence for annealing of dust in the outflows. Laboratory data show that annealing of silicates causes dramatic changes in their spectra, sharpening the principal features and introducing distinct structure associated with various vibrational resonances in the lattice (Jäger et al. 1998; Fabian et al. 2000). The spectra differ in detail according to the specific composition and mineralogy of the silicate: results for crystalline forsterite are used here for illustration (Figure 8.9 and Table 8.1). Incorporating the data into spectral models is challenging, in part because the profiles are sensitive to grain size, shape and temperature as well as mineralogy, and it can therefore be difficult to constrain the mass ratio of crystalline to amorphous silicates in a given source by modeling its SED. Those with simple, smooth spectra (e.g., red curve in Figure 8.9) are clearly dominated by amorphous silicates, but the converse is not necessarily true. Even stars with complex spectra often lack diagnostic features in the 21–28 μm range that would be indicative of crystalline silicates (e.g., blue curve in Figure 8.9; more examples in Golriz et al. 2014), perhaps in some cases due to self-absorption (Jones et al. 2012). Others show clear evidence of crystallization (e.g., review by Henning 2010). It has been suggested that the degree of crystallinity may depend on the mass-loss rate of the star, but this has not been confirmed: see Speck et al. (2008), Jones et al. (2012) and Liu et al. (2017) for further discussion.

Ice may form at sufficiently low temperatures in the outer envelopes of some O-rich stars (Section 8.1.3). The conditions needed for this to occur are most likely to arise during late phases of AGB evolution, as the mass-loss rate increases and the circumstellar envelope becomes optically thick: prime candidates are the OH/IR stars, distinguished observationally by OH maser line emission at radio wavelengths and continuum emission from cold dust in the infrared, with silicate features in absorption (e.g., Figure 8.6). The outer layers of the envelope are shielded from photospheric radiation, and a steep temperature gradient is established. H_2O molecules in the gas may then freeze onto the surfaces of refractory grains that formed previously in a hotter region of the outflow. Spectral features identified with H_2O-ice (Table 8.1) include the O–H stretching, bending and libration modes in absorption, and lower-energy lattice modes in emission (Boogert et al. 2015 and references therein).

It is informative to compare ice features observed in circumstellar outflows with those in interstellar environments. Examples are shown in Figure 8.10 of the absorption centered near 3.1 μm, identified with the O–H stretching mode of H_2O-ice, in the circumstellar envelope of OH 231.8+4.2 and in a prototypical molecular cloud. The profile observed in the OH/IR star is consistent with a model based on *pure* H_2O-ice mantles deposited onto silicate cores at a temperature of 80 K. Thus, as expected, they appear to be forming in substantially warmer conditions compared with ices in molecular clouds. Another difference is that the circumstellar ices are likely to be condensates of preformed gaseous H_2O monomers rather than the products of reactions between constituent atoms on grain surfaces (see Section 9.2.2). The circumstellar ice is

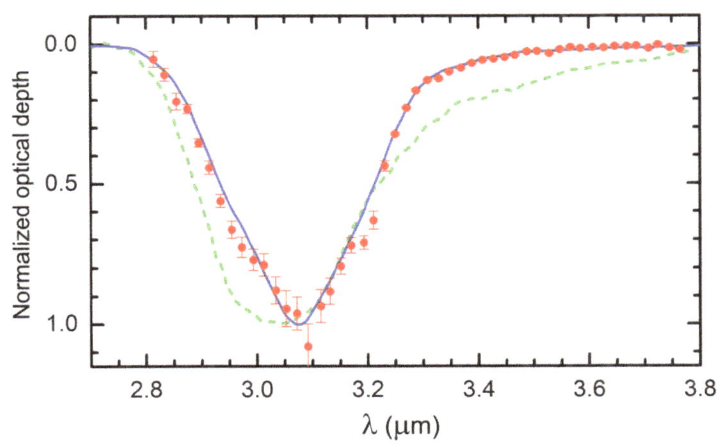

Figure 8.10. Circumstellar absorption centered at 3.1 μm toward OH 231.8+4.2 (red points), attributed to freeze-out of H_2O in its ejecta. The blue curve is a model fitted to the observed profile, assuming spherical grains consisting of H_2O-ice mantles of temperature 77 K and maximum radius 0.6 μm deposited onto silicate cores of radius 0.1 μm (Smith et al. 1988). Also shown for comparison is the profile of the interstellar ice feature observed in the Taurus dark cloud toward Elias 16 (green dashed curve: see also Figure 5.12).

at least partially crystalline, as confirmed by studies of its features at longer wavelengths (Boogert et al. 2015). Significantly, the circumstellar 3.1 μm profile lacks the extended long-wavelength wing characteristic of its interstellar and protostellar counterparts (Section 5.3.3). This difference might be compositional, or it might reflect a disparity in grain size. Circumstellar ice may be purer than interstellar ice, lacking additional absorbers such as ammonium hydrates and hydrocarbons that add opacity in the 3.2–3.6 μm spectral range; indeed, no carbon-bearing condensates are predicted in an oxygen-rich outflow (Section 8.1.2). A maximum radius $a \approx 0.6$ μm is suggested for ice-mantled grains toward OH 231.8+4.2 (Figure 8.10); those in dense molecular clouds may grow significantly larger, causing enhanced extinction by scattering that might account for the long-wavelength wing.

In contrast, the primary condensate in a C-rich outflow—carbon itself—is largely featureless in the infrared. Potential exceptions include lattice resonances in graphite (Draine 1984, 2016), but the predict features at 6.3 and 11.5 μm are very weak and have so far eluded detection. In any case, both condensation theory (Section 8.1.4) and radiative transfer models (Martin & Rogers 1987; Groenewegen 2012) suggest a preponderance of amorphous carbon over graphite. Vibrational modes in hydrocarbons are further candidates for detection, but cool stars do not emit sufficient UV flux to "pump" the familiar PAH emission spectrum (Section 6.2.2), and larger particles seem likely to have low hydrogen content (Chiar et al. 2013). Indeed, the only widely-observed circumstellar dust signature in cool carbon stars that lays claim to a secure identification is that of SiC near 11 μm (Table 8.1). This feature is most commonly seen in emission, but cases of its occurrence in net absorption have also been reported (Speck et al. 2009). An example of each is shown in Figure 8.11.

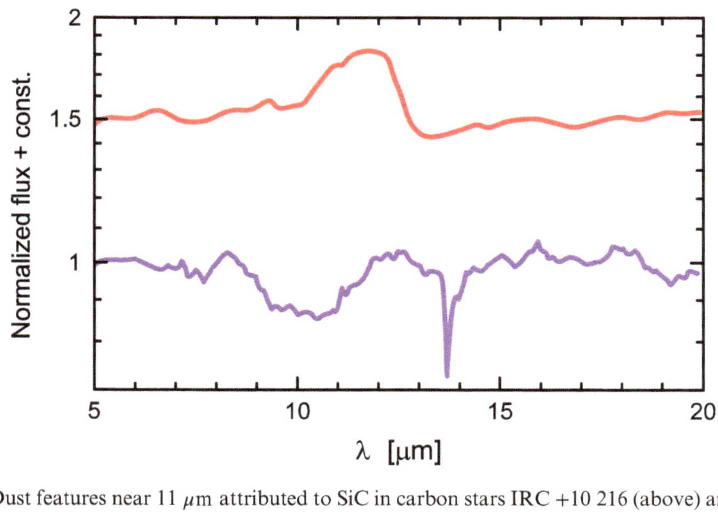

Figure 8.11. Dust features near 11 μm attributed to SiC in carbon stars IRC +10 216 (above) and IRAS 21318 +5631 (below). Each curve is the ratio of observed spectrum to fitted continuum (the upper curve displaced for display). The narrow 13.7 μm absorption in IRAS 21318+5631 is attributed to gas-phase C_2H_2. Based on spectra from the ISO archive analyzed by Thompson et al. (2006), Speck et al. (2009) and Messenger et al. (2013).

Detection of SiC in carbon stars is qualitatively consistent with condensation models for C-rich atmospheres (Section 8.1.4). However, extracting quantitative information on the nature and abundance of SiC in stardust from the observations is challenging for several reasons. Laboratory studies show that the position, width, strength and profile of the 11 μm feature are highly sensitive to the precise properties of the particles, including size, shape, purity and crystallinity (Mutschke et al. 1999). Condensation zones for SiC and solid carbon are partially overlapping in the phase diagram (Figure 8.5), so SiC-bearing particles in the outflows might assume structures ranging from pure SiC crystals to SiC seed particles coated with carbon or irregular mixtures of both materials. Thus, variations in physical conditions in the condensation zone may affect the structure of the particles produced, both from one source to another and over time in a specific source. In cases affected by self-absorption, the emitting grains may differ from the absorbing grains, not only in temperature but potentially in other ways that influence the profile, such as structure, purity, and size distribution. These factors may account for source-to-source differences in the position and width of the 11 μm feature, notably in the shift from net emission to net absorption (Figure 8.11). Because of such concerns, estimates of the mass ratio of SiC to generic carbon dust in stellar ejecta are uncertain. Results based on observations of stars in the Milky Way with approximately solar metal abundances typically range from 5% to 15% (Groenewegen et al. 1998), significantly lower than the predictions of AGB models constructed by Nanni et al. (2013).

As a star reaches the end of the AGB phase of its evolution, it sheds its outer layers to form a planetary nebula (Section 8.1.5). During the transition period, lasting perhaps a few thousand years, the color temperature of the stellar spectrum becomes progressively hotter as the core is gradually unveiled. Once the metamorphosis is complete, the circumstellar environment is exposed to intense, hard UV radiation emanating from the $T \gtrsim 50{,}000$ K core, now revealed as a white dwarf. The dramatic change in visual appearance is accompanied by corresponding changes in the spectrum. In carbon-rich objects, a signature development is the appearance of the PAH emission spectrum (Figure 8.12; see also Figure 6.7). Of course, the presence of PAHs in this environment is no surprise—their production is implicated in the growth of larger particles (Section 8.1.4). PAHs are not generally detectable in absorption in optically-thick circumstellar envelopes, the opacity arising in much larger grains; but on exposure to photons sufficiently energetic to induce thermal transients, free-flying PAHs immediately reveal their presence. See Section 6.2.2 for related discussion.

In addition to the PAH spectrum, Figure 8.12 exhibits emission features peaking at longer wavelengths that have proved difficult to identify. Of these, the broad band centered near 30 μm seems to be a common signature of stardust in C-rich environments, observed in AGB stars (see, e.g., Figure 8.7) as well as transition objects and PNe. An identification with MgS has been proposed but this is not well-supported by available evidence (see Messenger et al. 2013 and Zhang 2020 for detailed discussion). Its extreme width and star-to-star variations in shape suggest that the 30 μm "feature" is in fact a blend (Hrivnak et al. 2000), potentially with multiple carriers: Mishra et al. (2015, 2016) use two or three distinct Drude functions

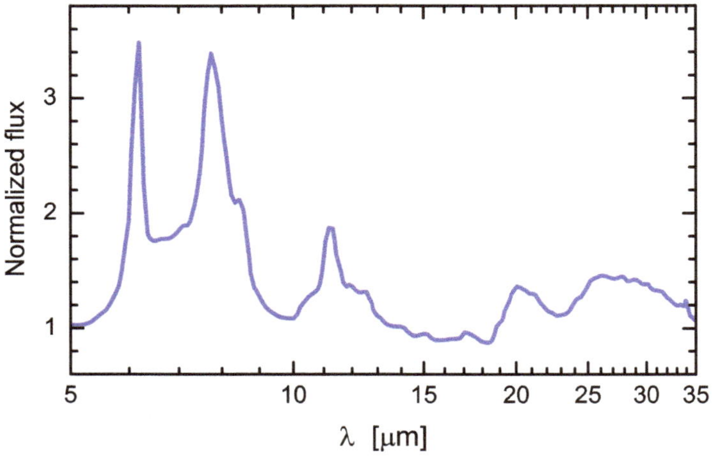

Figure 8.12. Circumstellar emission features toward IRAS 14429−4539, an example of a carbon-rich post-AGB star in the brief transitional period during which it evolves to become a white dwarf with a planetary nebula. The vertical axis is the ratio of observed flux to continuum fit from an analysis of archival ISO data by Mishra et al. (2016). The spectrum exhibits PAH features and unidentified peaks centered near 21 and 30 μm.

to model the observed profiles. Another unidentified feature, centered near 21 μm, is prominent *only* in the spectra of C-rich sources in transition between AGB and PN phases, where stellar color temperatures are intermediate between those of red giants and hot white dwarfs. Perhaps, in common with the PAH features, the carrier requires UV excitation to be observable, but is subsequently destroyed when the radiation field becomes too harsh.

As previously noted, not all carbon-rich sources display exclusively carbonaceous dust features: a small subset exhibits silicate emission features in their circumstellar envelopes (Little-Marenin 1986; Willems & de Jong 1986; Waters et al. 1998; Kwon & Suh 2014). It seems unlikely that these objects are true exceptions to the premise described in Section 8.1.2, which predicts that the gas-phase C/O ratio in the condensation zone controls the chemistry of dust production. In a single star, this ratio changes with time if the stellar atmosphere becomes carbon-enriched during successive episodes of dredge-up. Enrichment is associated with thermal pulses that induce temporary phases of rapid mass-loss (Zijlstra et al. 1992). The switch from O-rich to C-rich chemistry may therefore occur quite rapidly, creating a source of carbonaceous dust within an expanding envelope of silicate dust from earlier cycles of production. However, the expanding shells are expected to cool on timescales much shorter than the thermal pulse cycle (10^3–10^4 years; Iben & Renzini 1983), and it is then difficult to understand why the silicate features are observed in *emission*. But if the mass-losing star is a member of a binary system, its ejecta may be captured into the circumstellar envelope of the companion, which could then act as a reservoir of warm silicate dust formed at an earlier epoch. A further albeit less probable scenario is that both members of a binary system could be dust producers, one carbon-enriched, the other not.

8.2.3 Circumstellar Extinction

One important result arising from extinction by circumstellar dust has already been noted: an increase in infrared flux from the ejecta of certain stellar cataclysms (novae, core-collapse supernovae) may be accompanied by simultaneous dimming of the visible light, attributed to extinction, thereby providing corroborative evidence for the onset of dust formation. If the wavelength-dependence of the extinction can be measured, it may provide further insight into the nature of the dust. A specific question that can be addressed by such observations is the relationship between carbonaceous stardust and the absorber responsible for the interstellar UV extinction “bump” (Section 3.4), widely attributed to graphitic nanoparticles. Do the carrier particles (or their precursors) originate in stellar outflows?

A practical problem that constrains observing campaigns designed to address this question is the simple fact that most dust-producing stars are insignificant sources of ultraviolet radiation. This property is intrinsic to low-temperature photospheres, exacerbated by high extinction if the circumstellar envelopes are not optically thin. Because of this, only a small (and perhaps unrepresentative) subset of sources are feasible targets, specifically post-AGB stars (Figure 8.2) and members of the rare RCB class of eruptive variables. Observations designed to measure extinction in these objects were enabled by space telescope missions, most notably the International Ultraviolet Explorer, which was operational from 1978 to 1996 (little further progress seems to have been made since then). Extinction curves may be generated by methods analogous to those applied to studies of the ISM (Section 3.1), using either a model stellar atmosphere or observations of an unreddened star of matching spectral type as the basis for comparison. These methods are generally less reliable for studies of circumstellar extinction, however, as the targets typically have complex, non-standard photospheric spectra and are intrinsically variable. Care must also be taken to separate circumstellar and interstellar components of the measured extinction. However, if stellar variability results primarily from varying degrees of circumstellar extinction, as is generally the case for RCB-type stars, a comparison between spectra measured at bright and faint phases of the light curve may provide an extinction curve without reference to another object or model atmosphere, subject only to matching the stellar pulsational phase at high and low obscuration (e.g., Hecht et al. 1984). In serendipitous cases where a normal early-type star lies within or behind the circumstellar envelope of a dust-forming star, the former may be used as a probe of the extinction in the latter using the standard technique (e.g., Snow et al. 1987).

Examples of circumstellar UV extinction curves are illustrated in Figure 8.13 (displayed as extinction versus wavenumber, the conventional format for interstellar studies; Chapter 3). A broad peak centered near 4 μm^{-1} is evident in two of the three sources (see Hecht et al. 1984, Buss et al. 1989, and Drilling et al. 1997 for further examples). The mean wavelength of the peak is 240 ± 2 nm (Muci et al. 1994), substantially longer than that of the 217 nm interstellar feature. Both RY Sgr (an RCB-type star) and Abell 30 (a planetary nebula) are carbon-rich and hydrogen-deficient,

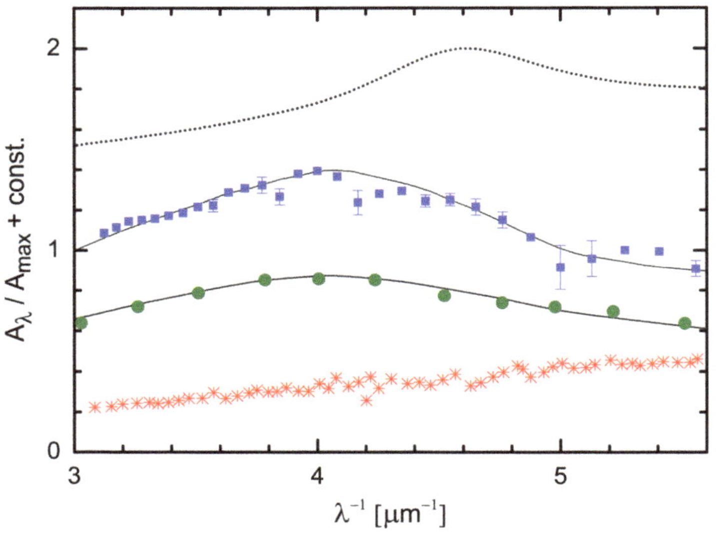

Figure 8.13. Examples of circumstellar ultraviolet extinction curves, compared with the mean for the diffuse ISM (dotted curve; Section 3.2.1). Vertical placement is arbitrary for display. RY Sgr (an RCB star, blue squares) and Abell 30 (a planetary nebula, green circles) both exhibit a broad absorption peak centered near 4 μm^{-1}, consistent with predicted profiles for absorption by sub-micrometer-sized amorphous carbon grains (solid curves). Extinction toward HR 4049 (also known as HD 89353, a post-AGB star, red asterisks) appears to be featureless. Based on data from Muci et al. (1994) and references therein.

as are other objects in which the 240 nm feature is detected. The most probable dust condensate in such an environment is amorphous or partially annealed carbon with low hydrogen content, and indeed the profile of the feature is well-matched by numerical models and laboratory simulations based on such particles (Hecht et al. 1984; Hecht 1991; Muci et al. 1994; Drilling et al. 1997). Grain radii yielding good fits are typically $a \sim 0.03$ μm, larger than those used to fit the interstellar feature (Figure 3.11). In contrast, the circumstellar UV extinction toward HR 4049 appears to be featureless (Figure 8.13, red). This carbon-rich post-AGB star may be atypical: it is a member of a close binary system with a circumbinary disk, in which the companion exerts a major influence on the physical and chemical evolution of the system (Malek & Cami 2014). Unlike stars that exhibit 240 nm features, HR 4049 is not depleted in hydrogen, a factor that suggests a plausible explanation: hydrogenation tends to suppress UV features in graphitic or amorphous carbon by localizing the electrons responsible for absorption (Hecht 1986).

The majority of stars for which circumstellar extinction curves have been studied are C-rich, but a few O-rich supergiants have also been investigated (Snow et al. 1987; Buss & Snow 1988; Seab & Snow 1989). To within the limits of the data, there is no evidence for either a mid-UV bump or a far-UV rise. The absence of an absorption feature is, of course, expected as dust forming in an O-rich environment should lack a carbonaceous component. In the best-studied case (α Sco), the extinction declines steadily with increasing wavenumber in the UV: the data are consistent with relatively large ($a \gtrsim 0.1$ μm) silicate grains (Seab & Snow 1989).

If these results are typical, the O-rich stardust entering the ISM will tend to populate the upper range of the size distribution.

8.3 Stardust in Meteorites

8.3.1 Background

Our solar system was born some 4.568 Gyr ago. Its age is known rather precisely by means of radiometric dating techniques applied to samples of primitive meteorites (e.g., Bouvier & Wadhwa 2010). These fragments of asteroidal parent bodies formed at early times in the Sun's protoplanetary disk, and their radiometric ages refer to the epoch at which constituent minerals condensed. Most solids from the interstellar birth cloud had been vaporized and homogenized prior to recondensation, thereby erasing all chemical and mineralogical fingerprints of their presolar state. Exotic particles identified in the meteorites are exceptions: a tiny but highly significant population of survivors from the presolar interstellar medium, some of which predate the solar system's origin by at least a billion years (Heck et al. 2020).

Discovery of this population simultaneously answered a perceptive question posed many years previously ("Interstellar Grains in Museums?"; Cameron 1973) and instigated a new field ("Astrophysics with Extraterrestrial Materials"; review by Nittler & Ciesla 2016). A new window on cosmic chemical evolution was opened by the ability to apply advanced analytical techniques to physical samples of extrasolar dust in the laboratory (see Zinner et al. 2011 for a review of the relevant technology). Results have major implications for our understanding of the nature and origin of the dust itself and the nucleosynthetic processes that created its constituent elements. The field has been well-served by excellent review papers: see Anders & Zinner (1993), Zinner (1998), Hoppe (2011), and Davis (2011) in addition to those cited above.

Examples of stardust grains from meteorites are shown in Figure 8.14. Identification depends primarily on detailed isotopic analyses of samples extracted from carbonaceous chondrites, the least altered remnants of the protoplanetary disk. Solids that condensed locally have elemental and isotopic abundances that closely match those in the Sun (Section 7.1.3), whereas extrasolar grains (Table 8.2) display isotopic patterns that deviate from solar values to an extent that is incompatible with an origin in the solar system. The analytical techniques not only identify the particles as extrasolar but also provide evidence on their birth environment. Nuclear reactions in stars (Section 7.1) generate distinct isotopic abundance patterns according to physical conditions and the timescales on which they operate. Insight is often gained from trace impurity elements, including trapped noble gases, as well as from those that compose the bulk of a specific solid. Different patterns are predicted for certain elements depending on whether the reactions responsible proceed slowly during an extended phase of stellar evolution, or rapidly during a cataclysmic event. It is thus possible to distinguish stardust formed in (e.g.) the outflow of an AGB star from that formed in a supernova explosion. An example of such a diagnostic pattern is illustrated in Figure 8.15, and some properties of the various forms of stardust are summarized in Table 8.2.

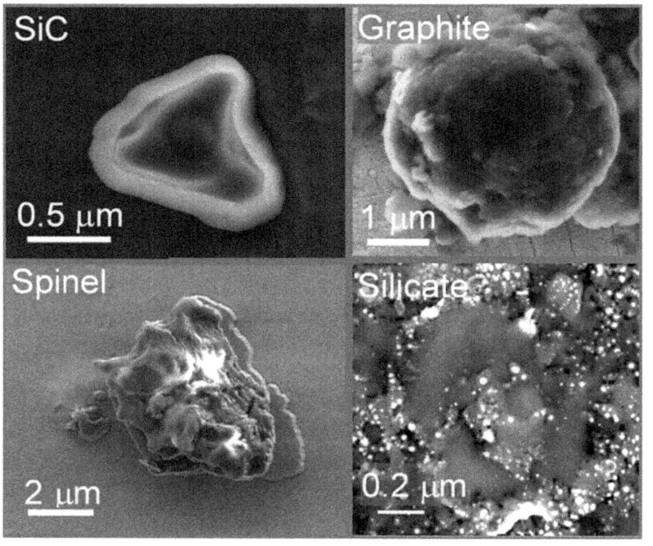

Figure 8.14. Examples of presolar grains identified in meteorites (Hoppe 2011). The grain labeled "silicate" (lower right) consists of an Al-rich core surrounded by a silicate mantle. The SiC grain originated in a supernova, the spinel ($MgAl_2O_4$) and silicate grains in AGB stars; the graphite grain could be of either AGB or supernova origin. Image credit: Max Planck Institute for Chemistry, courtesy of Peter Hoppe.

Table 8.2. Properties of Stardust Grains in Carbonaceous Chondrites.

Constituent	Grain Size (μm)	Abundance (ppm)	Stellar Sources
Silicates	$\lesssim 1$	200	AGB, SNe (9:1)
Oxides	0.1–2	50	AGB, SNe (9:1)
SiC	0.3–50	30	AGB, SNe (20:1)
Si_3N_4	$\lesssim 1$	0.002	SNe
Graphite	1–20	10	SNe, AGB (3:2)
Diamond	0.002	1400	SNe?
TiC	0.01–0.05	0.001	Similar to graphite

Notes: Abundances are in parts per million by mass relative to the bulk meteorite. Where two stellar sources are listed, a rough estimate of their relative numerical contributions is given. The presolar status of diamond is uncertain (see text). TiC is detected as subparticles within graphite rather than an independent population. Data are from Davis (2011) and references therein.

8.3.2 Composition and Origin: Carbon-rich Grains

The solar system is intrinsically oxygen-rich (C/O ≈ 0.5). Thus, we expect native mineral grains to consist mostly of oxygenated forms such as oxides and silicates, whereas carbonaceous forms such as graphite and silicon carbide are more likely to have originated elsewhere. This is largely confirmed by the investigations. Meteoritic

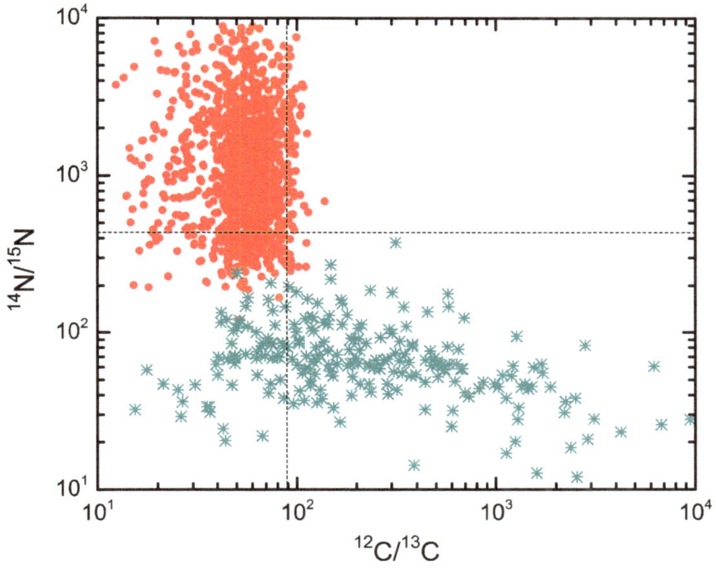

Figure 8.15. Stable isotope ratios for carbon versus nitrogen, comparing distributions for two classes of presolar silicon carbide grains: "mainstream" (red circles), and type X (asterisks), originating principally in AGB stars and core-collapse SNe, respectively. Dashed lines indicate mean solar values. Data are from the Presolar Grain Database (Stephan et al. 2020).

SiC, in particular, appears to be almost entirely extrasolar. Although by no means the most common, SiC is the most extensively studied of all forms of presolar dust, in part because its crystal structure accepts a number of trace elements in grains large enough to contain measurable quantities. They reveal orders-of-magnitude variations in isotopic abundances that allow the grains to be characterized into several groups (e.g., Davis 2011). Figure 8.15 compares stable isotope ratios for C and N in two such classes. The mainstream group, which accounts for about 90% of all SiC grains, exhibits a general pattern consistent with theoretical models and observed values for normal C-rich AGB stars of otherwise solar metallicity. In such stars, the CNO cycle in hydrogen-burning shells (Figure 8.3) controls the distribution of C versus N isotopes (e.g., Anders & Zinner 1993), such that grains from their outflows tend to cluster toward the upper left quadrant of Figure 8.15. The distribution of the rarer type X grains (1% to 2% of the sample) is dramatically different, with little or no overlap. Isotopic abundance patterns for these and other elements in type X SiC grains are consistent with an origin in the ejecta of core-collapse SNe. Similarly, the much rarer Si_3N_4 grains (Table 8.2) also exhibit a pattern indicative of an origin in SNe.

As well as establishing supernovae as the source, isotopic abundance patterns in X-type SiC grains also indicate a probable timescale for condensation (Liu et al. 2018). Grain formation in core-collapse SNe proceeds under conditions that evolve rapidly (Section 8.1.6). However, to make SiC in the outflow it is necessary to mix elements from different layers in the progenitor: Si and C are created by distinct processes, such that the Si-rich zone is internal to the C-rich zone at the time of detonation. How long did this take? An answer is suggested by the abundances of certain trace elements in the grains that are products of extinct, short-lived radioactive nuclides produced in the explosion—effectively radiometric chronometers. By measuring the abundances of isotopes such as ^{49}Ti (the decay product of ^{49}V with a half-life 330 days), it is possible to constrain the time of formation. Liu et al. (2018) deduce that X-type SiC grains formed at least 2 years after their parent stars exploded, a result that places their origins in the mid-to-late phase of dust nucleation in core-collapse SNe (Sarangi & Cherchneff 2013). About two-thirds of all presolar graphite grains also display signatures of a supernova source, with abundance patterns similar to those of the SiC particles, suggesting a contemporary origin in the ejecta.

Presolar graphite grains display distinctive internal structures (Bernatowicz et al. 1991, 1996, 2005). Onion-like concentric layers of graphite may encase a core composed of PAH clusters or amorphous carbon, suggesting a formation sequence in which cores condense in isolation and then acquire graphitic mantles by vapor deposition in a C-rich atmosphere. The grains frequently contain much smaller particles (subgrains) embedded in the structure. Subgrains composed of TiC are common to graphites originating in both AGB stars and SNe (the compositions of other subgrains are different, with additional refractory carbides predominant in the former, pure metals in the latter; Croat et al. 2003, 2005). TiC particles embedded in the mantles must have formed prior to mantle deposition: their presence (and the absence of embedded carbides with lower condensation temperatures, such as SiC) places constraints on the condensation sequence. The density required, both to grow $>1\ \mu$m-sized grains within reasonable timescales and to explain the efficient inclusion of carbide crystals in the structure, is a factor $\sim$100 higher than would typically occur in a spherically symmetric wind, implying the occurrence of local high-density clumps in the outflows.

Another carbon-bearing candidate that might originate in SNe is diamond. Nanometer-sized diamond crystals (nanodiamonds) are abundant in carbonaceous chondrites (Table 8.2) and were among the first meteoritic grains to be proposed as presolar (Lewis et al. 1987). However, their status has since been questioned (e.g., Dai et al. 2002). Because they are so tiny (a nanodiamond of average size contains about 1800 carbon atoms), detailed isotopic analyses of trace elements that might confirm their origins cannot be performed on individual particles, and studies of aggregates yield ambiguous results. Mean isotopic ratios for elements such as C and N are close to solar (Davis 2011), but those of certain trace elements such as Xe suggest an extrasolar origin for some unknown fraction of the particles (see Lewis et al. 2020 for detailed discussion). It is difficult to understand how nanodiamonds could form in large numbers in a gas of solar composition, and their status thus remains unclear.

8.3.3 Composition and Origin: Oxygen-rich Grains

Identifying oxygen-rich presolar grains in meteorites is more challenging because they are composed of compounds similar to those formed in the solar nebula.

Indeed, the techniques commonly used to isolate refractory carbon grains often involve dissolving the dominant silicate component in acid, a process akin to "burning down a haystack to find the needle" (e.g., Nittler & Ciesa 2016). The development of advanced ion microprobe facilities capable of isotopic mapping with high spatial resolution enabled a different approach. This yielded the first detections of presolar silicates, identified in interplanetary dust particles by their anomalous isotopic oxygen compositions (Messenger et al. 2003). Application of this technique to meteorite sections led to the recognition of silicates and oxides as the most abundant presolar constituents (Table 8.2) if nanodiamonds are disregarded. As with C-rich particles, detailed studies of isotopic patterns allow insight into their origins: for example, the abundance ratios $^{17}O/^{16}O$ and $^{18}O/^{16}O$ are highly diagnostic (e.g., Hoppe 2011; Nittler & Ciesa 2016). The results identify some 10% of the sample as originating in SNe. A small fraction (< 1%) may originate in novae, but the majority formed in the outflows of AGB stars.

Presolar silicate grains contain a wide range of compounds and structures, with no obvious preference for any specific phase. Examples include single crystals of Mg_2SiO_4 or $MgSiO_3$, non-stoichiometric amorphous grains, and compound grains containing multiple minerals, including silicate/oxide aggregates (Nittler & Ciesla 2016; Nguyen et al. 2016). Mg silicates are dominant but some Fe-bearing compounds are also present, suggestive of non-equilibrium condensation. The crystalline fraction is about 40%. Presolar oxide grains include Al_2O_3 (corundum) and $MgAl_2O_4$ (spinel), often in the form of single crystals with structures and compositions consistent with equilibrium condensation. Thus, AGB stars appear capable of producing a range of oxidized phases under both equilibrium and non-equilibrium conditions. Identification of Al_2O_3 corroborates observational evidence for a probable contribution by this mineral to the mid-infrared spectra of AGB stars (Section 8.2.2). Crystallinity may be related to interstellar residence times prior to inclusion in the meteorites, given that interstellar silicates appear to be largely amorphized (Section 5.2.1).

8.3.4 Particle Sizes

It is interesting to compare sizes represented in the presolar grain population (Figure 8.14; Table 8.2) with those of interstellar particles, inferred from their optical properties. Presolar silicates have radii typically in the range $0.1 \lesssim a \lesssim 0.2$ μm, with a maximum of about 1 μm, broadly consistent with the silicate components of models for interstellar extinction and polarization (Sections 3.6.1 and 4.2.8). In notable contrast, presolar graphite grains identified in meteorites are enormously large compared with the graphitic nanoparticles proposed as carriers of the 217 nm absorption bump (Section 3.4), and larger, even, than particles invoked to account for the general extinction at longer wavelengths. This might seem perplexing, but there are reasons to suppose that the apparent size distribution of graphite in meteorites is skewed toward large grains and therefore unrepresentative of the parent population. Smaller grains may have been destroyed selectively, prior to accretion into the meteorites, and further selection effects that tend to favor large grains may arise in the laboratory during the extraction process. Meteoritic graphite is, in fact, rare in terms of fractional mass—rarer, even, than SiC (Table 8.2), which is below detectable levels in the ISM (Section 5.2.2). The bulk of the C in carbonaceous chondrites is in organic refractory matter, which might have formed in part by surface reactions on graphite grains in the solar nebula (Barlow & Silk 1977), a process that would naturally tend to consume the smallest graphitic particles most efficiently (while leaving the more chemically-stable nanodiamonds intact). The discovery of large (>1 μm) particles of both graphite and SiC in meteorites is significant, nevertheless, for two reasons: (i) it demonstrates the occurrence of conditions in stellar outflows that enable growth to such sizes, as previously discussed; and (ii) it demonstrates that large grains do indeed exist in the ISM, even if they represent a mere “tip of the iceberg” with regard to the overall size distribution (see Sections 3.2.5, 3.6.1, and 8.4.2 for related discussions).

8.4 Stars as Sources of Interstellar Grains

Stardust undoubtedly contributes to the interstellar grain population. This is implied by observational evidence that evolved dust-forming stars may undergo rapid mass loss (see below), and confirmed by detections of presolar stardust in the solar system, discussed in the previous section. The mix of dust generated by a diverse population of stars includes both carbon-rich and oxygen-rich particles, in qualitative agreement with the need for both kinds of dust to exist in the ISM to account for its observed properties. The quantity of dust present in the ISM at any given time represents the balance between sources and sinks. This section attempts to assesses the significance of stardust as a source of interstellar dust.

8.4.1 Mass Loss

All mature stars lose mass to some degree, in the form of gaseous stellar winds driven by thermodynamic pressure. The flux of solar wind particles currently emanating from the Sun amounts to about 3×10^{-14} $M_\odot\,yr^{-1}$, equivalent to a mere 0.03% of the total mass over its entire main-sequence lifetime. The flux correlates with stellar luminosity (L_*) and radius (R_*), so the mass loss rates of Sun-like stars increase dramatically during the red giant phase. The rate $\dot{M} = dM/dt$ for a star of mass $M = M_*$ arising from a thermally driven gaseous wind may be related to stellar parameters by the approximation

$$\dot{M} \approx 4 \times 10^{-13} \frac{L_*}{gR_*} \tag{8.4}$$

where $g = GM_*/R_*^2$ is the gravitational acceleration at the stellar surface and all quantities are expressed in solar units (Reimers 1975; Dupree 1986). Equation (8.4) thus predicts mass loss rates of $\dot{M} \sim 10^{-8}$ $M_\odot\,yr^{-1}$ for a first-ascent red (M-type) giant and $\dot{M} \sim 5 \times 10^{-7}$ $M_\odot\,yr^{-1}$ for a red supergiant; however, these values are effectively lower limits for dust-producing stars because they do not take grain dynamics into account.

Cool stars with high mass loss rates invariably show evidence of dust in their envelopes, revealed most clearly by circumstellar infrared emission (Section 8.2.1). Grains nucleating in stellar atmospheres undergo outward acceleration due to radiation pressure, and transfer momentum to the gas by collisions. Empirical estimates of mass loss rates are based on observed expansion velocities in the gaseous outflows, in combination with measures of circumstellar mass, yielding results broadly consistent with detailed dynamical models that include the effect of radiation pressure on the dust (e.g., Bladh et al. 2019). Results for luminous stars with dusty envelopes show that mass loss rates increase with luminosity far more steeply than implied by Reimers' formula (see Höfner & Olofsson 2018 for a review). Values in the range 10^{-6}–10^{-4} $M_\odot\,yr^{-1}$ are typical of red supergiants (Jura & Kleinmann 1990) and both C-rich and O-rich AGB stars (Ramstedt et al. 2008; Eriksson et al. 2014; Bladh et al. 2019). A histogram for M-type stars over a range of evolutionary states (Figure 8.16) suggests a bimodal distribution, with peaks separated by two orders of magnitude. The highest rates may be attributed to dust-driven superwinds during the late phases of AGB evolution that precede ejection of a planetary nebula (Section 8.1.5). Such a star might lose a substantial fraction of its mass in a few thousand years.

That dust provides the impetus for such enhancements may be demonstrated by considering the balance between the forces of radiation pressure and gravity. Consider a spherical grain of mass m_d and radius a situated in a circumstellar shell

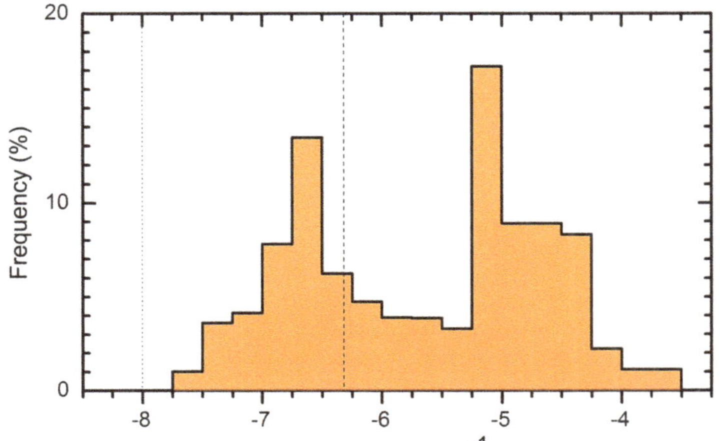

Figure 8.16. Histogram of observed mass loss rates in M-type luminous stars. The sample combines data for semi-regular variables, Miras, and OH/IR stars from Bladh et al. (2019) and references therein. Frequency as a percentage of the total corresponds to bins of width 0.25 dex in log(dM/dt). Vertical dotted and dashed lines correspond to values predicted by Reimers' formula (Equation (8.4)) for M-type giants and supergiants, respectively.

at a radial distance r from the center of a star. Let us assume that the circumstellar zone internal to the location of the grain is optically thin to stellar radiation, as expected near the inner boundary of the dust condensation region (Section 8.1.1). The outward force imposed by radiation pressure is then

$$F_{\rm pr} = \pi a^2 \langle Q_{\rm pr}\rangle \left(\frac{L_*}{4\pi r^2 c}\right) \tag{8.5}$$

where $Q_{\rm pr}$ is the efficiency factor for radiation pressure, defined in Section 2.4.1, and $\langle Q_{\rm pr}\rangle$ is its average value with respect to wavelength over the stellar spectrum. The opposing gravitational force is

$$F_{\rm gr} = \frac{GM_* m_{\rm d}}{r^2} \tag{8.6}$$

and we require $F_{\rm pr}/F_{\rm gr} > 1$ for outward acceleration. Combining Equations (8.5) and (8.6) and expressing $m_{\rm d}$ in terms of the density $\rho_{\rm s}$ of the solid grain material, this ratio is given by

$$\frac{F_{\rm pr}}{F_{\rm gr}} = \frac{3L_*}{16\pi GM_* c}\left\{\frac{\langle Q_{\rm pr}\rangle}{a\rho_{\rm s}}\right\} \tag{8.7}$$

which is independent of r and varies with grain properties in proportion to the bracketed term on the right. $Q_{\rm pr}$ may be calculated as a function of wavelength for a given grain model using Mie theory (Section 2.1). It is typically maximized when $2\pi a/\lambda \sim 1$, and greater for strongly absorbing grains than for dielectric grains of the same size (because $Q_{\rm abs}$ is the dominant contributor to $Q_{\rm pr}$ for plausible values of the scattering asymmetry parameter; Section 2.4.1). As an example, Martin (1978) estimates $\langle Q_{\rm pr}\rangle \approx 0.18$ for graphite and 0.003 for silicates, assuming spheres of constant radius $a = 0.05$ μm in the atmosphere of a red giant of luminosity 10^4 $L_\odot$ and mass 4 $M_\odot$. It may be shown that dust is driven outward from red giants of luminosity $L_* \gtrsim 10^3$ $L_\odot$ for a range of grain size and composition. In the above example, $F_{\rm pr}$ exceeds $F_{\rm gr}$ by a factor of ~2000 for graphite grains and by a factor of ~40 for silicate grains (Equation (8.7)).

The spectrum of the radiation field naturally varies with r when the dust shell becomes optically thick, shifting the peak from the near-infrared (typical of an M-type photosphere) to the mid-infrared at large distances from the star, as radiation is absorbed and reradiated by warm circumstellar dust. This change is unlikely to inhibit the outward flow of the particles, given that both carbon and silicate grains are efficient absorbers in the mid-infrared. The outward speeds of the grains are limited primarily by frictional drag exerted by collisions with the gas. Grain speeds cannot be measured directly but may become coupled with those of the gas in dense regions of the outflow. Typical gaseous wind speeds, determined by the Doppler method, are in the range 5–30 km s^{-1}, values that exceed the local escape velocity at sufficiently large distances from the star.

8.4.2 Predicting the Size Distribution

The terminal speed of a particle driven through a gas by radiation pressure depends on its size, such that large grains tend to overtake smaller ones, resulting in grain–grain collisions. Depending on their relative speeds, the impacts might be constructive (resulting in coagulation) or destructive (resulting in fragmentation). The influence of collisions on the size distribution of the particles in red giant winds and planetary nebulae is assessed by Biermann & Harwit (1980). Frequent collisions at relative speeds $\gtrsim 1$ km s^{-1} are predicted, resulting in the imposition of a power-law form for the size distribution, independent of the initial size distribution of the condensates. On the basis of fragmentation theory (originally applied to asteroids), Biermann and Harwit argue that the emergent grains follow a size distribution of the form $n(a) \propto a^{-3.5}$. This model suggests a physical basis for the origin of the interstellar grain-size distribution, deduced by fitting the extinction curve (Section 3.6.1).

A parameter relevant to calculations of interstellar extinction that must be reconciled with empirical evidence from stardust is the upper bound of the grain radius, $a_{\rm max}$. In early work, fixed upper bounds of $a_{\rm max} \sim 0.2$ (silicates) and ~ 1 μm (graphite) were typical of models generating acceptable fits to the diffuse-ISM extinction curve using the power-law form for the distribution function (Equation (3.30)). However, a smooth decline in $n(a)$ with increasing a is more physically plausible, and we have proof from studies of meteorites that some of the dust grains injected into the ISM from evolved stars are much larger than these limits (Section 8.3.4). A functional form that accommodates this is

$$n(a) \propto a^{-3.5} \exp(-a/a_0) \tag{8.8}$$

where the parameter a_0 specifies a characteristic size, such that power-law behavior is retained for $a \ll a_0$, and there is no specific cut-off for $a \gg a_0$. The observed properties of both C-rich and O-rich mass-losing red giants are consistent with a size distribution of this form with $0.10 \lesssim a_0 \lesssim 0.15$ μm (Jura 1994, 1996), similar to values obtained from fits to interstellar extinction (Kim et al. 1994).

8.4.3 Dust-to-Gas Ratios

In principle, the mass ratio of dust to gas ($Z_{\rm d}$) in a circumstellar outflow may be determined simply by comparing independent measures of the mass loss rates for dust and for gas. The mass of dust may be estimated from observations of far-infrared continuum emission, that of gas from millimeter-wave CO line emission (see Knapp 1985 and Olofsson et al. 1993 for technical details). Results are subject to substantial uncertainty because relevant parameters are either unknown or not tightly constrained. The terminal speed of the dust cannot be measured directly and is therefore assumed to be tied to that of the gas, and the reliability of this assumption is likely to depend on the evolutionary status of the star and its mass loss rate. Also, because the principal gaseous component is not generally accessible to direct measurement, the mass of gas is subject to uncertainties in the CO/H ratio. See the review by Höfner & Olofsson (2018) for further discussion and caveats.

Another important variable is the initial metallicity of the star. A general increase in Z_d with metallicity is expected, and this is confirmed by comparisons of data for AGB stars in the Milky Way and the Magellanic Clouds (van Loon 2000). Metallicity is fundamental to oxygen-rich dust production because elemental Mg, Si, and Fe are not synthesized on the AGB. Thus, an M-type star must rely on its initial endowment of these elements—i.e., on its metallicity—to make silicates and oxides, whereas a carbon star generates its own primary condensate. However, the dominant factor determining Z_d in carbon stars is the C/O ratio (Eriksson et al. 2014), and this depends on both initial metallicity and evolutionary status.

Observational estimates of dust-to-gas ratios in AGB outflows generally fall within the range $10^{-3} \lesssim Z_d \lesssim 10^{-2}$ (e.g., Knapp 1985; Olofsson et al. 1993; Justtanont et al. 1994; Ramstedt et al. 2008). Results for planetary nebulae are typically of the same order (Stasinska & Szczerba 1999). The upper end of this range extends to values that are comparable with results for the diffuse ISM (Sections 3.2.6 and 7.2.2) but higher than predictions based on models for dust production in either C-type and M-type stars ($Z_d \lesssim 3 \times 10^{-3}$; Eriksson et al. 2014; Bladh et al. 2019). Z_d falls well below theoretical upper limits for a given metallicity because significant proportions of elemental C and O remain in the gas.

8.4.4 Production and Destruction Timescales

The contribution of stardust to the overall dust content of a galaxy may be assessed by comparing the rate of production, summed over all relevant sources, with the rate of destruction in the interstellar medium. This is a complex task that depends on many parameters, some of which are not well constrained. The primary stellar sources are considered to be AGB stars and core-collapse supernovae, which correspond to the most productive phases of dust formation for intermediate and high mass stars, as previously discussed. Other types of star and other evolutionary phases are known to make dust, but their net contributions are considered to be negligible because their populations and/or production rates are relatively small. For example, the contribution from red supergiants is estimated to be no more than a few percent of that from AGB stars (Massey et al. 2005). Therefore, only AGB stars and core-collapse SNe are considered here.

The overall rate at which stardust is injected into the ISM of a galaxy may be stated as

$$[\dot{M}_G]_d = \sum_{\mathrm{AGB}} \dot{M}_d + \sum_{\mathrm{SNe}} \dot{M}_d \tag{8.9}$$

where $\dot{M}_d = \dot{M}\, Z_d$ is the mean mass loss rate for dust flowing from each individual star. The summation for supernovae may be approximated by $f\,\Delta m_d$, where f is the mean frequency of core-collapse events in the galaxy ($\sim$ 2 per century for those similar to the Milky Way), and Δm_d is the average dust mass injected into the ISM per event. The ideal way to evaluate the total for AGB stars would be to detect the entire population, estimate the dust-injection rate of each star individually, and sum the results. A complete census of dust-forming stars in the Milky Way is not possible because of source confusion and high extinction in the galactic disk, but can be attempted for nearby dwarf galaxies such as the Magellanic Clouds (e.g., Schneider et al. 2014). A more feasible approach is to combine estimates of the space densities of dust-forming stars with average values of $\dot{M}_d$ to evaluate the injection rate in a specific galactic region such as the solar neighborhood (Gehrz 1989).

Intricate models have been developed that link dust production to the stellar initial mass function (IMF), star formation history and evolving metallicity of a galaxy, with a view to understanding the significance of stardust not only in the Milky Way and its near neighbors but also in galaxies at high redshift. Examples from an extensive literature include Dwek (1998), Ferrarotti & Gail (2006), Zhukovska et al. (2008), Zhukovska & Henning (2013), Ventura et al. (2014), McKinnon et al. (2016, 2017), Ginolfi et al. (2018), and Nanni et al. (2020). The IMF specifies the relative numbers of stars born within a given mass range: for stars relevant to this discussion ($M_* \gtrsim 0.8$ M$_\odot$), the IMF varies as $M_*^{-\alpha}$ where $\alpha \approx 2.35$ (Salpeter 1955). The timescale for a star to reach its epoch of optimal dust production is of the same order as its main-sequence lifetime ($\sim 10^{10} M_*^{-2.5}$ years with M_* in solar units), and this ranges from a few million years for massive supernova progenitors to times approaching the age of the universe for the least massive stars considered. Thus, a period of intense star formation activity instigates a rapid onset of dust production by supernovae and the most massive AGB stars, consistent with observational evidence for dust in high-redshift galaxies. This is followed by sustained production on longer timescales as less massive stars reach the AGB. Dust yields are closely linked to metallicity, both spatially within a galaxy such as the Milky Way and from one galaxy to another. As carbon stars have relatively low mass ($\sim$ 2 M$_\odot$), C-rich grain production lags behind O-rich grain production if AGB stars are the primary source.

An important metric constrained by the models is the timescale for dust production in stars relative to the existing dust content of a galaxy. A calculation for the solar region of the Milky Way is informative. Consider a column of cross-sectional area A_G with its axis perpendicular to the galactic disk at the solar location. The stardust production rate per unit cross-section is then $[\dot{M}_G]_d/A_G$, summed for the two source types (Equation (8.9)). Taking their space density, mass loss rates, and dust-to-gas ratios into account, an estimate of the total from AGB stars is

$$\left[\frac{[\dot{M}_G]_d}{A_G}\right]_{AGB} \sim 8 \times 10^{-6}\ M_\odot\ kpc^{-2} yr^{-1} \tag{8.10}$$

(Jones 1997; Tielens 2012), to which O-rich and C-rich sources contribute approximately in the ratio 3:1. The global average rate for dust production in supernovae is

$$\left[\frac{[\dot{M}_G]_d}{A_G}\right]_{SNe} = \frac{f \Delta m_d}{\pi R_G^2} \lesssim 14 \times 10^{-6}\ M_\odot\ kpc^{-2} yr^{-1} \tag{8.11}$$

where $R_G \approx 15$ kpc is the radius of the (stellar) galactic disk, and the expected frequency is $f \approx 0.02$ yr^{-1}. By far the greatest source of uncertainty in the estimate for SNe is the mass of dust produced per event: the above calculation assumes $\Delta m_{\rm d} \lesssim 0.5\ {\rm M}_{\odot}$, consistent with relatively high estimates for SN 1987A and Cassiopeia A (Section 8.2.1) but allowing for the possibility that typical values may be significantly lower (Tielens 2012; Bocchio et al. 2016; Ginolfi et al. 2018). Context for these numbers is provided by an estimate of the mass of dust currently present in the ISM, enabling the timescale required to produce an equivalent mass to be calculated. Combining the observed column mass density of neutral (diffuse) gas in the solar neighborhood ($\sim 8\ {\rm M}_{\odot}\ {\rm pc}^{-2}$; Gilmore & Zeilik 2000) with a mean dust-to-gas ratio of $Z_{\rm d} \sim 5 \times 10^{-3}$ for the diffuse ISM (Table 7.2, assuming intermediate depletion), the column mass density of dust is $[M_{\rm G}]_{\rm d}/A_{\rm G} \sim 4 \times 10^{4}\ {\rm M}_{\odot}\ {\rm kpc}^{-2}$. The resultant timescale is then the ratio of this estimate and the summed production rates (Equations (8.10) and (8.11)):

$$\tau_{\rm sd} = \frac{[M_{\rm G}]_{\rm d}}{[\dot{M}_{\rm G}]_{\rm d}} \gtrsim 2\ {\rm Gyr}. \tag{8.12}$$

On reaching the ISM, stardust grains are "subject to a variety of indignities" (Seab 1988), of which sputtering in supernova-generated shock waves (Section 9.1.2) is the most severe. This occurs mainly in low-density phases of the ISM on timescales that may be estimated with reference to physical models describing the process. A global mean destruction rate is determined for an effective supernova frequency, adjusted to allow for the fact that SNe tend to occur in associations rather than in isolation (the same grain cannot be destroyed multiple times!). The corresponding timescale is then estimated in a manner analogous to the stardust calculation above (total mass divided by destruction rate), in effect the timescale ($\tau_{\rm sh}$) for shocks to destroy all the dust in a galaxy if no new dust were being produced.

The results have created a dilemma. It was concluded in early work that $\tau_{\rm sh} \lesssim 0.5$ Gyr for both silicate and carbon dust in our Galaxy (see Jones & Nuth 2011 for a review). If so, interstellar dust is destroyed by shocks far more rapidly than it is replenished by stellar mass loss ($\tau_{\rm sh} \ll \tau_{\rm sd}$). It is then challenging to account for the origin of the dust currently observed in the ISM, and it is also difficult to understand how large stardust grains, some of which predate the birth of the solar system by a billion years (Section 8.3), could have survived long enough to find their way into our meteorite collection (Slavin et al. 2004). More recent models, based on potentially more realistic assumptions, may have mitigated this discrepancy somewhat, suggesting that $\tau_{\rm sh} \sim \tau_{\rm sd}$ is within the range of uncertainty (Slavin et al. 2015; Priestley et al. 2020).

The proportion of all interstellar dust originating in stellar outflows remains an open question, but there is general agreement that a substantial fraction must be formed by interstellar rather than circumstellar processes. For example, limits on the abundance of SiC (Section 5.2.2) suggest that carbon stars may not be the dominant source of C-rich dust (Chen et al. 2022). Because densities are low, even in "dense" clouds, formation of solids in the ISM must proceed by gradual deposition of matter onto the surfaces of existing grains rather than by nucleation of new particles. We know from depletion data (Chapter 7) that gas and dust exchange material in the diffuse ISM at a density-dependent rate, and we know from studies of spectral features (Chapter 5) that grains acquire icy mantles in dense clouds. Feasible routes to carbonaceous dust production might thus include direct deposition of C atoms onto refractory dust, or photolytic processing of ice mantles.

An interstellar route to *silicate* dust production is harder to envisage. Abundance limits on Si and Mg require that these elements remain mostly sequestered in amorphous silicates to explain the strengths and profiles shapes of the infrared spectral features (Sections 5.2.1 and 7.3.2). Si, Mg, and Fe atoms released to the gas by destruction of stardust in shocks may subsequently reoxidize and reattach to grain surfaces, but it is unclear how they could assemble into silicate compounds consistent with the observed properties of the dust under interstellar conditions. Recombination in an annealed ice matrix has been proposed, supported by laboratory simulations (Rouillé et al. 2020), but the efficiency of this process is not yet known. A related issue is the need for the silicates to remain largely distinct from the carbon population to explain differences in their response to alignment (Section 4.2.7); this seems improbable if most of the dust is reprocessed in the ISM, as the Si-rich and C-rich components would tend to become blended.

To summarize, it seems hard to avoid the conclusion that circumstellar environments around O-rich stars are by far the most efficient and effective sites for silicate dust production (Section 8.1.3). It may be relevant to recall that such environments arise during the earliest as well as the latest stages of stellar evolution (the phase diagram in Figure 8.4 is applicable to both situations for solar metallicity). It was suggested many years ago that star formation might be a significant *source* of interstellar dust (Herbig 1970; Burke & Silk 1976). Silicates readily condense and grow in a protoplanetary disk, thereby forming minerals that we now find in interplanetary dust, in stony meteorites, and in the ground beneath our feet. As a young star approaches the main sequence, it generates a wind that disperses solid particles that failed to accrete into larger bodies, driven mainly by radiation pressure. Of course, we should not necessarily expect the star formation process to be a net producer of dust: it may consume at least as much as it creates, given that the condensible elements are likely to be substantially depleted into solids prior to collapse of the birth cloud. But it can alter the *quality* of the dust that gets recycled to the ISM, for example by vaporizing coatings of randomly accreted Si, Mg, and Fe atoms, and recondensing them into real silicates. Although largely ignored in subsequent decades, this is an idea worth reconsidering.

References

Abia, C., de Laverny, P., Cristallo, S., Kordopatis, G., & Straniero, O. 2020, A&A, 633, A135

Agúndez, M., Martínez, J. I., de Andres, P. L., Cernicharo, J., & Martín-Gago, J. A. 2020, A&A, 637, A59

Anders, E., & Zinner, E. 1993, Metic, 28, 490

Arendt, R. G., Dwek, E., Bouchet, P., et al. 2016, AJ, 151, 62

Barlow, M. J., & Silk, J. 1977, ApJ, 215, 800

Barshay, S. S., & Lewis, J. S. 1976, ARA&A, 14, 81

Begemann, B., Dorschner, J., Henning, T., et al. 1997, ApJ, 476, 199

Bernatowicz, T. J., Amari, S., Zinner, E. K., & Lewis, R. S. 1991, ApJ, 373, L73
Bernatowicz, T. J., Cowsik, R., Gibbons, P. C., et al. 1996, ApJ, 472, 760
Bernatowicz, T. J., Akande, O. W., Croat, T. K., & Cowsik, R. 2005, ApJ, 631, 988
Biermann, P., & Harwit, M. 1980, ApJ, 241, L105
Bladh, S., Liljegren, S., Höfner, S., Aringer, B., & Marigo, P. 2019, A&A, 626, A100
Bocchio, M., Marassi, S., Schneider, R., et al. 2016, A&A, 587, A157
Bode, M. F. 1988, in Dust in the Universe, ed. M. E. Bailey, & D. A. Williams (Cambridge: Cambridge Univ. Press), 73
Bode, M. F. 2010, AN, 331, 160
Boogert, A. C. A., Gerakines, P. A., & Whittet, D. C. B. 2015, ARA&A, 53, 541
Bouchet, P., Dwek, E., Danziger, J., et al. 2006, ApJ, 650, 212
Bouvier, A., & Wadhwa, M. 2010, NatGe, 3, 637
Burke, J. R., & Silk, J. 1976, ApJ, 210, 341
Buss, R. H., & Snow, T. P. 1988, ApJ, 335, 331
Buss, R. H., Lamers, H. J. G. L. M., & Snow, T. P. 1989, ApJ, 347, 977
Cameron, A. G. W. 1973, in IAU Symp. 52, Interstellar Dust & Related Topics, ed. J. M. Greenberg, & H. C. van de Hulst (Dordrecht: Reidel), 545
Chen, T., Xiao, C. Y., Li, A., & Zhou, C. T. 2022, MNRAS, 509, 5231
Chiar, J. E., Tielens, A. G. G. M., Adamson, A. J., & Ricca, A. 2013, ApJ, 770, 78
Cigan, P., Matsuura, M., Gomez, H. L., et al. 2019, ApJ, 886, 51
Clayton, D. D. 2013, ApJ, 762, 5
Croat, T. K., Bernatowicz, T., Amari, S., Messenger, S., & Stadermann, F. J. 2003, GeCoA, 67, 4705
Croat, T. K., Stadermann, F. J., & Bernatowicz, T. J. 2005, ApJ, 631, 976
Cox, N. L. J., Kerschbaum, F., van Marle, A.-J., et al. 2012, A&A, 537, A35
Dai, Z. R., Bradley, J. P., Joswiak, D. J., et al. 2002, Natur, 418, 157
Davis, A. M. 2011, PNAS, 108, 19142
De Looze, I., Barlow, M. J., Swinyard, B. M., et al. 2017, MNRAS, 465, 3309
De Looze, I., Barlow, M. J., Bandiera, R., et al. 2019, MNRAS, 488, 164
DePew, K., Speck, A., & Dijkstra, C. 2006, ApJ, 640, 971
Donn, B., & Nuth, J. A. 1985, ApJ, 288, 187
Dorschner, J., Begemann, B., Henning, T., Jäger, C., & Mutschke, H. 1995, A&A, 300, 503
Douvion, T., Lagage, P. O., Cesarcky, C. J., & Dwek, E. 2001, A&A, 373, 281
Draine, B. T. 1984, ApJ, 277, L71
Draine, B. T. 2016, ApJ, 831, 109
Drilling, J. S., Hecht, J. H., Clayton, G. C., et al. 1997, ApJ, 476, 865
Dupree, A. K. 1986, ARA&A, 24, 377
Dwek, E. 1998, ApJ, 501, 643
Dwek, E. 2016, ApJ, 825, 136
Dwek, E., & Arendt, R. G. 2015, ApJ, 810, 75
Dwek, E., Sarangi, A., & Arendt, R. G. 2019, ApJ, 871, L33
Eriksson, K., Nowotny, W., Höfner, S., Aringer, B., & Wachter, A. 2014, A&A, 566, A95
Fabian, D., Jäger, C., Henning, T., Dorschner, J., & Mutschke, H. 2000, A&A, 364, 282
Ferrarotti, A. S., & Gail, H.-P. 2006, A&A, 447, 553
Frenklach, M., & Feigelson, E. D. 1989, ApJ, 341, 372
Gail, H.-P., & Sedlmayr, E. 1986, A&A, 166, 225

Gail, H.-P., & Sedlmayr, E. 1987, in Physical Processes in Interstellar Clouds, ed. G. E. Morfill, & M. Scholer (Dordrecht: Reidel), 275
Gail, H.-P., Wetzel, S., Pucci, A., & Tamanai, A. 2013, A&A, 555, A119
Gail, H.-P., Scholz, M., & Pucci, A. 2016, A&A, 591, A17
Gehrz, R. D. 1989, in IAU Symp. 135, Interstellar Dust, ed. L. J. Allamandola, & A. G. G. M. Tielens (Dordrecht: Kluwer), 445
Gehrz, R. D., Truran, J. W., Williams, R. E., & Starrfield, S. 1998, PASP, 110, 3
Gilmore, G. F., & Zeilik, M. 2000, in Allen's Astrophysical Quantities, ed. A. N. Cox (4th ed.; New York: Springer), 471
Ginolfi, M., Graziani, L., Schneider, R., et al. 2018, MNRAS, 473, 4538
Golriz, S. S., Blommaert, J. A. D. L., Vanhollebeke, E., et al. 2014, MNRAS, 443, 3402
Groenewegen, M. A. T. 1997, A&A, 317, 503
Groenewegen, M. A. T. 2012, A&A, 543, A36
Groenewegen, M. A. T., & Sloan, G. C. 2018, A&A, 609, A114
Groenewegen, M. A. T., Whitelock, P. A., Smith, C. H., & Kerschbaum, F. 1998, MNRAS, 293, 18
Guélin, M., Patel, N. A., Bremer, M., et al. 2018, A&A, 610, A4
Guerrero, M. A., & Ortiz, R. 2020, MNRAS, 491, 680
Hecht, J. H. 1986, ApJ, 305, 817
Hecht, J. H. 1991, ApJ, 367, 635
Hecht, J. H., Holm, A. V., Donn, B., & Wu, C. C. 1984, ApJ, 280, 228
Heck, P. R., Greer, J., Kööp, L., et al. 2020, PNAS, 117, 1884
Henning, Th. 2010, ARA&A, 48, 21
Henning, Th., Begemann, B., Mutschke, H., & Dorschner, J. 1995, A&AS, 112, 143
Herbig, G. H. 1970, Mémoires de la Société Royale des Sciences de Liège, 19, 13
Höfner, S., & Olofsson, H. 2018, A&ARv, 26, 1
Hoppe, P. 2011, in 11th Symp. on Nuclei in the Cosmos, ed. K. Blaum, N. Christlieb, & G. Martinez-Pinedo, https://pos.sissa.it/100/021/
Hrivnak, B. J., Volk, K., & Kwok, S. 2000, ApJ, 535, 275
Iben, I., & Renzini, A. 1983, ARA&A, 21, 271
Jäger, C., Molster, F. J., Dorschner, J., et al. 1998, A&A, 339, 904
Johnson, J. A. 2019, Sci, 363, 474
Jones, A. P. 1997, in ASP Conf. Ser. 122, From Stardust to Planetesimals, ed. Y. J. Pendleton, & A. G. G. M. Tielens (San Francisco, CA: ASP), 97
Jones, A. P., & Nuth, J. A. 2011, A&A, 530, 44
Jones, O. C., Kemper, F., Sargent, B. A., et al. 2012, MNRAS, 427, 3209
Jura, M. 1994, ApJ, 434, 713
Jura, M. 1996, ApJ, 472, 806
Jura, M. 1999, ApJ, 515, 706
Jura, M., & Kleinmann, S. G. 1990, ApJS, 73, 769
Justtanont, K., Skinner, C. J., & Tielens, A. G. G. M. 1994, ApJ, 435, 852
Justtanont, K., Skinner, C. J., Tielens, A. G. G. M., Meixner, M., & Baas, F. 1996, ApJ, 456, 337
Kastner, J. H., Zijlstra, A., Balick, B., & Sahai, R. 2018, in ASP Conf. Ser. 517, Science with a Next Generation Very Large Array, ed. E. Murphy (San Francisco, CA: ASP), 395
Kastner, J. H., Moraga, P., Balick, B., Bublitz, J., Montez, R., Frank, A., & Blackman, E. 2022, ApJ, 927, 100
Kim, S. H., Martin, P. G., & Hendry, P. D. 1994, ApJ, 422, 164

Kimura, Y., Tanaka, K. K., Nozawa, T., Takeuchi, S., & Inatomi, Y. 2017, SciA, 3, e1601992
Knapp, G. R. 1985, ApJ, 293, 273
Kotak, R., Meikle, W. P. S., Farrah, D., et al. 2009, ApJ, 704, 306
Kwitter, K. B., & Henry, R. B. C. 2022, PASP, 134, 022001
Kwon, Y.-J., & Suh, K.-W. 2014, JKAS, 47, 123
Lewis, J. B., Floss, C., Isheim, D., et al. 2020, M&PS, 55, 1382
Lewis, R. S., Tang, M., Wacker, J. F., Anders, E., & Steel, E. 1987, Natur, 326, 160
Little-Marenin, I. R. 1986, ApJ, 307, L15
Liu, J., Jiang, B. W., Li, A., & Gao, J. 2017, MNRAS, 466, 1963
Liu, N., Nittler, L. R., Alexander, C. M. O'D., & Wang, J. 2018, SciA, 4, 1054
Lodders, K. 2003, ApJ, 591, 1220
Lodders, K., & Fegley, B. 1995, Metic, 30, 661
Malek, S. E., & Cami, J. 2014, ApJ, 780, 41
Martin, P. G. 1978, Cosmic Dust, its Impact on Astronomy (Oxford: Oxford Univ. Press)
Martin, P. G., & Rogers, C. 1987, ApJ, 322, 374
Massey, P., Plez, B., Levesque, E. M., et al. 2005, ApJ, 634, 1286
Matsuura, M., Dwek, E., Meixner, M., et al. 2011, Sci, 333, 1258
McKinnon, R., Torrey, P., & Vogelsberger, M. 2016, MNRAS, 457, 3775
McKinnon, R., Torrey, P., Vogelsberger, M., Hayward, C., & Marinacci, F. 2017, MNRAS, 468, 1505
Messenger, S., Keller, L. P., Stadermann, F. J., Walker, R. M., & Zinner, E. 2003, Sci, 300, 105
Messenger, S. J., Speck, A., & Volk, K. 2013, ApJ, 764, 142
Micelotta, E. R., Matsuura, M., & Sarangi, A. 2018, SSRv, 214, 53
Mishra, A., Li, A., & Jiang, B. W. 2015, ApJ, 802, 39
Mishra, A., Li, A., & Jiang, B. W. 2016, ApJ, 825, 68
Montiel, E. J., Clayton, G. C., Sugerman, B. E. K., et al. 2018, AJ, 156, 148
Muci, A. M., Blanco, A., Fonti, S., & Orofino, V. 1994, ApJ, 436, 831
Mutschke, H., Begemann, B., Dorschner, J., et al. 1998, A&A, 333, 188
Mutschke, H., Andersen, A. C., Clément, D., Henning, Th., & Peiter, G. 1999, A&A, 345, 187
Nanni, A., Bressan, A., Marigo, P., & Girardi, L. 2013, MNRAS, 434, 2390
Nanni, A., Burgarella, D., Theulé, P., Côté, B., & Hirashita, H. 2020, A&A, 641, A168
Nguyen, A. N., Keller, L. P., & Messenger, S. 2016, ApJ, 818, 51
Nittler, L. R., & Ciesla, F. 2016, ARA&A, 54, 53
Nuth, J. A. 1996, in The Cosmic Dust Connection, ed. J. M. Greenberg (Dordrecht: Kluwer), 205
Olofsson, H., Eriksson, K., Gustafsson, B., & Carltröm, U. 1993, ApJS, 87, 267
Onaka, T., de Jong, T., & Willems, F. J. 1989, A&A, 218, 169
Paquette, J. A., & Nuth, J. A. 2011, ApJ, 737, L6
Posch, T., Kerschbaum, F., Mutschke, H., Dorschner, J., & Jäger, C. 2002, A&A, 393, L7
Priestley, F. D., Chawner, H., Matsuura, M., et al. 2020, MNRAS, 500, 2543
Ramstedt, S., Schöier, F. L., Olofsson, H., & Lundgren, A. A. 2008, A&A, 487, 645
Reimers, D. 1975, Memoires of the Societe Royale des Sciences de Liege, 8, 369
Rietmeijer, F. J. M., Nuth, J. A., & Karner, J. M. 1999, ApJ, 527, 395
Rouillé, G., Jäger, C., & Henning, T. 2020, ApJ, 892, 96
Salpeter, E. E. 1955, ApJ, 121, 161
Salpeter, E. E. 1974, ApJ, 193, 579
Salpeter, E. E. 1977, ARA&A, 15, 267

Santoro, G., Martinez, L., Lauwaet, K., et al. 2020, ApJ, 895, 97
Sarangi, A., & Cherchneff, I. 2013, ApJ, 776, 107
Sarangi, A., Matsuura, M., & Micelotta, E. R. 2018, SSRv, 214, 63
Schneider, R., Valiante, R., Ventura, P., et al. 2014, MNRAS, 442, 1440
Seab, C. G. 1988, in Dust in the Universe, ed. M. E. Bailey, & D. A. Williams (Cambridge: Cambridge Univ. Press), 303
Seab, C. G., & Snow, T. P. 1989, ApJ, 347, 479
Shepard, L. M., & Speck, A. K. 2021, MNRAS, 506, 4750
Slavin, J. D., Jones, A. P., & Tielens, A. G. G. M. 2004, ApJ, 614, 796
Slavin, J. D., Dwek, E., & Jones, A. P. 2015, ApJ, 803, 7
Sloan, G. C., & Price, S. D. 1998, ApJS, 119, 141
Sloan, G. C., Kraemer, K. E., Goebel, J. H., & Price, S. D. 2003, ApJ, 594, 483
Smith, R. G., Sellgren, K., & Tokunaga, A. T. 1988, ApJ, 334, 209
Snow, T. P., Buss, R. H., Gilra, D. P., & Swings, J. P. 1987, ApJ, 321, 921
Speck, A. K., Whittington, A. G., & Tartar, J. B. 2008, ApJ, 687, L91
Speck, A. K., Corman, A. B., Wakeman, K., Wheeler, C. H., & Thompson, G. 2009, ApJ, 691, 1202
Stasinska, G., & Szczerba, R. 1999, A&A, 352, 297
Stephan, T., Bose, M., Boujibar, A., et al. 2020, LPSC, 51, 2140
Thompson, G. D., Corman, A. B., Speck, A. K., & Dijkstra, C. 2006, ApJ, 652, 1554
Tielens, A. G. G. M. 2008, ARA&A, 46, 289
Tielens, A. G. G. M. 2012, in IAU Symp. 284, The Spectral Energy Distribution of Galaxies, ed. R. J. Tuffs, & C. C. Popescu (Cambridge: Cambridge Univ. Press), 72
van Loon, J. Th. 2000, A&A, 354, 125
Ventura, P., Dell'Agli, F., Schneider, R., et al. 2014, MNRAS, 439, 977
Waters, L. B. F. M., Waelkens, C., van Winckel, H., et al. 1998, Natur, 391, 868
Whitney, B. A., Balm, S. P., & Clayton, G. C. 1993, in ASP Conf. Ser. 45, Luminous High-Latitude Stars, ed. D. D. Sasselov (San Francisco, CA: ASP), 115
Willems, F. J., & de Jong, T. 1986, A&A, 309, L39
Williams, B. J., Borkowski, K. J., Reynolds, S. P., et al. 2012, ApJ, 755, 3
Wooden, D. H., Rank, D. M., Bregman, J. D., et al. 1993, ApJS, 88, 477
Zhang, Y. 2020, ApJ, 898, 151
Zhukovska, S., & Henning, T. 2013, A&A, 555, A99
Zhukovska, S., Gail, H.-P., & Trieloff, M. 2008, A&A, 479, 453
Zijlstra, A. A., Loup, C., Waters, L. B. F. M., & de Jong, T. 1992, A&A, 265, L5
Zinner, E. 1998, AREPS, 26, 147
Zinner, E. K., Moynier, F., & Stroud, R. M. 2011, PNAS, 108, 19135

AAS | IOP Astronomy

Dust in the Galactic Environment (Third Edition)

D C B Whittet

Chapter 9

The Life Cycle of Dust—II. From the ISM to Protostars and Planets

This chapter tracks the life cycle of dust from its injection into the interstellar medium (ISM) by evolved stars (Chapter 8) to its amalgamation into newly-formed stars and planetary systems in molecular clouds, illustrated schematically in Figure 9.1. A vast range of physical conditions are encountered (Section 1.3), in which either constructive or destructive processes may occur. Intercloud regions of the ISM are permeated by energetic photons and shocks capable of destroying even the most hardy particles. Within an interstellar cloud, however, the grains are temporarily shielded from such extremes. Here they may undergo growth by coagulation and by deposition of surface coatings, at rates that increase with density. The exchange of matter between gas and dust regulates the chemical evolution of the cloud. The surfaces of the grains are not merely repositories for gas-phase atoms and molecules but also catalysts for production, opening new pathways to molecule formation that are not always possible in the gas (see Section 1.4.2 for an introductory overview). In the coldest, densest clouds, the grains accumulate surface volatiles as the result of both frosting from the gas and in-situ production of ices. Later, as the cloud evolves, the onset of star formation exposes the grains once more to stellar radiation and winds. This input of energy may drive chemical reactions toward greater complexity, and as the ices sublimate they contribute to the molecular content of the gas as well. By studying these processes we hope to gain insight into the evolution of matter that subsequently condenses to form planets around newly-born stars.

The sections of this chapter follow the sequence from diffuse environments to molecular clouds and star formation regions. Figure 9.1 represents the big-picture view, and various physical and chemical processes discussed in the text that may occur on or in the surface layers of an individual grain are illustrated schematically in Figure 9.2.

doi:10.1088/2514-3433/ac7204ch9

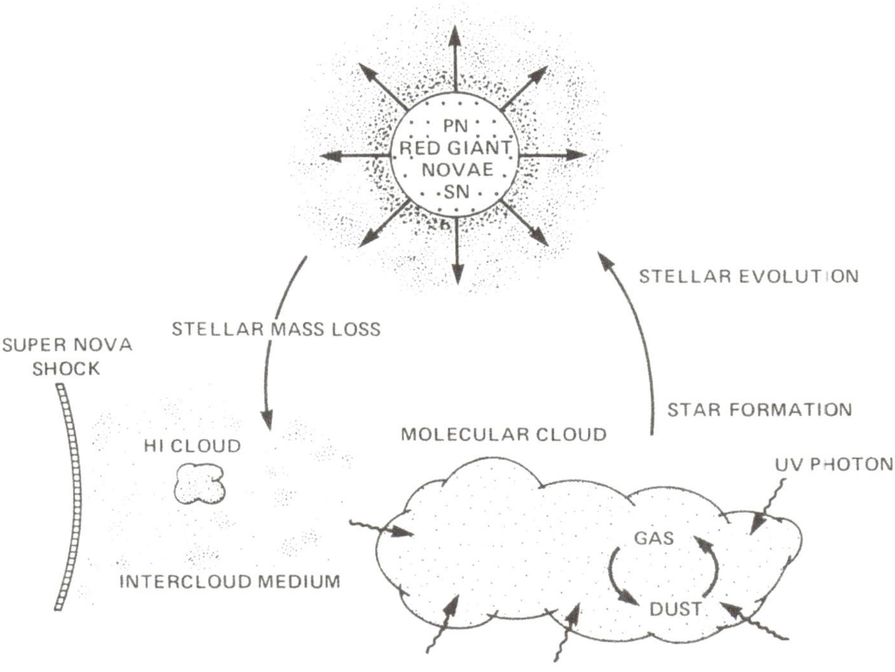

Figure 9.1. Schematic representation of the life cycle of cosmic dust. Grains condensing in the outflows of evolved stars (Chapter 8) are ejected into low-density phases of the ISM, where they are exposed to UV photons and supernova shocks. Gas and dust exchange material according to ambient conditions in diffuse (H I) and molecular clouds. Subsequent star formation and stellar evolution completes the cycle. Image credit: reproduced from Tielens & Allamandola (1987a) by permission of Kluwer Academic Publishers.

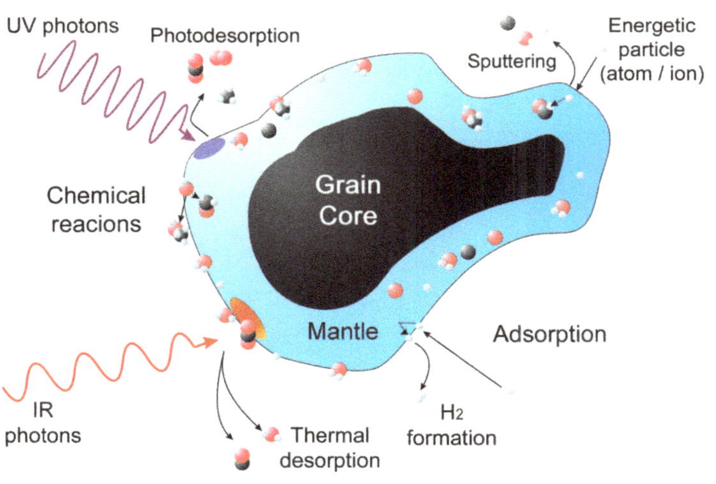

Figure 9.2. Schematic representation of surface processes. A generic core-mantle grain is illustrated, but such processes are similarly relevant to uncoated grains and aggregates. The grain exchanges matter with the gas by adsorption, desorption and sputtering. Small spheres represent individual atoms or ions: H (small light gray), C (dark gray), and O (light red).

9.1 Processes in the Diffuse ISM

The generic term "diffuse ISM" is used here, as in other chapters of this text, to imply regions where hydrogen gas is predominantly non-molecular (Section 1.3.2). Environments ranging from hot intercloud and warm diffuse gas to H I clouds are included (Table 1.1). Molecules are detected in H I clouds but their fractional abundances remain low. Grains exchange elements with the gas (Figure 9.2), as implied by density-dependent variations in the observed depletions (Section 7.2), but do not retain detectable ice mantles (Section 5.3.2). In this section, we consider the constructive and destructive process that operate on the grains in such environments and thereby control their size, structure and survival.

9.1.1 Grain–Grain Collisions

Collisions between grains regulate the size distribution of the dust. They may result in fragmentation or aggregation, depending on the inherent properties and impact speeds of the colliding particles. In general, collisions impose power-law behavior on the size distribution, as discussed in Section 8.4.2 in the context of stellar winds driven by radiation pressure. Collisions in the ISM arise primarily from turbulence driven by shocks, and again a power-law form is predicted, broadly consistent with the optical properties of the dust.

The dynamical interactions of colliding particles depend on many factors, including their relative speeds and trajectories, their size, composition and structure (e.g., monolithic grains or aggregates), and the presence of any net electrostatic charge (Dominik & Tielens 1997). The propensity of a grain to break up on impact is determined by the shear strength of the material: if the shock pressure of the impact exceeds the critical pressure for fracturing, the target grain will shatter (Jones et al. 1996). Models indicate that this critical pressure corresponds to a collision velocity $v_{\rm col} \sim 1$ km s^{-1} for typical particles, a threshold commonly exceeded in low-density phases of the ISM (but not in dark clouds). Kinetic energy imparted by a high-speed impact might also vaporize grain material, but the shattering threshold is much lower than the vaporization threshold. Thus, the effect of shattering is to alter the size distribution with little or no change in the dust-to-gas ratio. Fragmentation of graphitic grains, for example, may be an important source of carbon nanoparticles and PAH clusters (Tielens 2008) while conserving the overall abundance of C in dust.

The physics of grain aggregation is a topic of great interest and intensive experimental and theoretical research because it is, in essence, the initial step toward planet formation (Dominik et al. 2007). When two electrostatically neutral grains come into contact, they experience an attractive force equal, at a minimum, to the weak van der Waals force; depending on their compositions, stronger forces such as dipole-dipole interactions and metallic bonding may come into play. The impact causes deformation at the point of contact, and the outcome of the collision (adhesion or rebound) will depend on whether the attractive forces are sufficient to overcome elastic repulsion. For a given pair of grains of masses m_1 and m_2, this depends primarily on the impact energy, with coagulation occurring below some critical value $E_{crit} = \frac{1}{2}\mu v_{col}^2$, where $\mu = m_1 m_2/(m_1 + m_2)$ is the reduced mass. Coagulation is a feasible outcome of collisions between typical interstellar grains when $v_{col} \lesssim 0.1$ km s^{-1}, i.e., in the range likely to occur within dark clouds (Jura 1980; Ormel et al. 2009). Once formed, an aggregate may grow by capture of smaller grains with enhanced efficiency because some of the impact energy can be transferred into internal degrees of freedom of the aggregate.

These results imply a general pattern of growth and fragmentation as grains cycle in and out of interstellar clouds: the particles tend to coagulate within a cloud, and shatter in lower-density (intercloud) regions, resulting in an overall size distribution of the form $n(a) \propto a^{-q}$ with $q \approx 3.3$ (Jones et al. 1996). This behavior is consistent with the known optical properties of the dust, including variations in size dependent parameters such as R_V (the ratio of total to selective extinction; Section 3.3.1) and λ_{max} (the wavelength of maximum linear polarization; Section 4.2.6). The observations show a clear trend toward larger grains in denser environments, in agreement with models for growth by coagulation (Jura 1980; Wurm & Schnaiter 2002). This mechanism also explains an apparent anomaly in the observed ratio of hydrogen column density to extinction (N_H/A_V) in the dark cloud toward ρ Oph, which exceeds the mean value for the diffuse ISM by a factor of 2: subgrains in an aggregate tend to "hide behind" one another, reducing the opacity in the line of sight compared with an equivalent mass of independent grains (Jura 1980; Kim & Martin 1996).

The structure of aggregates produced by the coagulation process depends on the size and structure of the incident particles (Dorschner & Henning 1995; see also Dominik & Tielens 1997 for detailed discussion of the relevant physics). Compact, quasi-spherical aggregates tend to be favored, especially for particles toward the lower and upper ranges of the size distribution. As the smallest grains have the largest surface to volume energy ratios, they tend to form compact clusters that minimize their surface area. For the largest grains, energy transfer during a collision becomes comparable with internal binding energy, causing loose aggregates to shatter and thereby favoring the most compact structures by natural selection. In the size range between these extremes, loose fractal aggregates are more likely to occur. The growth of clusters may proceed through sequential addition of individual subgrains to the cluster (particle-cluster aggregation) or through coagulation of clusters (cluster-cluster aggregation). Some examples from computer simulations are shown in Figure 9.3.

9.1.2 Sputtering

Sputtering occurs if a solid is impacted by gas-phase atoms or ions that carry sufficient energy to eject atoms from the lattice, resulting in surface erosion (Figure 9.2). Sputtering in supernova-generated shock waves is considered to be the primary destruction mechanism for refractory dust in the ISM. A shock wave is an irreversible, supersonic, pressure-driven disturbance that leads to impulsive heating of the shocked gas. The shock slows gradually as it expands into the ambient medium, and may propagate through large volumes of low-density gas before its speed falls below the local sound speed (see Draine & McKee 1993 for a

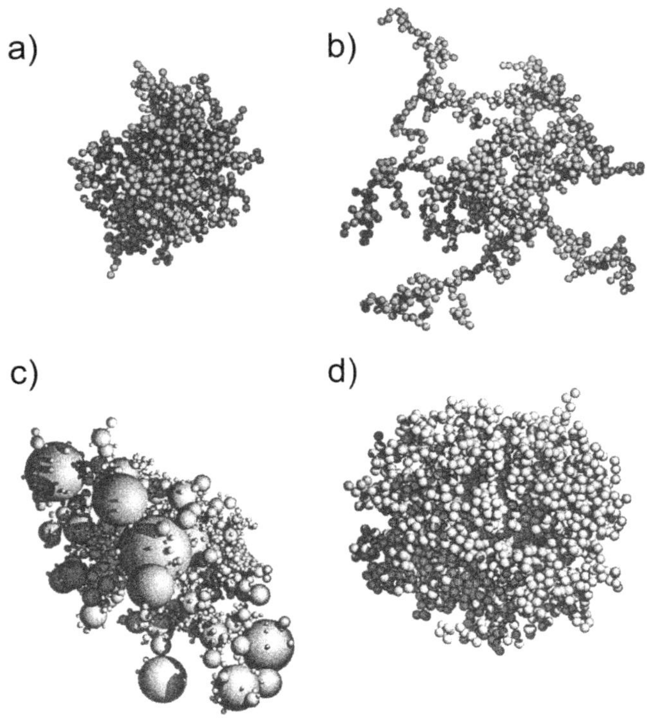

Figure 9.3. Computer simulations showing examples of particles formed by grain–grain coagulation. (a) A compact particle produced by particle-cluster aggregation of 1024 identical small spheres. (b) A loose fractal particle produced by cluster-cluster aggregation of 1024 identical small spheres. (c) A composite particle produced by particle-cluster aggregation of 2001 spheres with sizes following a power-law distribution with exponent −3.15. (d) A compact particle produced by particle-cluster aggregation of 2000 identical small spheres onto one larger, spherical seed particle. Image credit: reproduced with permission from Astronomy and Astrophysics Review (Dorschner & Henning 1995).

review of the theory of interstellar shocks). Sputtering of a dust grain may be thermal, arising from the kinetic motion of atoms or ions in heated gas, or non-thermal in cases where gas and dust are in rapid relative motion. The atoms within the lattice structure of a refractory solid typically have binding energies ~ 5 eV, so gas temperatures in excess of $\sim 10^5$ K are required for thermal sputtering to be effective. When a grain encounters a shock, its velocity does not change as quickly as that of the gas, resulting in differential motion. If the grain bears electric charge, it gyrates about magnetic field lines and may undergo betatron acceleration. Either of these effects may lead to non-thermal sputtering and enhanced rates of grain–grain collisions. Numerical simulations show that thermal sputtering is important at the highest shock speeds ($v > 150$ km s^{-1}), and non-thermal sputtering is dominant in the range 50–150 km s^{-1}, with similar results for both carbon and silicate grains (Jones et al. 1994). A typically encounter may remove a surface layer ~ 3 nm thick, causing total dissipation of the smallest grains (Tielens 2012). Shock-driven grain–grain collisions modify the size distribution by shattering, thereby increasing the surface area exposed to erosion.

Sputtering of refractory dust occurs most efficiently in the warm diffuse phase of the ISM (Section 1.3.1), as may be illustrated by a simple calculation. Suppose that a shock wave of velocity v_0 traveling in a medium of number density n_0 encounters a cloud of number density $n_1 \gg n_0$. The cloud is subjected to a sudden increase in ambient pressure that drives the shock into the cloud with velocity

$$v_1 = (n_0/n_1)^{\frac{1}{2}} v_0 \tag{9.1}$$

(Draine 1989). This reduction in shock speed as the square root of the relative density implies that shocks capable of destroying grains in the intercloud medium are decelerated to such an extent that they do not generally destroy grains inside clouds. Assuming $v_0 = 500$ km s^{-1}, and adopting values of $n_0 = 5 \times 10^3$ m^{-3} and $n_1 = 3 \times 10^7$ m^{-3}, typical of the intercloud medium and a diffuse H I cloud, respectively (Table 1.1), Equation (9.1) yields $v_1 = 6$ km s^{-1} (and lower values for denser clouds). However, shock speeds $\gtrsim 50$ km s^{-1} are needed to destroy refractory dust. Thus, destruction by shocks is efficient in the intercloud medium (where the grain number density is low) and generally inoperative in clouds (where the grain number density is high). It follows that ablation is most rapid in the intermediate warm phase, the densest environment in which shock speeds remain above the threshold. Note that ices, which have binding energies $\lesssim 0.5$ eV, are removed easily by low velocity shocks (as well as by photodesorption) in diffuse phases of the ISM, but are protected in dense clouds.

Sputtering of atoms from dust in the diffuse warm phase is presumably balanced by reattachment onto dust in denser regions of the ISM. This cycle is, indeed, implied by the observed pattern of gas-phase depletions for elements expected to make up the bulk of the dust mass (Sections 7.2 and 7.3). Relatively low depletion indices ($F_* \lesssim 0.5$), implying more atoms in the gas, occur in lines of sight with relatively low mean gas densities, i.e., $\langle n_H \rangle \lesssim 3 \times 10^5$ m^{-3} (see Equation (7.8)). This density level corresponds to environments in which most of the matter is in the warm diffuse phase (Spitzer 1985), i.e., to conditions where sputtering is most efficient. Conversely, larger depletions correspond to denser environments where grains are largely protected from shocks and adsorption rates are enhanced.

Further evidence for the influence of shocks on the depletion pattern arises when cloud dynamics are taken into account. High-velocity clouds (HVCs) are characterized by exceptionally large Doppler shifts of interstellar gas-phase absorption lines, measured in high-resolution UV–visual spectra of background stars: results indicate relative speeds of up to about 100 km s^{-1} (e.g., Cowie et al. 1979; Welty et al. 2002). HVCs are presumed to have been accelerated by supernova shocks or energetic stellar winds in regions of vigorous recent star formation. Notable examples are found toward the Orion OB association, in which the structure known as Barnard's Loop is attributed to a supernova event in the past 2 Myr. The observed depletions of condensible elements in HVCs tend to be systematically lower than in otherwise similar clouds, and to correlate with velocity, such that fewer atoms remain in the dust in faster-moving clouds. The data suggest that ~ 10% of the Si, Al and Fe (and essentially all of the C) that was originally locked up in dust in the pre-shock medium may have been returned to the gas phase by shocks.

As previously noted, sputtering in shocks is the primary mechanism responsible for destruction of refractory dust in the ISM. Average lifetimes are notoriously difficult to estimate, however, which is unfortunate because the results are fundamental to our understanding of grain ecology in galaxies. Numerical models yield values that differ by over an order of magnitude (ranging from 0.1 to 3 Gyr). If lifetimes are relatively short, most of the dust now present in the ISM must have formed in the ISM, because stellar outflows cannot replenish it quickly enough to explain its current mass. However, it is questionable whether interstellar processes can create grain materials that satisfy all observational constraints—silicates, in particular, seem problematic. This question remains unresolved at the time of writing. See Section 8.4.4 for more detailed discussion and references.

9.1.3 Metamorphosis

Energetic processes that fall below the threshold for widespread destruction of the dust may nevertheless induce important structural changes in both carbon and silicate grain materials. Energy is required to induce a change of state, whether it be to crystallize an initially disordered solid or to disrupt the structure of a crystal. Crystallinity affects the optical properties of the particles and is therefore constrained by the observations. Potential sources of energy include UV photons, high-speed grain–grain collisions, and cosmic rays.

Carbon is of great interest in this context because of its chemical and structural versatility. Known or suspected ingredients of the ISM range in scale from macromolecules and nanoparticles to 20 μm-sized presolar stardust grains, and in composition from amorphous soot to graphite, PAHs and diamond. The ternary diagram shown in Figure 9.4 provides a useful illustration of this versatility, mapping the various states of amorphous carbon with respect to bonding prevalence (graphitic versus tetrahedral) and degree of hydrogenation. PAH molecules are aromatic structures with sp^2 bonding and varying hydrogenation (e.g., Figure 6.6), so their loci are along the base of the triangle. PAHs are known to be produced in stellar outflows (their emission spectrum is observed in C-rich planetary nebulae) but may also be formed in the ISM, e.g., by collisional fragmentation of carbonaceous dust. The initial state of the carbonaceous dust is likely to be predominantly amorphous—a mix of randomly-connected units with linear, aromatic and tetrahedral bonding—regardless of whether it originates primarily in stellar outflows or by interstellar processes. It may then be subject to changes of state as it evolves in the ISM, which we may attempt to quantify.

Astronomical observations that enable the extant state of the dust to be characterized by placement on the ternary diagram are challenging, as they require high-quality spectra of several intrinsically-weak infrared absorption features (Section 5.2.3). The best available data yield an average for the line of sight toward

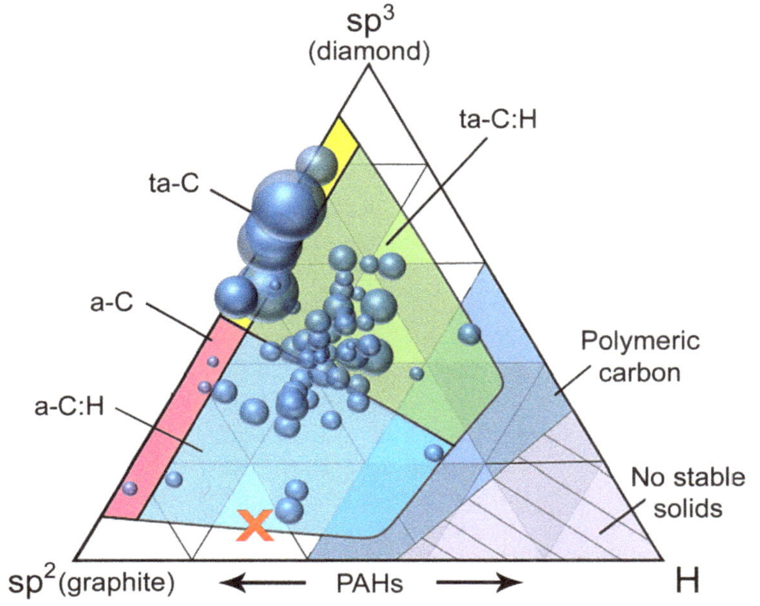

Figure 9.4. Ternary diagram mapping the nature of amorphous carbon (a-C) films (adapted from Ohtake et al. 2021). The films may be hydrogenated (a-C:H) and/or predominantly tetrahedrally bonded (ta-C). Apexes of the triangle (clockwise from lower left) correspond to pure graphite, diamond and hydrogen; each side represents relative proportions of these components. Blue circles show the distribution of laboratory-deposited samples, with sizes of the symbols scaled according to hardness. The red cross is the mean locus of interstellar dust deduced from observations of the galactic center (Chiar et al. 2013).

the center of our Galaxy (Chiar et al. 2013; red cross in Figure 9.4). This result shows that the dust is on average highly aromatic (sp^3/sp^2 ratio ~ 10%) and only modestly hydrogenated (H/C ~ 30%). The implication of this finding is that the bulk of the carbon contained within a typical grain is aromatic, with partial hydrogenation attributed to adsorption of H-atoms onto surface layers. Collisional fragmentation of such a grain is thus a logical source of free-flying interstellar PAHs.

The strength and profile of the 217 nm extinction bump imply that about 20% of all available carbon resides in graphitic nanoparticles with low H content (Section 3.4.5). Such particles should thus occupy an area toward the lower left apex in Figure 9.4. Let us assume, based on the above discussion, that interstellar dust has a general preference for sp^2 carbon bonding, and that collisions naturally produce an abundance of particles of appropriate size. A suitable population of 217 nm absorbers may then arise from modest changes in internal structure, accompanied by dehydrogenation. This transition may be driven by UV photon absorption, as proposed by Sorrell (1990). An incident photon induces heating, and if the resulting increase in temperature is of sufficient amplitude a phase change occurs, rearranging the lattice into a more ordered graphitic state. Graphitization may occur on timescales of about 100 Myr, shorter than the predicted lifetimes of the grains

(Jones et al. 1994). Simultaneously, irradiation may cause hydrogen to desorb from the grains. Laboratory simulations provide empirical support for this mechanism (Mennella et al. 1996; Gadallah et al. 2011): the sp^3/sp^2 and H/C ratios of the samples both decrease systematically with increasing UV exposure.

The diamond content of interstellar dust is enigmatic and more difficult to quantify. The discovery of nanodiamonds in meteorites of ostensibly extrasolar origin (Section 8.3.2) suggests that such particles exist in the ISM ("diamonds in the sky"; Tielens et al. 1987a). Surface H atoms bonded to this form of carbon produce characteristic spectral signatures in the infrared, and these have been detected in certain protoplanetary disks, but confirmation of their presence in the ISM is more challenging and elusive (see Jones 2022 for a review). Isotopic analyses imply that some fraction of the meteoritic diamond originated in stellar outflows, but energetic shock-driven grain–grain collisions in the ISM are also potential sources. Diamond is the thermodynamically favored form of solid carbon at high temperature and pressure. Graphite may be converted to diamond under such conditions, a phase change with a high activation energy (7.5 eV per C atom) because complete rearrangement of the crystal lattice is required. Modeling suggests that diamond nanocrystals ranging up to $\sim$ 10 nm in size may form in high-speed impacts, with an equilibrium abundance amounting to about 5% of the available carbon (Tielens et al. 1987b). Of course, such collisions must also affect the silicate dust, but in this case a melting-resolidification cycle is likely to produce glassy amorphous phases rather than crystals.

The effect of cosmic-ray exposure on the dust differs radically in comparison to the mechanisms considered so far, as it is regulated by individual encounters between projectile and target atoms rather than by a more general input of energy into the grain. The kinetic energy of a cosmic-ray particle is many orders of magnitude greater than that of an atom or ion traveling at thermal speeds. It can thus penetrate into a grain and cause internal damage, displacing atoms from the lattice. Exposure thus tends to destroy long-range order, reversing the crystallization process. This mechanism has been invoked to explain rather stringent upper limits on the crystallinity of interstellar silicate dust, in contrast to significant degrees of crystallization detected in circumstellar outflows and in meteoritic stardust (Bringa et al. 2007; Henning 2010). The mean timescale required to amorphize a population of initially crystalline silicates is estimated to be $\sim$ 70 Myr, much less than grain lifetimes. Cosmic-ray exposure also provides a means of estimating the interstellar residence times for presolar grains prior to their inclusion in meteorites, yielding results typically up to 300 Myr for large SiC particles (Heck et al. 2020).

9.1.4 Surface Reactions and Molecule Formation

Atoms in a cool gas impinging on a dust grain may become attached to the surface, and may participate in chemical reactions with other adsorbed species (Figure 9.2). Attachment may be physical or chemical, and the binding energies are markedly different in each case. Physical attachment is maintained by van der Waals forces with binding energies <0.1 eV, the surface acting merely as a passive substrate.

The surfaces of real solids terminate in atoms or ions in states of low symmetry and low coordination, and these tend to be chemically active. The surfaces of interstellar grains are likely to be highly disordered, containing numerous lattice defects and impurities that enhance their chemical reactivity (Duley & Williams 1981, 1986). Amorphous carbon dust may have a high concentration of active surface sites arising from the presence of C atoms with "dangling" bonds, and these tend to become saturated with hydrogen in a diffuse cloud. Similarly, silicate and oxide grains have active surfaces arising from the presence of oxygen ions in low coordination sites, and these also tend to attach hydrogen. The binding energies for chemical adsorption are typically in the range from 1 to 5 eV, at least an order of magnitude higher than those for physical adsorption.

Figure 9.5 illustrates potential energy diagrams for attachment of atoms and molecules onto a surface. Note that the forces for physical adsorption (P) have longer range than those for chemical bonding (C), and two distinct potential wells therefore occur: a particle approaching the surface first encounters a shallow well associated with physical adsorption forces, followed by a deeper well associated with chemical bonding. Chemisorption of accreting atoms is limited by the availability of active sites, and a surface in which all such sites are occupied may accrete further atoms into an outer layer by physical adsorption. The situation is different for accreting molecules, as chemisorption is often inhibited by the presence of an activation energy barrier (A) arising from the need to break or modify a molecular bond. Chemisorption of molecules is thus generally unimportant in interstellar clouds.

The rate at which a dust grain adsorbs atoms or molecules from its environment may be expressed as a function of the kinetic temperature and density of the gas.

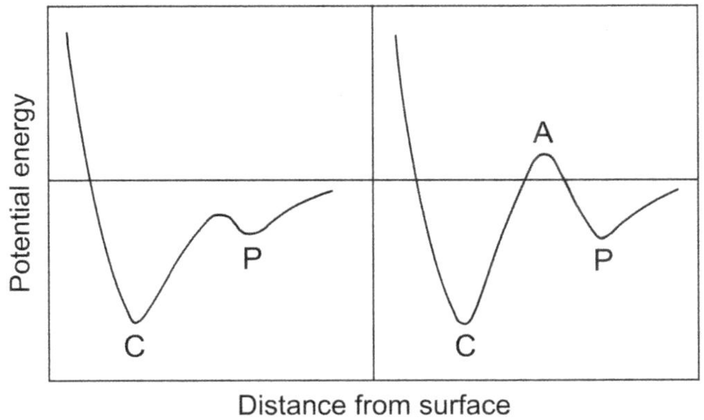

Figure 9.5. Schematic potential energy diagrams for the adsorption of an atom (left) and a molecule (right) onto the surface of a solid (adapted from Tielens & Allamandola 1987b). In each case, a particle approaching the surface (from right to left) encounters a relatively shallow negative potential well (P) associated with physical adsorption forces, and a deeper well (C) associated with chemical bonding. In the molecular case, chemical bonding is inhibited by an activation energy barrier (A).

Assuming a Maxwellian distribution of velocities, the mean speed of a particle of mass m in a gas of temperature $T_{\rm gas}$ is

$$\bar{v} = \left(\frac{8kT_{\rm gas}}{\pi m}\right)^{\frac{1}{2}}. \tag{9.2}$$

In diffuse clouds we are concerned primarily with the adsorption of atomic hydrogen, and assuming $T_{\rm gas} \approx 80$ K (Table 1.1), Equation (9.2) gives $\bar{v}_{\rm H} \approx 1.4$ km s^{-1}. Kinetic motion leads to collisions between the gas atoms and the grain surface. Considering a spherical grain of radius a, immersed in a gas of number density $n_{\rm H}$, the number of collisions per second is $\pi a^2 n_{\rm H} \bar{v}_{\rm H}$. An impinging atom may be adsorbed or returned to the gas; the probability of adsorption is denoted by ξ, the sticking coefficient, which has a value in the range $0 \leqslant \xi \leqslant 1$. The number of atoms adsorbed per second is therefore $\xi \pi a^2 n_{\rm H} \bar{v}_{\rm H}$, and the timescale (mean free time) for adsorption events is the reciprocal of this quantity:

$$t_{\rm ad} = \frac{1}{\xi \pi a^2 n_{\rm H} \bar{v}_{\rm H}}. \tag{9.3}$$

The sticking coefficient depends on the nature of the surface and on the temperatures of the gas and dust (Leitch-Devlin & Williams 1985; Chaabouni et al. 2012). In a hot gas the mean kinetic energy of the impinging atoms exceeds the adsorption energy, and collisions tend to be elastic with low probability of attachment ($\xi \ll 1$). However, theoretical calculations and experimental investigations suggest that $\xi \to 1$ for adsorption of atoms onto amorphous carbon or silicate grains in cool interstellar clouds. Equation (9.3) then yields $t_{\rm ad} \sim 800$ s for grains of radius $a \sim 0.1$ μm in diffuse clouds of density $n_{\rm H} \sim 3 \times 10^7$ m^{-3}.

Adsorbed atoms are subject to thermal evaporation. The mean residence time depends on the dust temperature $T_{\rm d}$ and the binding energy ε, and is given by

$$t_{\rm ev} = t_0 \exp\left(\frac{\varepsilon}{kT_{\rm d}}\right) \tag{9.4}$$

where t_0 is the oscillation time period of the adsorbed species in the direction perpendicular to the surface (typically 10^{-12} s). For chemical binding ($\varepsilon \sim 3$ eV), we have $\varepsilon \gg kT_{\rm d}$ for any reasonable value of dust temperature ($T_{\rm d} \lesssim 500$ K); thus, a chemisorbed atom is stable with respect to thermal evaporation because $t_{\rm ev}$ is long compared with cloud lifetimes. For physical binding, $t_{\rm ev}$ is much shorter and extremely sensitive to temperature: if, for example, we set $\varepsilon = 0.03$ eV, then values of 1.3×10^3 s and 1.1×10^{-7} s are predicted by Equation (9.4) for $T_{\rm d} = 10$ K and 30 K, respectively.

The accretion of atoms onto grains is of primary importance in astrochemistry because it provides the only plausible mechanism for the efficient production of molecular hydrogen (Hollenbach & Salpeter 1971). The process may be written symbolically as

$$\mathrm{H} + \mathrm{H} + \mathrm{grain} \to \mathrm{H_2} + \mathrm{grain}^* \tag{9.5}$$

where the asterisk indicates that the grain has absorbed energy. H atoms cannot recombine directly by two-body encounters in the gas phase because there is no feasible means to release binding energy. The surface of a grain provides both a substrate on which the atoms congregate and a heat reservoir that absorbs excess energy. Three distinct steps are involved (Figure 9.2): (i) adsorption of H atoms onto the surface, as described above; (ii) migration of an atom from one surface site to another until a second atom is encountered; and (iii) recombination, leading to ejection of the resulting H_2 molecule from the grain.

The mobility of an atom attached to a dust grain is limited by the energy binding it to the surface, and this varies with position because of the discrete nature of the lattice. The potential barrier against thermal diffusion to an adjacent site is typically of the same order as the surface binding energy for chemical adsorption, much less for physical adsorption. Migration proceeds efficiently by quantum mechanical tunneling and thermal hopping (Cazaux & Tielens 2004), on timescales short compared with timescales for evaporation at low temperature ($T_d < 30$ K). Thus, migrating physically-adsorbed atoms rapidly seek out and fill any vacant sites for chemisorption on the grain. H_2 formation at low temperature is most likely to involve an encounter between a migrating physically-adsorbed atom and a trapped chemisorbed atom. Chemisorbed atoms become mobile at higher temperatures, enabling molecule formation to continue when physisorbed atoms are no longer stable against evaporation. A binding energy of 4.48 eV is released on each recombination, some fraction of which is imparted to the newly-formed molecule as rotational excitation and translational motion, ejecting it from the surface; any excess energy is absorbed by the grain as heat.

The chemical complexity of diffuse clouds is limited by the ambient flux of UV photons, which dissociates H_2 and other simple molecules on timescales of a few hundred years. A rather rapid transition to a richer chemistry occurs as progressively denser environments are considered, because increasing rates of molecule formation are accompanied by declining rates of photodissociation.

9.2 Processes in Dark Clouds

9.2.1 Gas-phase Chemistry

Models that attempt to account for the rich inventory of molecules observed in interstellar clouds rely on networks of chemical reactions, including those that occur in the gas phase and on the surfaces of dust grains (see Herbst 2014 for a review). The most fundamental role of dust in gas-phase chemistry is to provide a source of H_2, as discussed in Section 9.1.4 above. Once H_2 is available in the gas, new routes to molecule formation are enabled that do not depend on the presence of a substrate. Molecules heavier than H_2 that either form on or attach to grain surfaces in dense clouds are mostly retained in the solid phase as icy mantles (Figure 9.2), until an energy source is available to remove them.

Under normal interstellar conditions only binary (two-body) reactions need be considered in the gas, as the probability of a three-body interaction is extremely low even in the densest clouds. Also, in general, only exothermic reactions lacking activation energy barriers are likely to occur at a significant rate in cold, dark clouds. Exothermic chemical-exchange reactions, in which two reactants produce two (or more) products, are energetically favored because the excess energy is transferred to kinetic energy of the products. As an example, consider the reaction

$$O^+ + H_2 \rightarrow OH^+ + H \tag{9.6}$$

in which an H atom is exchanged between the reactants to form a new molecule. If, as in this case, one of the reactants is ionized, the electric field of the ion polarizes the electrostatic charge of the neutral reactant, leading to a net Coulomb attraction. The reaction rate is thus enhanced, typically by a factor of 100, with a corresponding increase in the resultant abundance of the product. Such ion–molecule reactions play an important and, indeed, dominant role in gas-phase astrochemistry (e.g., Snow & Bierbaum 2008). A supply of ions is maintained by cosmic rays in dark clouds shielded from the external radiation field.

Reaction networks based on this principle can account for many of the observed interstellar molecules. For example, reaction (9.6) initiates a route to water via the sequence

$$\begin{gathered} OH^+ + H_2 \rightarrow H_2O^+ + H \\ H_2O^+ + H_2 \rightarrow H_3O^+ + H \\ H_3O^+ + e^- \rightarrow H_2O + H \end{gathered} \tag{9.7}$$

and an exactly analogous sequence starting with N^+ and involving one additional step produces NH_3. Carbon hydrogenation is initiated in a different manner because the equivalent first step ($C^+ + H_2 \rightarrow CH^+ + H$) is endothermic and therefore inhibited in interstellar clouds. A feasible alternative route depends on prior formation of the highly reactive H_3^+ ion:

$$\begin{gathered} H_2^+ + H_2 \rightarrow H_3^+ + H \\ C + H_3^+ \rightarrow CH^+ + H_2 \end{gathered} \tag{9.8}$$

with further H-additions analogous to the sequence in Equation (9.7) enabling production of CH_4. Recombination may occur at any stage in these sequences to form the neutral radicals OH, NH, CH, NH_2, CH_2, and CH_3. Carbon also undergoes oxidation to produce CO, the most abundant interstellar molecule after H_2, which can form in a number of ways, including by the ion–molecule reaction

$$C^+ + OH \rightarrow CO + H^+ \tag{9.9}$$

and the neutral reaction

$$CH + O \rightarrow CO + H. \tag{9.10}$$

Once created, CO is resilient because of its relatively large binding energy. In diffuse clouds, CO is destroyed principally by UV photodissociation; in dense clouds, the most probable sink for gas phase CO is adsorption onto grains.

These and a host of other reactions have been used in chemical models that predict gas-phase abundances in accord with the observations for the majority of known species. See the review papers by Snow & Bierbaum (2008), Herbst & van Dishoeck (2009), and Herbst (2014) for further details and in-depth discussions.

9.2.2 Mantle Growth and Surface Chemistry

The exchange of atoms and molecules between dust and gas favors adsorption over desorption within the shielded environment of a cold, dense cloud, allowing accumulation of surface coatings (mantles) on the grains. This section assesses the physical process of mantle growth, and considers the role of surface chemistry as the primary route to production of species observed to be present in the mantles.

Consider a spherical grain of radius a immersed in a gas containing particles of mass m and number density n. The grain adsorbs particles at a rate $\xi\pi a^2 n\bar{v}$, where ξ is the sticking coefficient, and $\bar{v}$ is the mean particle speed defined in Equation (9.2). The rate of increase in the mass of the dust grain is thus

$$\frac{\mathrm{d}m_\mathrm{d}}{\mathrm{d}t} = \xi\pi a^2 n\left(8kT_\mathrm{gas}m/\pi\right)^{\frac{1}{2}}. \tag{9.11}$$

If adsorption leads to the formation of a mantle containing solids of mean density ρ_s, the radius of the grain increases at a rate

$$\begin{aligned}\frac{\mathrm{d}a}{\mathrm{d}t} &= \frac{1}{4\pi a^2\rho_\mathrm{s}}\frac{\mathrm{d}m_\mathrm{d}}{\mathrm{d}t}\\ &= 0.4\,\xi\, n\rho_\mathrm{s}^{-1}\left(kT_\mathrm{gas}m\right)^{\frac{1}{2}}.\end{aligned} \tag{9.12}$$

If the quantities on which the rate depends remain constant with respect to time, integration gives the radius at time t simply as

$$a(t) = a_0 + \Delta a(t) \tag{9.13}$$

where a_0 is the radius at $t = 0$, i.e., the core radius, and

$$\Delta a(t) = \left[0.4\,\xi\, n\rho_\mathrm{s}^{-1}\left(kT_\mathrm{gas}m\right)^{\frac{1}{2}}\right]t \tag{9.14}$$

is the thickness of the mantle. Note that $\Delta a(t)$ is independent of a: if this simple treatment is valid, all grains in the same environment are predicted to acquire mantles of equal thickness, irrespective of their initial size, leading to a shift of the distribution function to larger sizes while preserving its functional form. However, thermal transients may result in selective desorption of ice mantles from the smallest grains in a radiation field, and coagulation of mantled grains may lead to further growth.

If desorption rates remain low in dark clouds, the timescale for growth of a mantle to a given thickness Δa is specified by a rearrangement of Equation (9.14):

$$t_\mathrm{m} = \frac{2.5\rho_\mathrm{s}\Delta a}{\xi n\left(kT_\mathrm{gas}m\right)^{\frac{1}{2}}}. \tag{9.15}$$

H_2O is the most abundant constituent of the mantles (Section 5.3), and the relationship in Equation (9.15) may be used to test whether it accumulates primarily by frosting (direct freeze-out of gaseous H_2O molecules formed in the gas) or by surface reactions between adsorbed atoms (Jones & Williams 1984). Models based on gas-phase chemical networks for molecules other than H_2 predict an equilibrium gas-phase abundance $n_{H_2O}/n_H \sim 10^{-7}$ in a cloud of density $n_H = 10^9$ m^{-3}, temperature $T_{gas} = 15$ K, and solar elemental abundances. Assuming that adsorption is efficient ($\xi \approx 1$), the time needed to accumulate a thin layer ($\Delta a \approx 0.02$ μm) of amorphous ice ($\rho_s \approx 750$ kg m^{-3}), consistent with observational evidence (Boogert et al. 2015), is $t_m \sim 5$ Gyr! As cloud lifetimes are <50 Myr, this calculation clearly demonstrates that the observed ice mantles *cannot* be explained by frosting of H_2O molecules; most of the available gaseous oxygen resides in O and CO for any reasonable chemical model (e.g., Herbst & Leung 1986; Hollenbach et al. 2009). If we repeat the above calculation for atomic O, and assume that surface reactions result in efficient hydrogenation of O to H_2O, the timescale is reduced to $t_m \sim 2$ Myr. This result is entirely plausible because it is shorter than typical age estimates for clouds in which ice is observed. It must therefore be inferred that H_2O in the ice mantles is produced almost entirely by surface reactions. This conclusion also holds true for most other known ice constituents; only in the case of CO is frosting likely to be important.

The composition of the ice mixture is sensitive to environment and regulated by the fraction of hydrogen in molecular form ($f(H_2)$; Section 1.3.2). To illustrate this, consider the evolution of a starless dark cloud formed by gradual contraction of diffuse gas and dust. Initially, $f(H_2)$ is close to zero; it remains low in the outer layers of the cloud, but increases internally with density as H_2 production becomes more rapid and self-shielding protects against photodissociation, until eventually $f(H_2) \rightarrow 1$ inside the cloud. At early times, O, C, and N atoms attaching to a grain surface encounter (and react with) H atoms on timescales much shorter than those on which they react with each other. Whereas newly-formed H_2 molecules are expected to desorb from the grain on release of binding energy, the heavier products OH, CH, and NH are more likely to remain attached to the surface and undergo further hydrogenation reactions such as OH + H $\rightarrow$ H_2O. Surface chemistry during this initial phase thus leads to efficient production of H_2O and other hydrogenated molecules on the grain (Tielens & Hagen 1982; Cuppen & Herbst 2007; Dulieu et al. 2010).

In isolation, a physisorbed H_2O molecule may be removed from the grain by absorption of an infrared ($\lambda \sim 3$ μm) photon corresponding to the vibrational resonance frequency of the molecule (Williams et al. 1992). However, once a complete monolayer is formed, the ice becomes more stable because the energy required to remove a molecule from bulk ice exceeds that of the resonant photons, and further growth may then be rapid. As the cloud evolves and H $\rightarrow$ H_2 conversion approaches completion, surface hydrogenation reactions become far less frequent. A qualitatively different mantle material is then predicted, in which the abundances of molecules such as CO, CO_2, O_2, and N_2 may exceed that of H_2O. As CO is already abundant in the gas, its presence in the ice arises largely from frosting. CO_2 forms by oxidation of CO (see below), and O_2 and N_2 may form by recombination of adsorbed O and N atoms. The outcome is a hierarchical structure in which an H-poor layer forms on top of an H-rich layer, as illustrated schematically in Figure 5.14. In practice, the transition between majority species near the interface is likely be gradual rather than abrupt.

This prediction of hierarchical growth provides a basis for interpreting the observations reviewed in Section 5.3. Studies of laboratory analogs show that H-rich and H-poor ices are distinguished by clear differences in their vibrational spectra, and for species such as CO and CO_2 these disparities correlate with the mean dipole moment of the mixture that contains them. The observations indicate the presence of distinct polar and apolar ices, dominated by H_2O and CO, respectively, with an admixture of other molecules. The polar mantles include NH_3, CO, and CO_2 as minority species. The apolar mantles contain CO, CO_2, and presumably also O_2 and N_2, although these homonuclear molecules are not available to direct spectroscopic detection. The general scenario outlined above seems reasonably well established, although many details of the surface chemistry are not yet fully understood.

9.2.3 The Surface Chemistry of CO

Surface reactions that utilize CO to produce larger molecules are of special significance in our quest to understand the origin of the ices and their potential for further chemical evolution. One evident outcome is oxidation to CO_2, a widely-observed species associated with both early and late phases of mantle deposition, as discussed above. Another is hydrogenation to HCO, a precursor of organic molecules such as H_2CO and CH_3OH detected in the ices. These simple organics may themselves be gateways to a more complex chemistry.

Unlike CO, CO_2 is not predicted to have an appreciable abundance in the gas phase, as confirmed by observations (van Dishoeck et al. 1996), so frosting cannot account for the presence of CO_2 in the ices. Its abundance in a variety of environments (Table 5.3) must therefore be attributed to surface reactions. The possibilities include

$$\begin{aligned} CO + O &\rightarrow CO_2 \\ CO + OH &\rightarrow CO_2 + H \end{aligned} \tag{9.16}$$

but both of these reactions possess activation energy barriers that may inhibit CO_2 production at low temperature. If a flux of UV photons is available, laboratory simulations show that CO_2 is easily produced in H_2O:CO ice mixtures by the sequence

$$\begin{aligned} H_2O + h\nu &\rightarrow OH^* + H \\ CO + OH^* &\rightarrow CO_2 + H \end{aligned} \tag{9.17}$$

where the asterisk denotes an excited state (d'Hendecourt et al. 1986; Sandford et al. 1988). In an interstellar cloud, the external radiation field may induce CO_2 formation in its outer layers, and cosmic-ray excitation of H_2 generates a weak UV field that might contribute at greater optical depths. However, photon excitation alone cannot explain the profusion of CO_2 in the shielded environments of dark cloud cores (Whittet et al. 1998; Gerakines et al. 1999). The existence of a small energy barrier may not be a major impediment to CO_2 production on grains at low temperature if surface diffusion is efficient (Ruffle & Herbst 2001). Formation rates are highly sensitive to temperature, such that small thermal transients arising from infrared photon absorption by small grains may enhance production (Pauly & Garrod 2016). Extensive laboratory investigations carried out under simulated interstellar conditions confirm that significant CO_2 production may occur in the absence of UV radiation (Roser et al. 2001; Ioppolo et al. 2011; Minissale et al. 2013).

CO_2 is a robust molecule that does not readily react to form other species at interstellar temperatures, and its production is therefore effectively a chemical dead end. In contrast, a potential route to the formation of organic molecules is initiated by hydrogenation:

$$CO \overset{H}{\rightarrow} HCO \overset{H}{\rightarrow} H_2CO \overset{H}{\rightarrow} CH_3O \overset{H}{\rightarrow} CH_3OH. \tag{9.18}$$

The first and third reactions in this sequence possess small activation energy barriers, but again laboratory simulations show that they may proceed with reasonable efficiency (Watanabe & Kouchi 2002; Fuchs et al. 2009). The stable molecules CO, H_2CO, and CH_3OH are thus expected to co-exist in the ices, consistent with observations, whereas the intermediate radicals react more rapidly and have lower concentrations. Alternative routes in the sequence may form additional species such as HNCO (HCO + N) and HCOOH (H_2CO + O).

Oxidation and hydrogenation reactions compete for the available CO. The question of which is dominant may be explored by comparing the abundances and spatial distributions of the resultant species. As H_2CO is a relatively minor product ($\lesssim$6% relative to H_2O), the comparison between CO_2 and CH_3OH is most informative. CO_2 is detected in polar and apolar ices and in both dark clouds and star formation regions, with an extreme abundance range of approximately 10%–50%. CH_3OH is less widely distributed, mostly restricted to denser environments, with abundances up to about 30%. Both observations and astrochemical models indicate that CH_3OH production is efficient during the second phase of mantle growth, after an H_2O-rich layer is already in place (Whittet et al. 2011; Punanova et al. 2022). Prior to this, the first step in the sequence (9.18) may occur, but further hydrogenation is limited by the efficiency of the barrier-less reaction

$$HCO + O \rightarrow CO_2 + H \tag{9.19}$$

so again the product is CO_2. At later times and in denser regions, freeze-out of CO is accompanied by a continuing supply of accreting H atoms, maintained by cosmic-ray dissociation of gas-phase H_2 deep within the cloud. At this stage, a substantial fraction of the elemental O available for accretion onto dust is expected to be tied up in CO and other molecules (e.g., Hollenbach et al. 2009), and lower grain temperatures may reduce the mobility of the remaining atomic O that adsorbs onto the surface, to a point where reaction (9.19) becomes irrelevant. A dearth of atomic O also removes a major sink for atomic H. Thus, a critical point may be reached when CO becomes more abundant than atomic O in the accreting gas. CO processing is then dominated by the hydrogenation sequence, such that CH_3OH may be produced in concentrations comparable with or greater than those of H_2O and CO_2, mixed with unaltered excess CO. Subsequent production of more complex molecules is likely to occur when stars form in the cores, creating energy sources to induce further processing of the ices or their sublimation products in the gas. We return to this topic in Section 9.3.

9.2.4 Depletion from the Gas

The growth of mantles on dust must of course result in a corresponding loss of elements from the gas, and this too can be assessed by observation. The timescale on which a gaseous atom or molecule with mean thermal speed $\overline{v}$ becomes depleted from the gas by adsorption onto grains of cross-sectional areas σ_d and number density n_d is given by

$$t_{dep} = \frac{1}{\xi n_d \sigma_d \overline{v}} \approx \frac{10^{25}}{\xi n_H \overline{v}} \tag{9.20}$$

where the approximation is based on the estimated mean projected area of grains per H atom (Spitzer 1978, pp. 161–162). Consider again the case of CO, an important ice constituent, but also widely used as a generic tracer of molecular material in the gas. Using Equation (9.2) to estimate $\overline{v}_{CO}$ in a molecular cloud of temperature 10 K, and assuming that sticking is efficient ($\xi \approx 1$), Equation (9.20) yields $t_{dep} \lesssim 3$ Myr for densities $n_H \gtrsim 10^9$ m^{-3}. Hence, CO is predicted to become almost entirely depleted from gas onto dust in dense regions over timescales short compared with cloud lifetimes, unless desorption is also efficient. Estimates for other species yield similar results. Calculations of cosmic-ray-induced desorption rates show that they are insufficient to prevent rapid freeze-out of CO during the late phase of mantle deposition in which the CO-rich ices are formed (Silsbee et al. 2021). The implication is that a substantial fraction of the molecular content of a dense core becomes undetectable by rotational spectroscopy, which sees only the gaseous component. This prediction is, indeed, confirmed by observations. Distinct depletion zones, revealed by central emission-line intensity minima, have been mapped for CO and other gas-phase molecules in dense, starless cores (e.g., Caselli et al. 1999; Fuente et al. 2019; Sabatini et al. 2019).

Quantitative information on the distribution of CO between gas and dust in dense clouds is important to our understanding of surface processes, and also necessary to clarify the applicability limits of CO gas as a tracer of H_2. Determinations for individual lines of sight can be made by observing gas-phase spectral lines in the same sources used to study the ices. Ideally, infrared *absorption* lines (Section 5.1.1) should be studied to ensure that gas and ice are measured in identical columns. This is challenging, however, because the gas-phase lines require very high spectral resolution: current technology limits targets to bright protostellar objects that seem unlikely to provide results representative of dark clouds. Millimeter-wave observations are therefore used to enumerate CO column densities in cold molecular gas.

Emission lines of the common isotopic form of CO are generally saturated, so rarer forms are observed and the results scaled using standard abundance ratios. Systematic errors arise if the volume of space being sampled is not identical for gaseous and solid phases: differences in beam size may introduce errors if the density changes across the line of sight, and any material behind the source contributes to the molecular emission but not to the solid-state absorption. Background field stars behind nearby dark clouds that lie in directions away from the disk of the Milky Way are least likely to be affected by such errors.

The Taurus dark cloud is a good example of a favorably-placed target. Correlations of CO column density versus extinction for this cloud are shown in Figure 9.6. The gas-phase data (left frame) are from millimeter-wave observations of various isotopes, with all results normalized to the common isotopic form. A steep initial rise is followed by a decline in slope with increasing A_V, represented by the fitted curve. This change is attributed to adsorption of CO onto dust, as suggested by the right-hand frame, which plots the summation

$$\sum N(\mathrm{CO}) = N(\mathrm{CO})_{\mathrm{gas}} + N(\mathrm{CO})_{\mathrm{ice}} + N(\mathrm{CO_2})_{\mathrm{ice}} \quad (9.21)$$

with the ice data based on infrared observations. The relation is now approximately linear, consistent with the conclusion that the "missing" CO has simply migrated to the dust and become partially oxidized to CO_2. This result is consistent with observations of gas-phase CO in other dark clouds at low to moderate extinctions. A correlation of the form

$$N(\mathrm{CO}) \approx 1.12 \times 10^{21}(A_V - 2) \quad (9.22)$$

proposed by Kainulainen et al. (2006) is shown in Figure 9.6 (dotted line in each frame). The linear fit to the Taurus data for total N(CO) is consistent with

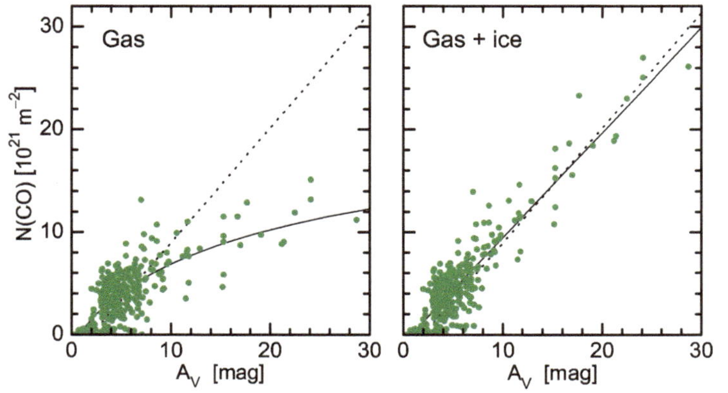

Figure 9.6. Correlations of CO column density with visual extinction in the Taurus dark cloud (Whittet et al. 2010). Each point represents the line of sight to a background star observed through the cloud. The left frame plots gaseous CO data only, and the solid curve is a sigmoidal fit to the points. The right frame plots the summation of CO gas, CO ice, and CO_2 ice for the same sample; the solid line is a linear fit. The dotted line in each frame extrapolates a relation proposed by Kainulainen et al. (2006) for $A_V \lesssim 10$ in dark clouds.

extrapolation of this relation to higher A_V. The positive intercept at $A_V \approx 2$ on the extinction axis suggests a threshold necessary for the rate of formation in the gas to exceed the rate of photodissociation. At the highest extinctions covered by the data set, ~ 60% of the CO is depleted into solids.

9.2.5 The Oxygen Crisis Revisited

A notable problem arising from studies of gas-phase depletions in the diffuse ISM concerns the distribution of elemental oxygen between candidate solids (Section 7.3.3). A substantial fraction of the depleted O in the densest diffuse clouds ($A_V \sim 1$) cannot be accounted for by silicates and oxides, leaving open the question of its chemical form. In dense clouds, ices and gas-phase CO sequester O in sufficiently shielded regions ($A_V \gtrsim 2$), as discussed above. This section extends the discussion of "missing" oxygen to the dark-cloud environment.

We begin by assuming that "protosolar" elemental abundances are applicable to clouds in the solar neighborhood of our Galaxy (Section 7.1.6). The same standard is adopted in studies of depletion by Jenkins (2009) and Ritchey et al. (2018), and it yields an oxygen abundance coincident with that observed in OB stars (O/H = 575 ppm; Nieva & Przybilla 2012). As abundant metallic elements such as Si, Mg, and Fe are already highly depleted in diffuse clouds, there is little scope for additional depletion of O into oxidized minerals in denser regions; assuming that the metals are optimally oxidized, silicates and other oxides account for at most ~ 170 ppm (30% of the total), as illustrated in Figure 9.7 (green slice).

In dense clouds, O is incorporated into gas-phase molecules, predominantly CO, and into ices. Observations of the Taurus region (Figures 5.11 and 9.6) provide the most extensive information on the uptake of O by these carriers. In a given line of sight, the column density of elemental O detected in the ices is

$$N_{\rm O} = N({\rm H_2O})_{\rm ice} + N({\rm CO})_{\rm ice} + 2N({\rm CO_2})_{\rm ice} \quad (9.23)$$

(other O-bearing ices are below detection limits in this cloud). The corresponding hydrogen column density is

$$N_{\rm H} \approx 2.1 \times 10^{25}\,(A_V - A_V^0) \quad (9.24)$$

where A_V^0 is the extinction threshold (Section 5.3.2). Analysis of data from Whittet et al. (2010) and references therein yields the proportions shown in Figure 9.7 for ices and CO gas at $A_V \sim 25$ (light and dark blue segments).

But so far we have accounted for only ~ 60% of the oxygen! Where is the rest? The line of sight to a background star passes through not only dense material but also the diffuse outer layers of the cloud, in which gas-phase atomic O remains ubiquitous. Spectroscopy of the diffuse ISM indicates that about 320 ppm of O remains in the gas at the highest levels of depletion accessible to observation. This value is consistent with the boundary level assumed by Hollenbach et al. (2009) in a model for internal dense-cloud chemistry. As expected, the model predicts that the abundance of atomic O drops rapidly with increasing extinction inside the cloud as it is used to form gas-phase CO and ices (predicted abundances of other gaseous O-

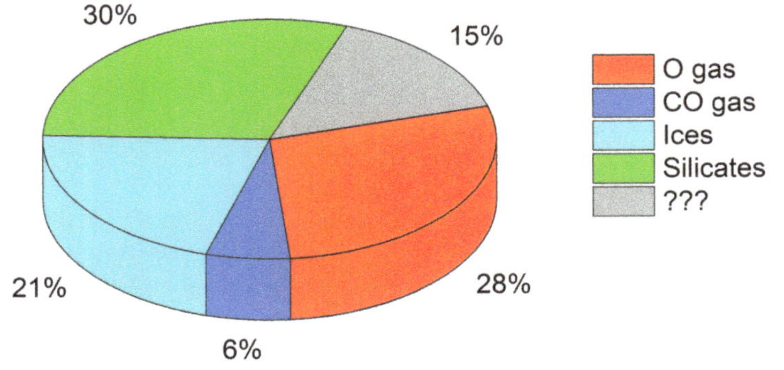

Figure 9.7. A pie chart of the distribution of elemental oxygen in a prototypical dark cloud. Percentages are relative to an assumed abundance O/H = 575 ppm. See text for details.

bearing species such as OH, H_2O, and O_2 are negligible). Gas-phase CO and ices together account for about 160 ppm of oxygen in the Taurus cloud. The segment of the pie allocated to gaseous atomic O in Figure 9.7 (red) assumes that all the free oxygen not consumed by CO and ice formation remains in atomic form, so it is *effectively an upper limit.*

These results show that the oxygen crisis identified in the diffuse ISM persists in the dense ISM. Even with these optimistic assumptions, a significant fraction ($\gtrsim$15%) remains unaccounted for by available data. It might be argued that the reference abundance is too high, but the anomaly in the diffuse-ISM data stems from the slope of the trend in depletion, not its absolute value. It might also be argued that the standard N_H/A_V ratio, based on observations of the diffuse ISM, is subject to change in dense clouds; however, if coagulation is affecting the optical properties of the dust (Section 9.1.1), the standard ratio is expected to yield underestimates of N_H and hence *overestimates* of the proportionate CO gas and ice abundances in Figure 9.7. A further possibility is that (undetected) O_2 may be abundant in the ices, but even if it were as common as (say) CO_2 it would be insufficient to close the gap. See Section 7.3.3 for further discussion and proposed solutions.

Finally, it may be informative to consider whether carbon is also anomalous. Depletions in the diffuse ISM are less well constrained for this element, as discussed in Section 7.2.1. The assumed reference abundance is 288 ppm (Table 7.1), and of this, somewhere between 150 and 200 ppm must be in PAHs and refractory particles (graphitic and amorphous carbon) to explain the observed properties of the dust (Section 7.3.1). The undepleted gas is mostly ionized (C II) in the diffuse ISM, trending to neutral C I and CO with increasing density in more shielded environments. Using the same data set as before, the summation of observed abundances for carbon in gaseous CO and ices (CO + CO_2) in the cloud amounts to about 70 ppm. We can therefore account for $\sim$ 240 ppm in the solid phase, leaving a shortfall of $\sim$ 50 ppm that is readily explained by the presence of ionized and neutral C atoms in the outer layers of the cloud. Thus, there does not seem to be an equivalent problem for carbon.

9.3 Processes in Star Formation Regions

Previous sections of this chapter examine astrophysical processes occurring in diffuse and molecular phases of the ISM prior to the onset of star formation. The focus now shifts to the role of dust in regions of active star formation. Stars are born within dense molecular clouds. The radiation fields and winds of newborn stars alter and ultimately dissipate their parent clouds. The clouds are typically far more massive than the stellar populations that they spawn: during each generation of star birth, a minor fraction of associated interstellar material is converted into stars and planetary systems, the remainder is recycled to the ISM.

9.3.1 Birth Aggregates: From Quiescent Clouds to Starbursts

Stars do not generally form in isolation. A birth cloud will typically undergo fragmentation into individual prestellar condensations as a result of the interplay of gravity, magnetic fields and turbulence (McKee & Ostriker 2007). A family of stellar siblings is produced that might contain anywhere from 10 to 10^6 members, depending on the initial mass and morphology of the birth cloud.[1] The number of stars in the group has a profound effect on its environment, controlling both the ambient radiation field and the dynamical evolution of the aggregate. The influence of each individual star depends on its mass. High-mass stars are by far the most significant because of their luminosities, their strong stellar winds, and the speed at which they evolve. Stars of mass $\gtrsim 10\ M_\odot$ may reach maturity while their lower-mass siblings are still forming: examples may be seen in Figure 9.8. Indeed, it is possible for the most massive stars ($\gtrsim 25\ M_\odot$) to evolve through their entire life-cycles and become supernovae before star formation in other regions of the same birth cloud is complete.

The relative number of stars forming in a given stellar mass range is described empirically by the initial mass function (Section A.3). The key result in this context is simply that low-mass stars always greatly outnumber high-mass stars in any given sample. Because of this, the size of the group is crucial: the probability that a small group with less than a hundred members will contain even one star of mass $>10\ M_\odot$ is negligible, whereas a much larger group will contain at least a few. A notable distinction between "small" and "large" groups occurs as the membership exceeds a few hundred: the internal UV radiation field generated by young stars rises steeply with membership at this cross-over point (Adams & Myers 2001). Because of this, we can consider star formation in two distinct modes: low-mass and high-mass. Low-mass star formation regions such as Lupus 3 (Figure 1.3), are relatively quiescent and often filamentary in structure rather than centrally condensed (Bergin & Tafalla 2007). They contain low and intermediate-mass YSOs that illuminate reflection nebulae and heat their circumstellar envelopes, but lack massive stars that become powerful emitters of UV radiation. High-mass star formation (e.g., Figure 9.8) occurs in giant molecular clouds and produces both low-mass and massive stars (Motte et al. 2018). The energy output of a massive star may have a

[1] The median membership of star clusters in the solar region of the Milky Way is ~300 (Adams 2010).

Figure 9.8. Example of a massive star formation region, NGC 602 in the Small Magellanic Cloud. Luminous stars in the central cluster are gradually dissipating the natal molecular cloud, while others are still forming within the cone-shaped structures above, below and to the right of center. Image credit: NASA/ESA/Hubble Space Telescope.

dramatic effect on not only its own circumstellar envelope but also those of its siblings and portions of the birth cloud that have not yet condensed to form stars.

9.3.2 Thermal Processing

The observed properties of dust in a molecular cloud prior to the onset of star formation are consistent with core-mantle particles (e.g., Figure 5.14), the mantles formed by sequential deposition of distinct layers containing amorphous ices of differing volatilities. Star formation induces both physical and chemical changes in these layers. Warming may result in crystallization, segregation, distillation, and partial or complete sublimation, and may activate chemical reactions. Spectroscopic observations in the infrared (Chapter 5) provide metrics for assessing such changes, including (i) evidence of ice-phase abundance variations that correlate with volatility, attributed to sublimation; (ii) evidence of temperature-dependent variations in profile shape, attributed to structural changes in the solids; and (iii) the detection of features identified with new products. We consider these in turn.

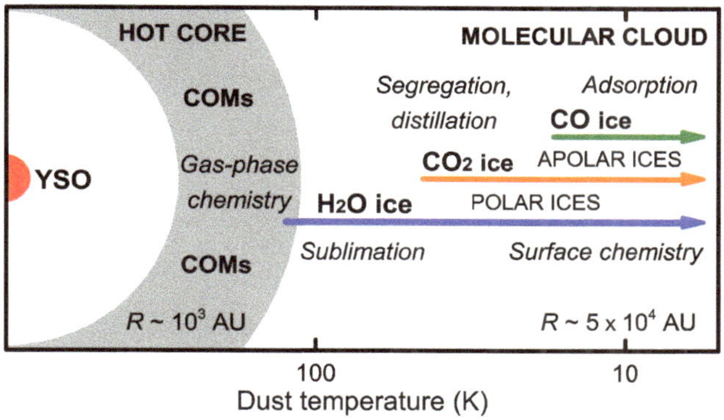

Figure 9.9. Schematic illustration of environments around a massive young stellar object (YSO) embedded in a dense molecular cloud. Radiation from the YSO imposes a circumstellar temperature gradient which controls the nature of the processes acting in different zones. Colored arrows indicate approximate stability limits of common ices. Approximate radial distances are indicated for the gaseous hot core around the star (gray) and the cold outer zone within which the apolar ices are sublimated. Complex organic molecules (COMs) in the hot core may result in part from chemical processing of ice sublimation products.

Sublimation zones around a massive young star are illustrated schematically in Figure 9.9. UV radiation from the star is absorbed and re-emitted in the circumstellar envelope, and the resulting infrared flux warms the surrounding dust. Of species commonly observed in the ices, CO and H_2O are, respectively, the most and least volatile, with CO_2 an intermediate case. Warming removes the CO-dominated (apolar) layers at dust temperatures above about 20 K, whereas the H_2O-dominated (polar) ices may survive closer to the source at temperatures up to least 100 K. In general, the line of sight to a YSO will contain both warm and cold dust, but the average CO/H_2O-ice ratio is significantly less toward high-mass YSOs compared with both lower-mass YSOs and field stars (Table 5.3). The spectra of high-mass YSOs also reveal dramatic variations in the distribution of CO between ice and gas (Figure 9.10), implying source-to-source differences in levels of sublimation (van Dishoeck & Blake 1998). However, some CO is retained as a minority constituent in the H_2O-dominated polar ices at temperatures up to $\sim$ 100 K (Schmitt et al. 1989). Distinct polar and apolar components are seen in the profile of the solid CO feature, and their relative strengths vary with the degree of thermal processing, as discussed in Section 5.3.4.

The ices that accumulate on grains in dark clouds are initially amorphous, as demonstrated by their observed absorption profiles (Section 5.3.3). This is to be expected because they generally form at temperatures well below the melting or sublimation points of the relevant molecules. Amorphous solids contain a distribution of molecular environments that produce absorption features with broad, Gaussian profiles. As the ice is warmed, the molecules rearrange themselves into more energetically-favorable orientations, resulting in narrower, more sharply peaked profiles characteristic of crystalline solids. In the case of H_2O-ice, for

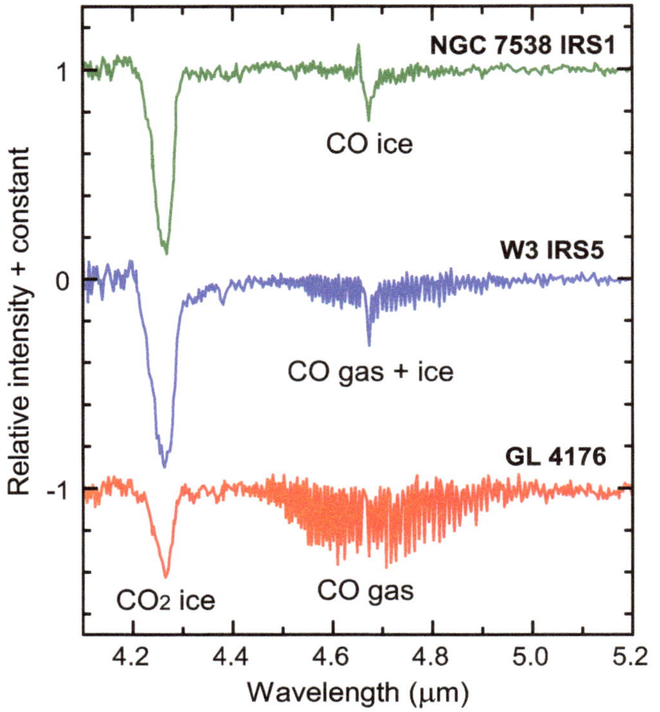

Figure 9.10. Comparison of 4–5 μm spectra for three massive YSOs, illustrating different levels of apolar ice sublimation in their circumstellar environments. Features of solid and gaseous CO centered near 4.67 μm are detected with differing relative strengths. CO remains frozen on grains toward NGC 7538 IRS1, but has been partially and completely vaporized toward W3 IRS5 and GL 4176, respectively. CO_2, which is less volatile than CO, produces absorption centered at 4.27 μm in all three spectra. Figure based on archival data obtained by the short-wavelength spectrometer of the Infrared Space Observatory.

example, heating from 10 K to 80 K results in a narrowing of the 3 μm profile and a shift of the peak to longer wavelength (see the inset graph in Figure 5.12). A great deal of laboratory work has been carried out to investigate the effect of temperature on different ice mixtures that might be representative of those present in the ISM, and the results are vital to our interpretation of the observational data (see Boogert et al. 2015 for a review).

The profile of the CO_2 bending-mode absorption near 15 μm is particularly sensitive to changes in the ices induced by thermal processing (Ehrenfreund et al. 1999; Pontoppidan et al. 2008). The symmetry of the linear molecule is broken in its pure solid form, resulting in a double-peaked structure. Laboratory simulations show that this structure is weak or absent in amorphous ice mixtures, and that subpeaks centered near 15.12 and 15.25 μm appear as the ice is warmed. Representative spectra that suggest an evolutionary sequence are shown in Figure 9.11. In the dark-cloud environment, represented by the line of sight to a field star (top), the profile is smooth and consistent with a superposition of polar

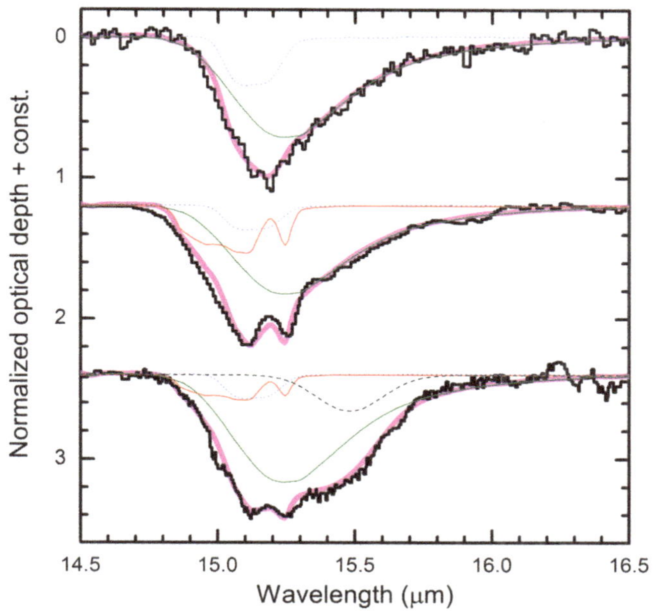

Figure 9.11. Evidence for thermal evolution of the ices. Spectra of the CO_2 bending-mode absorption toward the field star Elias 16 (top), the low-mass YSO RNO 91 (middle), and the high-mass YSO W33A (bottom) are shown in black, and fitted with models derived from laboratory data (magenta curves). Each fit is the sum of two or more components: polar H_2O:CO_2 ice (solid green), apolar CO:CO_2 ice (dotted blue), pure CO_2 (solid red), and CO_2:CH_3OH (dashed gray). Data are from Boogert et al. (2015) and references therein.

and apolar amorphous ices; but the YSO profiles require additional components. Heating may increase the purity of CO_2-bearing ices two ways. In an apolar CO:CO_2 mixture, it may arise by distillation, i.e., by sublimation of CO at $T_d \gtrsim 20$ K, while CO_2 is retained in solid form. At somewhat higher temperatures, the rearrangement of bonds accompanying crystallization of H_2O in the polar ices causes segregation, forming clusters of pure CO_2 within the ice matrix.

A broadening of the CO_2 band, consistent with the presence of an additional component centered near 15.45 μm, is also evident in the spectra of some YSOs. An example is shown in Figure 9.11 (bottom spectrum). This is attributed to the effect of intermolecular interactions between CO_2 and CH_3OH in the ices. Competing oxidation and hydrogenation reactions during CO freeze-out may lead to simultaneous production of CO_2 and CH_3OH during the late stages of mantle growth (Section 9.2.3). On subsequent heating, sublimation of CO and structural changes in the remaining ices may lead to the formation of complexes in which the O atoms of CH_3OH link with the C atoms of CO_2 to form stable intermolecular bonds (Lewis acid-base pairs; Ehrenfreund et al. 1998, 1999; Dartois et al. 1999). The CO_2 bending mode in these complexes matches the additional component seen in the observed spectrum.

9.3.3 Energetic Processing

Thermal processes described above result from the cumulative heating of particles exposed to infrared radiation from stars that remain embedded in placental gas and dust. As a young star evolves and circumstellar material gradually disperses, higher energy photons and particle winds permeate the surrounding medium in regions of high-mass star formation. Unless an ice-mantled grain is very small, the energy imparted when it absorbs a UV photon will not thermalize throughout the particle, instead it produces a local hot spot. Photon energies of a few eV may break chemical bonds and convert saturated molecules such as H_2O, NH_3, and CH_4 into radicals. These changes occur regardless of temperature; but if the ice is sufficiently warm, the radicals may be free to migrate and react with other species to form new products. For example, CO_2 may form in this manner by the reaction sequence in Equation (9.17). Energetic ion impacts may induce similar changes. Laboratory simulations demonstrate that a wide variety of species are formed by exposure of simple ice mixtures known to be present in the ISM to energetic photon irradiation or ion bombardment. The products include complex organic molecules (COMs) and kerogen-like polymers (e.g., Agarwal et al. 1985; Briggs et al. 1992; Bernstein et al. 2002; Muñoz Caro & Dartois 2009; Muñoz Caro et al. 2014; Urso et al. 2020).

The results of these laboratory investigations have proven to be especially relevant to studies of primitive bodies in our solar system, such as comets, asteroids, and icy moons, providing realistic analogs of constituents and elucidating the mechanisms that produced them. So far, however, they have had rather less direct impact on studies of interstellar solids. It is highly plausible that energetic processing occurs in star formation regions, but observational confirmation is hard to find. A spectral feature centered near 4.62 μm and attributed to molecules containing C≡N bonds (originally named "XCN" and subsequently attributed to the cyanate (OCN^-) ion; Table 5.2) was once considered to be a probable example, but more recent research casts doubt on this interpretation (see Boogert et al. 2015 for a review). CN-bearing species are, indeed, produced in the laboratory by energetic processing of interstellar ice analogs (e.g., Pendleton et al. 1999), but the spatial distribution of the 4.62 μm absorber suggests that it is formed at least in part by cold surface chemistry in dark clouds (van Broekhuizen et al. 2005). OCN^- production might be linked to that of the NH_4^+ ion. If the attribution of the 6.85 μm feature to NH_4^+ is correct, it must also be produced in cold, quiescent regions devoid of energetic sources because the feature is observed in dark clouds (see Section 5.3.5 for related discussion). OCN^- and NH_4^+ ions may form simultaneously at low temperature by acid-base reactions between NH_3 and HNCO in the ices.

Other potential products of energetic processing include refractory forms of solid carbon. Carbonaceous dust is a vital ingredient of all realistic models for interstellar grains (Section 3.6.2), and production in the outflows of carbon stars appears to be insufficient to account for the required abundance (Section 8.4.4), implying the need for an additional source. Organic refractory matter (ORM) has been widely discussed as a candidate (e.g., Greenberg et al. 1995; Li & Greenberg 1997). It is proposed that grains returned to the diffuse ISM from star formation regions retain mantles in which the ices have been converted into ORM by photolysis. This proposal is supported by laboratory simulations, but again observational confirmation is equivocal at best. If ORM is produced efficiently in the environments of embedded YSOs, we should expect to see evidence for its presence in their spectra. The 3.4 μm feature associated with C–H bonds in ORM (see Figure 5.9) has a potential counterpart in spectra of diffuse-ISM sources but not in those of YSOs (Pendleton & Allamandola 2002; Shenoy et al. 2003). Conversely, ORM has a stronger spectral signature near 6 μm attributed to C=O bonds that may contribute to absorption in this spectral region toward YSOs (Gibb & Whittet 2002) but is weak or absent in the diffuse ISM (Chiar et al. 2013). If ORM is a significant product of energetic processing near young stars, it must be promptly reduced to amorphous carbon as it cycles into lower-density regions of the ISM. Surface C–H bonds on small amorphous carbon grains may then yield satisfactory fits to the 3 and 6 μm regions of the spectrum simultaneously (Section 5.2.3).

9.3.4 Chemistry in Hot Cores and Corinos

During its early evolution, a massive YSO transitions from a collapsing protostar that emits its luminous energy almost entirely in the infrared to an incandescent main-sequence star that emits copiously in the ultraviolet (see Motte et al. 2018 for a review). Eventually, it will ionize the surrounding gas to form an H II region (Section 1.3.3). During the transition phase between these states, lasting some 10^5 years, the density and temperature of the molecular envelope may reach $\sim 10^{12}$–10^{14} m^{-3} and 100–300 K, significantly elevated in comparison to the surrounding molecular cloud (e.g., van Dishoeck & Blake 1998; Lahuis & van Dishoeck 2000; Gibb et al. 2000; Shimonishi et al. 2016). Physical conditions in these so-called hot cores (Figure 9.9) present opportunities for chemical evolution not possible in colder dense clouds. Ices that formed by surface chemistry on dust are vaporized and hence available for further processing in the gas, feeding reactions that produce molecules of greater complexity. Analogous regions, termed hot corinos, are detected around lower-mass YSOs (e.g., Bottinelli et al. 2004; Ceccarelli 2006; Imai et al. 2016). Whereas hot cores are generally large (radii $\sim$ 10^3–10^4 AU) in comparison to planetary systems, hot corinos ($\sim 10^2$ AU) are comparable in size to protoplanetary disks and may therefore influence their chemical content. Species commonly observed by radio techniques in hot cores and corinos include a wide range of COMs, of which $HCOOCH_3$ (methyl formate), $HOCH_2CHO$ (glycolaldehyde), CH_3OCH_3 (dimethyl ether), CH_2CHCN (vinyl cyanide), and CH_3CH_2CN (ethyl cyanide) are just a few examples (see Herbst & van Dishoeck 2009, and Jørgensen et al. 2020 for reviews).

It is of great interest to understand how chemistry evolves in these unique environments. Three distinct phases are considered in chemical modeling. Hot-core formation is preceded by a cold phase in which reactions both in the gas and on the dust proceed at low temperature ($T \lesssim 20$ K) in the parent cloud, as discussed in Sections 9.2.1–9.2.3. This is followed by a warm-up phase, in which the most volatile constituents of the grain mantles begin to sublimate and species remaining on the dust become more mobile and may react more easily. Finally, in the hot-core phase, essentially all of the ices have sublimated and chemistry proceeds entirely in the gas. Important products of gaseous and/or surface reactions in the cold phase include CO and its hydrogenation products (Equation (9.18)), together with H_2O, NH_3, and CH_4. Such compounds may react to generate molecules of greater complexity as the temperature increases. For example, radicals resulting from CO hydrogenation may combine in the ice mantles during warm-up to produce methyl formate and glycolaldehyde (Herbst 2014):

$$\begin{aligned} HCO + CH_3O &\rightarrow HCOOCH_3 \\ HCO + CH_2OH &\rightarrow HOCH_2CHO. \end{aligned} \tag{9.25}$$

When the hot-core phase is reached, additional species form by means of chemical networks fed by the products of prior gaseous and surface chemistry. Whereas ion–molecule reactions tend to dominate gas-phase chemistry at low temperature, many neutral-neutral reactions become relevant in hot cores as elevated temperatures and wind-driven shocks provide the means to overcome activation energies and drive endothermic reactions. Abundances of the products vary significantly from one source to another, suggesting that differences in physical conditions and/or evolutionary status affect the chemistry. The origin of these differences and the chemical networks that underpin them are not yet fully understood.

Some relatively simple molecules observed to be abundant in hot cores are undetected in the ices. In some cases, this may arise from differences in sensitivity: the spectral signatures of trace constituents in solids are generally much harder to detect than those in the gas; in other cases, the differences seem surprising and might inform our efforts to understand both cold and hot-core chemistries. Nitrogen compounds provide a notable example. NH_3 is a significant ice constituent and its abundance in the gas may arise in large part from sublimation, but other compounds are more puzzling. A clear distinction can be made between nitriles (generic formula R–C≡N) and equivalent oxygen-bearing molecules (cyanates, R–O–C≡N). Gas-phase HCN and heavier nitriles such as HC_3N and CH_3CN are widely observed and abundant in hot-cores, but species containing cyanate groups are relatively rare (Herbst & van Dishoeck 2009). If nitriles are a significant constituent of the ices they should produce detectable absorptions in the 4.35–4.55 μm region of the infrared spectrum, but none have been seen to date. Instead, a feature observed at 4.62 μm (very weak in most sources) is attributed to the OCN^- ion, as discussed in Section 9.3.3 above. The preponderance of nitriles in hot cores must therefore arise from gas-phase chemistry. This may be instigated by the ion–molecule reaction

$$C^+ + NH_3 = HCNH^+ + H \tag{9.26}$$

followed by dissociative recombination of $HCNH^+$ to form HCN (Charnley et al. 2001). The lack of detectable nitrile compounds in the ices may be explained by their relatively low gas-phase abundances in the parent molecular cloud (no significant frosting) and preferential production of NH_3 and N_2 from adsorbed N atoms on grain surfaces.

9.3.5 From Cores to Disks

Pre-main-sequence evolution may be characterized by distinct phases, as illustrated schematically in Figure 9.12 for a single star of moderate mass. Gravitational collapse of a dense core in a dark cloud forms a dust-embedded protostar, internally heated by release of gravitational energy as it accretes material. The protostar inevitably inherits some non-zero net rotation from the parent core, such that conservation of angular momentum causes its envelope to flatten into a rotating equatorial disk as collapse proceeds. Meanwhile, the temperature of the protostellar nucleus rises until thermonuclear reactions are ignited, and the young star begins to drive a wind. The wind is rapidly decelerated when it encounters the disk, but travels more freely in other directions, blowing cavities in the polar regions of the envelope to establish a bipolar outflow. During the early phases of collapse (Core → Class 0 → Class I), the young star remains deeply enshrouded in dense molecular gas and dust, and most of its luminous energy emerges in the mid to far-infrared. Eventually, the remaining material in the extended envelope either falls into the disk or is swept up and ejected by the outflow. The YSO then becomes a T Tauri-type pre-main-sequence star (Class II → Class III) with a protoplanetary disk.

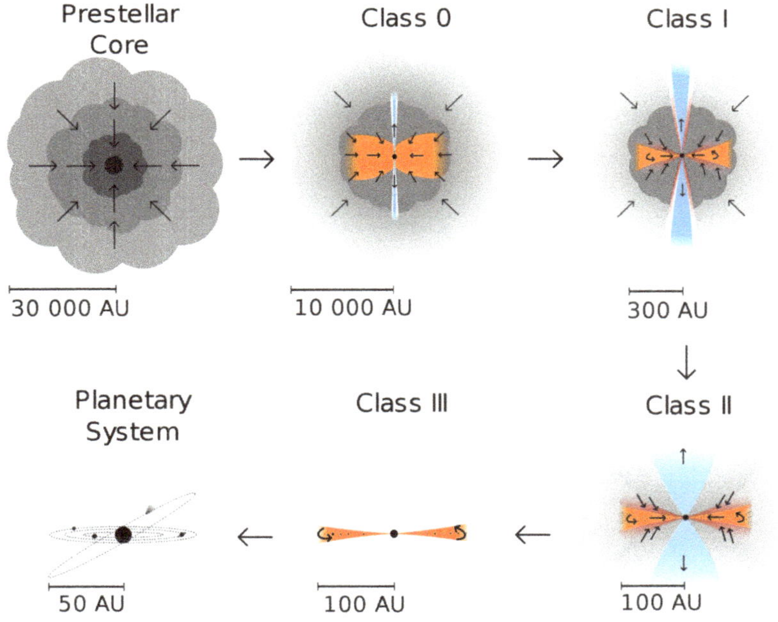

Figure 9.12. Schematic illustration of sequential phases in the early evolution of a single star of moderate mass, from initial collapse of a prestellar core to arrival on the main sequence (clockwise from top left). Loci of the accretion disk and bipolar jets are shown in orange and blue, respectively. Classes are based on the slope of the spectral energy distribution (see Equation (9.27) and related text). Image credit: Magnus Vilhelm Persson (Creative Commons BY 4.0 license).

This sequence of events considers the evolution of a single star, but a substantial fraction of all stars are known to be members of binary or multiple systems. As random capture is dynamically unfeasible, their status as such must be predetermined by the birth environment. A crucial parameter appears to be the angular momentum of the system, which may determine whether collapse results in a stable disk around a single protostar or a fragmented disk with more than one major mass concentration (Bodenheimer et al. 1993; Larson 2003). This issue is not considered further here: as one of the goals of this chapter is to examine what can be learned about the evolution of matter incorporated into potentially habitable planetary systems, it is natural to focus on single stars (e.g., Ballantyne et al. 2021).

YSOs that have reached the T Tauri phase may display visible photospheres, and many such stars have been detected and studied. Depending on its locus and orientation, however, a T Tauri star may remain hidden at visible wavelengths by foreground extinction in the parent cloud, or by extinction in the protoplanetary disk if our viewing angle is close to its equatorial plane. In the latter case, visible light from the star may sometimes be detected indirectly, via reflection, in which starlight is scattered into our line of sight by ambient dust lying out of the plane of the disk (e.g., Weintraub et al. 1995). The bipolar jets driven by the wind from the star are collimated and accelerated by the stellar magnetic field, resulting in shock waves that may generate optical emission nebulae (Herbig–Haro objects) in the surrounding ISM. The jets provide indirect but important insights into the nature of the accretion and mass-loss processes that govern star formation (Reipurth & Bally 2001). Mass is being lost from the system while material in the envelope continues to accumulate into the disk, which becomes flared in shape (Kenyon & Hartmann 1987; Rich et al. 2021). The motions of gas and dust are coupled by collisional friction. Doppler analyses of molecular line emissions show that the general rotation of the disk becomes Keplerian, i.e., the tangential speed at a radial distance r from the star in the midplane approximates to an $r^{-1/2}$ law, consistent with planetary motion (Sargent & Beckwith 1987; Hogerheijde et al. 2002; Pinte et al. 2016).

The fate of dust falling into the protoplanetary disk of a solar-mass star depends on a number of factors, including its volatility, location and accretion rate (Chick & Cassen 1997). In the hot innermost zone of the disk ($r \lesssim 1$ AU) all solids are vaporized, whereas at large distances from the star ($r \gtrsim 30$ AU) interstellar dust and ices may survive essentially unaltered. Between these extremes, vaporization of the volatiles and modification of the more robust particles is to be expected. Infall creates an accretion shock that compresses the gas and results in frictional heating of the dust. Heating may be sufficient to anneal and partially crystallize the silicates, consistent with spectral signatures of crystallinity detected in the disks (Jäger et al. 1998; Olofsson et al. 2009; Oliveira et al. 2011). The observations indicate that significant radial mixing must occur, as crystalline silicates are detected further from the star than their probably points of origin. Ices sublimated during infall may recondense in cooler regions of the midplane ($r \gtrsim 3$ AU), where they are shielded from stellar radiation by the disk itself, creating a "snowline" beyond which ices as well as minerals are available for growth into larger bodies.

Evolution of a typical YSO through these various stages is accompanied by a progressive thinning of the circumstellar environment. Initially, mass loss is largely limited to the bipolar jets. Eventually, after the envelope that fed the disk has dispersed, the disk itself begins to thin out. Timescales for disk dissipation are a few Myr (Williams & Cieza 2011). Relevant processes include orbital kinematics and radiation pressure. Absorption and re-emission of radiation by a particle in the inner disk may cause its orbit to decay by Poynting–Robertson drag, arising from the fact that the absorbed radiation is incident from a specific direction whereas the emitted radiation is isotropic; depending on the size, optical properties and dynamics of the particle, this may cause it to spiral into the star. Radiation pressure has the opposite effect, introducing a force that accelerates particles away from the star (Section 2.4.1). In both cases, the amplitude of the effect depends on particle size. Detailed calculations by Artymowicz (1988) suggest that radiation pressure becomes the dominant process as the disk becomes optically thin: particles of a given composition and with dimensions less than some critical size may then be expelled efficiently. For any realistic grain composition, this critical size is $a \sim 1\ \mu$m. If the size distribution of the orbiting particles were to remain the same as their interstellar precursors, radiation pressure might lead to almost complete dissipation of the disk. However, growth by grain–grain coagulation (Section 9.1.1) produces clusters with dimensions $>10\ \mu$m, i.e., well above the critical size, on timescales $\lesssim 10^4$ years (Suttner & Yorke 2001). In effect, this is the initial step toward planet formation.

The evolutionary stages described above and illustrated in Figure 9.12 are manifested to an external observer by systematic differences in the spectral energy distributions (SEDs) of the sources. With SEDs plotted in $\log \lambda F_\lambda$ versus $\log \lambda$ format (e.g., Figure 9.13), a classification scheme in common use is based on the mean slope at wavelengths from near to mid-infrared ($2 < \lambda < 25\ \mu$m):

$$\alpha_{\rm IR} = \frac{{\rm d}(\log \lambda F_\lambda)}{{\rm d}(\log \lambda)} \tag{9.27}$$

(e.g., Wilking 1989). During the earliest stages of gravitational core collapse ($T_{\rm d} \sim 15$ K), a source emits almost nothing in this spectral range, so $\alpha_{\rm IR}$ is indeterminate (see Figure 6.3, red curve); protostars in this phase are designated "Class 0" (André et al. 1993, 2000). The youngest objects detected in the mid-infrared, designated Class I, have warmer envelopes characterized by SEDs that increase toward longer wavelength ($\alpha_{\rm IR} > 0$). As an embedded YSO evolves, the photospheric contribution increases, dust emission becomes less dominant, and the slope becomes negative. Class II sources have slopes in the range $0 > \alpha_{\rm IR} > -2$ and are associated with T Tauri stars or their more massive counterparts, the Herbig Ae/Be stars. Note, however, that a Class II star with an *edge-on* disk may mimic a Class I source because the photosphere remains hidden (see Crapsi et al. 2008 for detailed discussion). Finally, stars close to or on the zero-age main sequence have spectral energy distributions resembling blackbodies at photospheric temperatures, for which

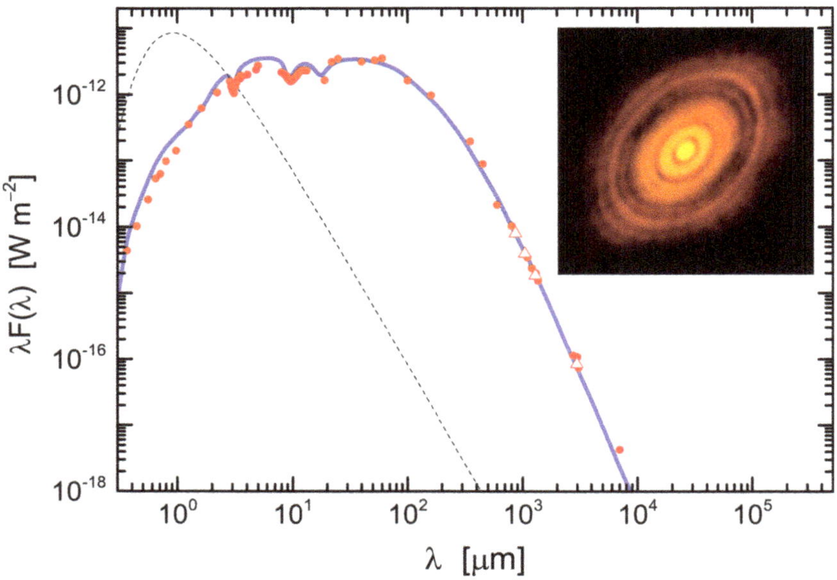

Figure 9.13. Spectral energy distribution of the young stellar object HL Tauri and an image of its protoplanetary disk at $\lambda = 1.28$ mm. Flux data are from Pinte et al. (2016) and references therein: ALMA observations (triangles) and a compilation of previous literature (red points). The dashed curve is an approximate (blackbody) representation of the stellar continuum that would be seen if no dust were present. A model fit to the SED (blue curve) assumes an inclined, toroidal disk in an envelope of mass 0.1 $M_\odot$, containing ices, silicates and amorphous carbon dust (Men'shchikov et al. 1999). Inset image credit: ALMA (ESO/NAOJ/NRAO).

the infrared slope approaches the value predicted by the Rayleigh–Jeans approximation ($\alpha_{IR} = -3$). These objects, designated Class III, have active chromospheres and are often strong X-ray emitters.

The SED of a well-studied pre-main-sequence star is illustrated in Figure 9.13. HL Tauri is similar in mass to the Sun and has an estimated age $\sim 10^5$ years. The SED is almost flat in the 2–25 μm range, indicating that the star is in transition between the embedded (Class I) and T Tauri (Class II) phases. Models fitted to the flux data are consistent with the presence of an inclined circumstellar disk (Men'shchikov et al. 1999; Robitaille et al. 2007), as spectacularly confirmed by imaging of the submillimeter dust emission (Figure 9.13 inset, discussed further below). The SEDs of young stars often exhibit structure arising from dust-related emission or absorption features: the presence of both ice and silicate dust in the line of sight to HL Tau is indicated by absorption features at 3, 10 and 19 μm, superposed on the continuum in Figure 9.13. In general, the observed spectra of such sources may contain contributions from dust both within the circumstellar disk/envelope and in the foreground molecular cloud and diffuse ISM. Models suggest that ices toward HL Tauri are present in the cold outer regions of the envelope (Furlan et al. 2008). As silicates may appear in net emission or net absorption, depending on their distribution with respect to ambient conditions along the line of sight, a superposition of components is required to match the observed profiles (Bowey & Adamson 2001). See also Section 6.2.1 for related discussion.

The sensitivities and resolutions available with facilities such as ALMA are providing new insights into the nature of protoplanetary disks. The disks around even relatively nearby YSOs such as HL Tauri ($d \approx 140$ pc) subtend angles no more than about 1 arcsecond on the sky; but using interferometric techniques, line emission from the gas and continuum emission and scattered radiation from the dust may be mapped at millimeter and submillimeter wavelengths at angular resolutions sufficient to resolve structure in the disks. These observations offer deep insights on how planetary systems form (see Andrews 2020 for a review). The disk of HL Tauri (Figure 9.13) is about 120 AU in radius (roughly four times the mean orbital radius of Neptune around the Sun), and inclined at an angle of about 47° to our line of sight. The spatial resolution of the image (3.5 AU) is sufficient to reveal detailed structure, including a pattern of concentric bright rings separated by darker "gaps." The gaps are not empty but depleted in dust by a factor of about 10 relative to the rings. This may be attributed to accretion of small grains into much larger aggregates (Pinte et al. 2016). The darkest gaps are consistent with the removal of up to 40 Earth-masses of material. Accretion on this scale could be sufficient to construct large planetary embryos capable of gas capture, and thus to form giant planets. Detailed modeling gives insight into the distribution of grain sizes and suggests that grain-growth is most efficient in central regions of the disk. The distributions of gas-phase molecules such as H_2O, NH_3, CH_3OH, and CO may be used to identify snowlines, setting radial limits on their presence in solid versus gaseous form according to their individual volatilities, and thus distinguishing accretion zones that form icy versus rocky aggregates. See Carrasco-González et al. (2019) and Rich et al. (2021) for further discussion and examples. For comprehensive reviews of theoretical and experimental work on the physics of grain growth in protoplanetary disks, see Blum & Wurm (2008) and Blum (2010, 2018).

Observations of polarized radiation from young stars and their envelopes are also vital. Results may elucidate the properties of the dust grains, especially their sizes, the orientation and strength of the local magnetic field, and the geometry of the system (e.g., Weintraub et al. 2000; Cho & Lazarian 2007; Andrews 2020). Polarization may be introduced in different ways. If the line of sight contains non-spherical grains that are being systematically aligned (e.g., by a magnetic field or a supersonic wind; Section 4.3), linear polarization in the transmitted beam will result from dichroic extinction, as discussed in Section 2.2.1, and corresponding polarization with orthogonal position angle will appear in the thermal emission from the dust at longer wavelengths (Section 2.3.5). If starlight is scattered into the line of sight, it will be polarized even if the grains are spherical or have no net alignment. Scattered light from a star embedded in a uniform dusty medium produces a centrosymmetric pattern of linear polarization vectors, modified in the case of a YSO by the disk-structure of the circumstellar envelope. Polarizations produced by dichroism and by scattering are expected have different spatial distributions, arising from their differing dependencies on magnetic field direction and the direction of incident radiation, which allows the possibility to disentangle their contributions. However, the interpretation of such observations is complex and challenging (e.g., Bertrang & Wolf 2017; Stephens et al. 2017).

If the dust grains are spherical, only linear polarization is produced when unpolarized radiation is scattered. However, circular polarization will result if radiation that is already linearly polarized is scattered, or if unpolarized radiation is scattered by aligned non-spherical particles. Either of these situations is likely to arise in YSO envelopes, e.g., when radiation already linearly polarized by a scattering event is scattered again (multiple scattering), or when target grains similar to those responsible for polarization in the ISM are aligned prior to scattering. Observations have revealed remarkably high degrees of near-infrared circular polarization toward many YSOs (Kwon et al. 2014 and references therein). The highest levels ($\sim$ 20%), detected in certain high-mass star formation regions such as OMC–1, vastly exceed those measured in the ISM (Section 4.2.5). The observed distributions typically display quadrupolar structure centered on the source: its architecture and amplitude can be understood if non-spherical grains are being aligned in circumstellar environments (Whitney & Wolff 2002). The models suggest that the net circular polarization observed toward a specific source is highly dependent on viewing angle.

9.3.6 Debris Disks

A protoplanetary disk around a Sun-like star may have a typical mass $\sim 0.1\ M_\odot$ prior to the onset of planet formation (see Franceschi et al. 2022 for caveats and uncertainties). The current mass of our planetary system is $\sim 1.4 \times 10^{-3}\ M_\odot$, most of it in the giant planets; only a tiny fraction ($\sim$ 1 part in 10^{11}) is in small particles such as micrometeoroids and interplanetary dust responsible for the zodiacal light. Protoplanetary disks are expected to dissipate and thin-out during the T Tauri phase, as matter is lost from the system and the remaining material is incorporated into larger bodies, as discussed in Section 9.3.5 above. Is this process complete by the time the star reaches maturity? In many cases it is clearly not, as evidenced by detections of dusty disks around main sequence stars (Aumann et al. 1984; Smith & Terrile 1984). These so-called "debris" disks occur in about 20% of single main-sequence stars with spectral types in the range A to K, their estimated ages ranging from about 10 Myr to a few Gyr (see Hughes et al. 2018 and Wyatt 2022 for reviews). Well-known examples include the bright stars Vega, Fomalhaut, and β Pictoris.

The primary means of detecting debris disks is the presence of far-infrared flux in excess of that predicted from a stellar photosphere, caused by thermal emission from circumstellar particles. Their spectral energy distributions imply dust temperatures typically in the range from $\sim$ 50 to 150 K. With sufficient resolution, the brightest disks may be imaged in the thermal infrared; in a few cases, scattered starlight is also detected at shorter wavelengths. Radial distributions of the disks typically fall in the range 30–150 AU, which overlaps the locus of the Kuiper Belt (30–50 AU) in our solar system. Searches for line emissions suggest that the disks generally have very low gas content (e.g., Rebollido et al. 2022). The amplitude of infrared excess emission enables the total mass of dust to be estimated: results for the most massive disks amount to $\sim 10^{-6}$ $M_{\odot}$, or 0.3 Earth masses (Holland et al. 2017). However, the small-particle approximation inherent in dust-mass calculations based on far-infrared to millimeter-wave observations (Section 2.3.3) is invalid for large ($a \gtrsim 500$ μm) particles presumed to be present in debris disks, hence the dust estimates place only lower limits on total disk masses.

There can be little doubt that the debris disks are, in most if not all cases, the remnants of protoplanetary accretion disks. In some instances, exoplanets with orbits approximately coplanar with the debris disk have been detected in the same system. Some disks also exhibit structure that might result from the gravitational influence of unseen planets. However, the timescale for dissipation of small particles from a protoplanetary disk (Section 9.3.5) is short compared with the age range of stars with debris disks, by orders of magnitude in many cases. Hence, the disks cannot represent a permanent population of unaccreted small grains. Instead, the emitters are presumed to be aggregates (planetesimals) that have grown sufficiently to be immune from expulsion by radiation pressure and winds. The aggregates are subject to collisions, however, leading to fragmentation and orbital scattering, which may result in accretion onto planets (or the star itself) or to expulsion from the system. Hence, the disks are expected to dissipate slowly over time. This scenario is consistent with observations showing that the incidence of debris disks, and the amplitudes of their emission when present, tend to decline with stellar age (Rieke et al. 2005; Holland et al. 2017). Debris disks may thus represent more massive analogs of the Kuiper Belt; currently available sensitivities are insufficient to detect debris disks of mass equivalent to that of the Kuiper Belt around other stars.

9.3.7 The Astrochemical Heritage of the Solar System

I conclude this chapter with an assessment of links between astrochemical processes, described in the preceding sections, and the nature of the raw materials available to build planets in protoplanetary disks. The focus is on what may be learned about the content and early evolution of our own planetary system—with the expectation that similar circumstances may arise elsewhere. Primitive bodies such as asteroids and comets are remnants of the solar nebula, and they retain chemical memories of the events and processes that forged them. Information is gathered from astronomical observations, space rendezvous missions, and laboratory analyses of physical samples. Results offer unique insights into the composition of the solar nebula and its astrochemical heritage. See Brownlee (2014), van Dishoeck et al. (2014), Altwegg et al. (2019), and Jørgensen et al. (2020) for related reviews.

A logical starting point is the carbonaceous chondrite class of meteorites, physical samples of great significance discussed in different contexts elsewhere in this text. These stony meteorites are considered to be collisional fragments of asteroids, emanating from the outer regions of the main asteroid belt, and to represent the least altered samples available from that population. The currently accepted value for the age of the solar system is based on radiometric dating of these samples. On average, their elemental abundances are essentially solar for all but the lightest condensible elements, and because of this they provide an important reference standard (Section 7.1.3). On closer examination, their elemental and isotopic abundances have been found to retain diagnostic information that opens new windows on cosmic evolution. A small but significant fraction of the subparticles contained within the aggregates are identified unambiguously as presolar in origin, on the basis of their isotopic compositions, as discussed in detail in Section 8.3.

Isotopic abundance patterns in carbonaceous chondrites also provide clues regarding the nature of the solar birth aggregate. As discussed in Section 9.3.1, the circumstances of its birth may have a major influence on the early evolution of a solar-mass star and its protoplanetary disk. As a mature, isolated star, the Sun is now evidently far removed from its birthplace, but the weight of evidence suggests that it formed in a relatively large group, with a membership of several hundred to a few thousand stars (Adams 2010; Gounelle & Meynet 2012). Support for this conclusion arises in part from abundance data for elements produced by decay of short-lived radioactive isotopes in the meteorites. Examples include ^{26}Al and ^{60}Fe, which decay to ^{26}Mg and ^{60}Ni with half-lives of 0.6 and 1.5 Myr, respectively, timescales short compared with the pre-main-sequence lifetime of a solar-mass star. The most likely explanation is that the parent isotopes were injected into the solar nebula by a contemporary supernova. This is a reasonably probable outcome if the Sun formed alongside massive stars in a large group, an extremely unlikely coincidence if it did not. Large groups often become gravitationally-bound star clusters, but the Sun's current isolation by no means contradicts the evidence for a populous birth environment; all but the densest star clusters are expected to disperse on timescales shorter than the current age of the Sun.

The architecture of the solar system provides additional clues. Possible consequences of a populous birth aggregate include exposure to hard radiation fields, energetic winds and shocks, and gravitational disruption of the protoplanetary disk caused by close encounters with more massive siblings. The orderly structure of the planets' present-day orbits around the Sun (concentric, almost circular, almost coplanar, with prograde motion) argues against any major disruption of the protoplanetary disk arising from a close encounter. However, the distribution of mass declines very rapidly at solar distances beyond the current orbit of Neptune, suggesting that the radial extent of the protoplanetary disk may have been limited to about 40 AU, perhaps by photo-evaporation or erosion driven by an external source (see Figure 9.8). It thus seems reasonable to surmise that the photochemical evolution of the solar nebula may have been influenced by external as well as internal irradiation.

Presolar grains identified in meteorites are composed of refractory solids (Table 8.2) for the practical reason that they are best equipped to survive both the vagaries of space and the laboratory methods that may be used to extract them. The provenances of other materials are generally more difficult to establish and often based on analyses of bulk extracts rather than individual grains. Essentially all solids were vaporized and recondensed in the innermost zone of the solar nebula, where the terrestrial planets formed. Subsequent radial mixing is evidenced by results from the Stardust mission, which identified high-temperature nebular condensates in a comet that must have condensed well beyond the snowline (Brownlee 2014). Conversely, icy planetesimals scattered into eccentric orbits transport volatiles into the inner solar system. Both carbonaceous chondrites and interplanetary dust particles (IDPs) contain organic matter, but with differing chemical histories: the meteorites are largely composed of hydrous minerals, the presence of which implies processing in aqueous environments on the parent asteroids, whereas many IDPs are anhydrous and most likely of cometary origin, released when ices sublimate without passing through a liquid phase. Nevertheless, these diverse materials display a common isotopic signature, deuterium enhancement, the "Ariadne's Thread" that connects the pieces of the puzzle (Ceccarelli et al. 2014).

Approximately one in every 50,000 hydrogen atoms in the cosmos is deuterated, i.e., its nucleus contains a neutron as well as a proton. The factor 2 ratio in mass affects the chemical kinetics and may result in very substantial levels of fractionation. Specifically, high molecular D/H ratios have been recognized as a signature of chemical processes operating in the low-temperature environments of dark clouds and prestellar cores (e.g., Tielens 2013). Molecular hydrogen may be deuterated in the gas phase by the ion–molecule reaction

$$D^+ + H_2 \rightarrow HD + H^+ \tag{9.28}$$

and the important H_3^+ ion (Section 9.2.1) is then deuterated by the exothermic reaction

$$H_3^+ + HD \rightarrow H_2D^+ + H_2. \tag{9.29}$$

As the reverse of reaction (9.29) is endothermic, the abundance ratio H_2D^+/H_3^+ becomes enhanced systematically at low temperatures, and carbon-bearing molecules formed from reaction sequences that begin with H_3^+ will inherit this enhancement. The observed D/H ratios of some gas-phase interstellar molecules exceed the cosmic value by orders of magnitude. Examples include the important precursor molecules H_2CO and CH_3OH, for which even doubly-deuterated forms have been detected (Parise et al. 2006). Surface chemistry is also implicated. The atomic D/H ratio of the gas becomes enhanced by dissociative recombination of deuterated H_3^+ as molecule formation proceeds, and thus H_2O and other ices that form by surface hydrogenation are expected to inherit this enhancement.[2]

Figure 9.14 presents an overview of available data on deuterium fractionation in the solar system. The baseline (yellow band) is set by the average "cosmic" value of $D/H \approx 2 \times 10^{-5}$ applicable to unfractionated solar nebular gas. D/H values for Jupiter and Saturn, suggest that most of their mass accumulated from the same reservoir, whereas Uranus and Neptune acquired significantly higher proportions of fractionated material. The Earth's oceans are enhanced by a factor of almost 10

[2] Observational confirmation should be possible by means of infrared spectroscopy: ices containing HDO exhibit an O–D vibrational feature at 4.1 μm, but this has so far eluded unambiguous detection. See Urso et al. (2018) for related discussion and laboratory work.

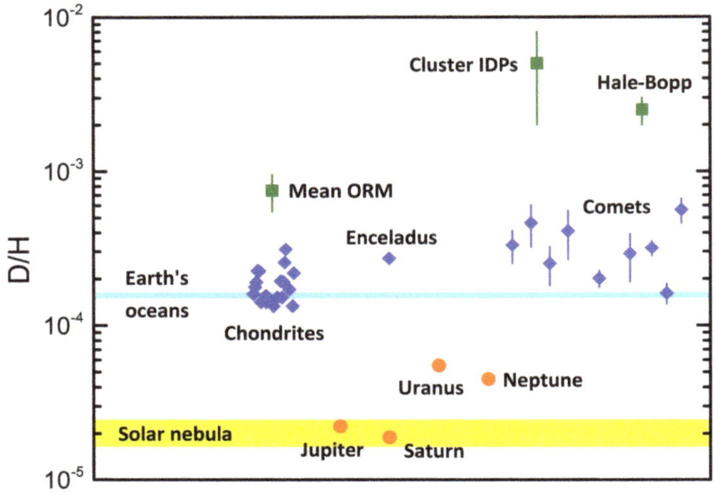

Figure 9.14. D/H ratios of various bodies and reservoirs in the solar system. The yellow band and cyan line indicate levels in unfractionated solar nebular gas and Earth's ocean water, respectively. Data are displayed roughly in order of increasing solar formation distance from left to right. Orange circles: HD/H_2 in the atmospheres of the giant planets. Blue diamonds: HDO/H_2O in chondrites, Saturn's moon Enceladus, and various comets. Green squares: organic D/H ratios, including means for organic refractory matter in carbonaceous chondrites and cometary "cluster" IDPs, and an observed value for HCN in comet Hale-Bopp. Based on data from Altwegg et al. (2015), Cleeves et al. (2016) and references therein.

relative to the baseline value. A general pattern is evident in the data for smaller bodies. H_2O in hydrated asteroidal minerals, cometary ices, and the icy moon Enceladus shows D/H enhancements similar to or greater than the terrestrial value, lending support to the hypothesis that the early Earth acquired some or all of its water by accretion of such bodies (e.g., Izidoro et al. 2013; Alexander et al. 2018). Enhancements are also characteristic of organic matter from both asteroidal and cometary sources. Among the largest are those found in fragile (cluster) IDPs thought to emanate from comets, their D/H ratios comparable with those observed in dense molecular clouds and prestellar cores (Messenger 2002). Models suggest that both the deuterium enhancement of H_2O and the differential fractionation of organics relative to H_2O may be inherited, at least in part, from chemical processes in the parent cloud (Cleeves et al. 2016; Furuya et al. 2017).

The provenance of the comets available for observation and analysis is diverse. With regard to orbital characteristics, the sample includes objects from the disk population and long-period comets from the Oort cloud. However, both of these groups are presumed to have originated in the protoplanetary disk, with subsequent redistribution imposed by gravitational perturbations. Comet-to-comet variations in the D/H ratio (Figure 9.14) provide a guide to points of origin irrespective of present-day orbital characteristics (Altwegg et al. 2019). The distribution most likely reflects differing degrees of mixing between solar nebular condensates and deuterium-enriched ices inherited from the parent cloud, and it implies that comets formed over a large radial distance range in the disk. The thermal history of a comet may also be assessed by comparing the abundances of ices with differing volatilities, for example those richest in CO are likely to be the most pristine.

A summary of molecular abundance data for comets appears in Table 5.3, alongside data for interstellar/protostellar ices in both dark clouds and regions of low and high-mass star formation. Comparisons between data for comets and low-mass YSOs seem most likely to be informative (Öberg et al. 2011; Boogert et al. 2015). However, it should be noted that the line-of-sight average toward a YSO will generally sample ices in the extended envelope of the source and/or the parent cloud, with little or no contribution from the protoplanetary disk in which comets are presumed to form. The results suggest some broad similarities, but also significant differences. Oxides of carbon (CO, CO_2) tend to dominate over hydrogenated forms in both samples; but although the abundance distributions overlap, many comets are relatively depleted in both CO and CO_2 compared with low-mass YSOs. The cometary abundances of CH_4 and (especially) NH_3 are surprisingly low, given that both are expected to form simultaneously with H_2O in the ISM during the hydrogen-dominated phase of mantle growth (Sections 5.3 and 9.2.2). Conversely, many species that remain undetected in interstellar ices have been identified at low levels in comets, with abundances $\lesssim 1\%$ relative to H_2O (Mumma & Charnley 2011; Altwegg et al. 2019). Notable examples include N_2, O_2, C_2H_2, C_2H_6, HCN, and H_2S. However, the possibility that these species might be present at similar levels in interstellar or protostellar ices is not ruled out by the observations.

The levels of chemical complexity detected in asteroidal and cometary material extend to prebiotic compounds, including amino acids, nucleobases and sugars. The study of amino acids in soluble organic extracts from carbonaceous chondrites is a field with an extensive history (e.g., Kvenvolden et al. 1970; Cronin & Chang 1993; Pizzarello & Huang 2005; Pizzarello 2012). They are identified as intrinsic to the meteorites and not terrestrial contaminants in various ways, most notably by their chemical diversity (over 70 different amino acids have been identified, with no particular preference for the 20 utilized by terrestrial biology), and by isotopic signatures such as higher D/H ratios. More recently, the Stardust and Rosetta missions have enabled detection of glycine, the simplest amino acid, in comets (Elsila et al. 2009; Altwegg et al. 2016). Nucleobases and sugars have also been identified in meteorites and comets (e.g., Callahan et al. 2011; Furukawa et al. 2019; Zellner et al. 2020). These discoveries support the view that the astrochemical heritage and early evolution of a solar system may yield resources necessary for life to emerge on a suitable planet (e.g., Ehrenfreund & Charnley 2000; Caselli & Ceccarelli 2012; Whittet 2017).

A specific connection between meteoritic amino acids and the origin of terrestrial life is suggested by the discovery that these molecules carry a chiral asymmetry. With the exception of glycine, all amino acids essential to life are chiral, i.e., they have left-handed or right-handed forms (L and D enantiomers) that are mirror images of each other. Terrestrial life is homochiral: nucleic acids use only the right-handed form of the appropriate sugar (ribose in RNA, deoxyribose in DNA); proteins and enzymes use only left-handed amino acids. The origin of terrestrial homochirality is one of the "Holy Grail" questions in all science. It seems suggestive that amino acids extracted from carbonaceous chondrites exhibit an excess of the L form, consistent with the chiral preference of terrestrial biology (see Pizzarello 2016 for a review). The possibility that this might result from terrestrial contamination is ruled out for reasons noted above. The degree of asymmetry varies from one meteorite to another and can be as high as 60%. The simple fact that meteorites carry intact organics to the Earth today demonstrates the plausibility of delivery, and accretion rates must have been far higher at early times when the Earth was acquiring its water. To date, this is the only confirmed case of chiral asymmetry in organic matter formed outside of the Earth's biosphere.

How did the asymmetry arise? A correlation has been found between its amplitude and the degree of aqueous processing of the parent body (Glavin et al. 2010). Some crystalline minerals (e.g., calcite) are intrinsically chiral, i.e., the arrangement of surface atoms on a given crystal face is not superimposable by its mirror image, and this property can be inherited by organic polymers that form on the surface. However, if one surface favors L over D enantiomers, its mirror image (occurring with equal probability) will favor D over L, so the overall chirality of the system is unchanged. It is more reasonable to suppose that chemical processing on the asteroids amplified some pre-existing excess. It is difficult to formulate a satisfactory model that will *create* a net system-wide excess by any process that is entirely internal to the solar system, because there is no obvious way to break the symmetry.

The problem might be solved by invoking an external factor. One possibility is exposure of the solar birth aggregate to circularly polarized radiation of a given handedness from an external source. As previously discussed, there is reason to suppose that the Sun formed in a group that may have contained massive, luminous stars, analogs of which are known to be sources of circularly polarized radiation (Fukue et al. 2010, Meinert et al. 2011, and references therein). The radiation emerging in a given direction has an equal probability of being left-handed or right-handed, but a neighboring protostar may be exposed preferentially to just one, depending on its location. Experimental work and model calculations show that when an assemblage of chiral organic molecules that is initially racemic (equal numbers of L and D) is exposed to circularly polarized UV radiation, the balance of L and D is altered by asymmetric photochemistry and photolysis (e.g., de Marcellus 2011; Da Pieve et al. 2014). One form is favored over the other, depending on the handedness of the radiation.

References

Adams, F. C. 2010, ARA&A, 48, 47
Adams, F. C., & Myers, P. C. 2001, ApJ, 553, 744
Agarwal, V. K., Schutte, W., Greenberg, J. M., et al. 1985, OLEB, 16, 21
Alexander, C. M. O'D., McKeegan, K. D., & Altwegg, K. 2018, SSRv, 214, 36
Altwegg, K., Balsiger, H., Bar-Nun, A., et al. 2015, Sci, 347, 1261952
Altwegg, K., Balsiger, H., Bar-Nun, A., et al. 2016, SciA, 2, e1600285

Altwegg, K., Balsiger, H., & Fuselier, S. A. 2019, ARA&A, 57, 113
André, P., Ward-Thompson, D., & Barsony, M. 1993, ApJ, 406, 122
André, P., Ward-Thompson, D., & Barsony, M. 2000, in Protostars & Planets IV, ed. V. Mannings, et al. (Tucson, AZ: Univ. Arizona Press), 59
Andrews, S. M. 2020, ARA&A, 58, 483
Artymowicz, P. 1988, ApJ, 335, L79
Aumann, H. H., Gillett, F. C., Beichman, C. A., et al. 1984, ApJ, 278, L23
Ballantyne, H. A., Espaas, T., Norgrove, B. Z., et al. 2021, MNRAS, 507, 4507
Bergin, E. A., & Tafalla, M. 2007, ARA&A, 45, 339
Bernstein, M. P., Dworkin, J. P., Sandford, S. A., Cooper, G. W., & Allamandola, L. J. 2002, Natur, 416, 401
Bertrang, G. H.-M., & Wolf, S. 2017, MNRAS, 469, 2869
Blum, J. 2010, RAA, 10, 1199
Blum, J. 2018, SSRv, 214, 52
Blum, J., & Wurm, G. 2008, ARA&A, 46, 21
Bodenheimer, P., Ruzmaikina, T., & Mathieu, R. D. 1993, in Protostars & Planets III, ed. E. H. Levy, et al. (Tucson, AZ: Univ. Arizona Press), 367
Boogert, A. C. A., Gerakines, P. A., & Whittet, D. C. B. 2015, ARA&A, 53, 541
Bottinelli, S., Ceccarelli, C., Neri, R., et al. 2004, ApJ, 617, L69
Bowey, J. E., & Adamson, A. J. 2001, MNRAS, 320, 131
Briggs, R., Ertem, G., Ferris, J. P., et al. 1992, OLEB, 22, 287
Bringa, E. M., Kucheyev, S. O., Loeffler, M. J., et al. 2007, ApJ, 662, 372
Brownlee, D. 2014, AREPS, 42, 179
Callahan, M. P., Smith, K. E., Cleaves, H. J., et al. 2011, PNAS, 108, 13995
Carrasco-González, C., Sierra, A., Flock, M., et al. 2019, ApJ, 883, 71
Caselli, P., & Ceccarelli, C. 2012, A&ARv, 20, 56
Caselli, P., Walmsley, C. M., Tafalla, M., Dore, L., & Myers, P. C. 1999, ApJ, 523, L165
Cazaux, S., & Tielens, A. G. G. M. 2004, ApJ, 604, 222
Ceccarelli, C. 2006, in IAU Symp. 231, Astrochemistry: Recent Successes and Current Challenges, ed. D. C. Lis, G. A. Blake, & E. Herbst (Cambridge: Cambridge Univ. Press), 1
Ceccarelli, C., Caselli, P., Bockelee-Morvan, D., et al. 2014, in Protostars and Planets VI, ed. H. Beuther, et al. (Tucson, AZ: Univ. Arizona Press), 859
Chaabouni, H., Bergeron, H., Baouche, S., et al. 2012, A&A, 538, A128
Charnley, S. B., Rodgers, S. D., & Ehrenfreund, P. 2001, A&A, 378, 1024
Chiar, J. E., Tielens, A. G. G. M., Adamson, A. J., & Ricca, A. 2013, ApJ, 770, 78
Chick, K. M., & Cassen, P. 1997, ApJ, 477, 398
Cho, J., & Lazarian, A. 2007, ApJ, 669, 1085
Cleeves, L. I., Bergin, E. A., & Alexander, C. M. O'D. 2016, ApJ, 819, 13
Cowie, L. L., Songaila, A., & York, D. G. 1979, ApJ, 230, 469
Crapsi, A., van Dishoeck, E. F., Hogerheijde, M. R., Pontoppidan, K. M., & Dullemond, C. P. 2008, A&A, 486, 245
Cronin, J. R., & Chang, S. 1993, in The Chemistry of Life's Origins, ed. J. M. Greenberg, et al. (Dordrecht: Kluwer), 209
Cuppen, H. M., & Herbst, E. 2007, ApJ, 668, 294
Da Pieve, F., Avendaño-Franco, G., De Proft, F., & Geerlings, P. 2014, MNRAS, 440, 494
Dartois, E., Demyk, K., d'Hendecourt, L., & Ehrenfreund, P. 1999, A&A, 351, 1066

de Marcellus, P., Meinert, C., Nuevo, M., et al. 2011, ApJ, 727, L27
d'Hendecourt, L. B., Allamandola, L. J., Grim, R. J. A., & Greenberg, J. M. 1986, A&A, 158, 119
Dominik, C., & Tielens, A. G. G. M. 1997, ApJ, 480, 647
Dominik, C., Blum, J., Cuzzi, J., & Wurm, G. 2007, in Protostars and Planets V, ed. B. Reipurth, D. Jewitt, & K. Keil (Tucson, AZ: Univ. Arizona Press), 783
Dorschner, J., & Henning, T. 1995, A&ARv, 6, 271
Draine, B. T. 1989, in Evolution of Interstellar Dust and Related Topics, ed. A. Bonetti (Amsterdam: North-Holland), 103
Draine, B. T., & McKee, C. F. 1993, ARA&A, 31, 373
Duley, W. W., & Williams, D. A. 1981, MNRAS, 196, 269
Duley, W. W., & Williams, D. A. 1986, MNRAS, 223, 177
Dulieu, F., Amiaud, L., Congiu, E., et al. 2010, A&A, 512, 30
Ehrenfreund, P., & Charnley, S. B. 2000, ARA&A, 38, 427
Ehrenfreund, P., Dartois, E., Demyk, K., & d'Hendecourt, L. 1998, A&A, 339, L17
Ehrenfreund, P., Kerkhof, O., Schutte, W. A., et al. 1999, A&A, 350, 240
Elsila, J. E., Glavin, D. P., & Dworkin, J. P. 2009, M&PS, 44, 1323
Franceschi, R., Birnstiel, T., Henning, T., et al. 2022, A&A, 657, A74
Fuchs, G. W., Cuppen, H. M., Ioppolo, S., et al. 2009, A&A, 505, 629
Fuente, A., Navarro, D. G., Caselli, P., et al. 2019, A&A, 624, A105
Fukue, T., Tamura, M., Kandori, R., et al. 2010, OLEB, 40, 335
Furlan, E., McClure, M., Calvet, N., et al. 2008, ApJS, 176, 184
Furukawa, Y., Chikaraishi, Y., Ohkouchi, N., et al. 2019, PNAS, 116, 24440
Furuya, K., Drozdovskaya, M. N., Visser, R., et al. 2017, A&A, 599, A40
Gadallah, K. A. K., Mutschke, H., & Jäger, C. 2011, A&A, 528, A56
Gerakines, P. A., Whittet, D. C. B., Ehrenfreund, P., et al. 1999, ApJ, 522, 357
Gibb, E. L., & Whittet, D. C. B. 2002, ApJ, 566, L113
Gibb, E. L., Nummelin, A., Irvine, W. M., Whittet, D. C. B., & Bergman, P. 2000, ApJ, 545, 309
Glavin, D. P., Callahan, M. P., Dworkin, J. P., & Elsila, J. E. 2010, M&PS, 45, 1948
Gounelle, M., & Meynet, G. 2012, A&A, 545, A4
Greenberg, J. M., Li, A., Mendoza-Gomez, C. X., et al. 1995, ApJ, 455, L177
Heck, P. R., Greer, J., Kööp, L., et al. 2020, PNAS, 117, 1884
Henning, T. 2010, ARA&A, 48, 21
Herbst, E. 2014, PCCP, 16, 3344
Herbst, E., & Leung, C. M. 1986, MNRAS, 222, 689
Herbst, E., & van Dishoeck, E. F. 2009, ARA&A, 47, 427
Hogerheijde, M. R., Boogert, A. C. A., van Dishoeck, E. F., & Blake, G. A. 2002, in The Origins of Stars and Planets: The VLT View, ed. J. Alves, & M. McCaughrean (New York: Springer), 53
Holland, W. S., Matthews, B. C., Kennedy, G. M., et al. 2017, MNRAS, 470, 3606
Hollenbach, D. J., & Salpeter, E. E. 1971, ApJ, 163, 155
Hollenbach, D., Kaufman, M. J., Bergin, E. A., & Melnick, G. J. 2009, ApJ, 690, 1497
Hughes, A. M., Duchêne, G., & Matthews, B. C. 2018, ARA&A, 56, 541
Imai, M., Sakai, N., Oya, Y., et al. 2016, ApJ, 830, L37
Ioppolo, S., van Boheemen, Y., Cuppen, H. M., van Dishoeck, E. F., & Linnartz, H. 2011, MNRAS, 413, 2281
Izidoro, A., de Souza Torres, K., Winter, O. C., & Haghighipour, N. 2013, ApJ, 767, 54

Jäger, C., Molster, F. J., Dorschner, J., et al. 1998, A&A, 339, 904
Jenkins, E. B. 2009, ApJ, 700, 1299
Jones, A. P. 2022, A&A, 657, A127
Jones, A. P., & Williams, D. A. 1984, MNRAS, 209, 955
Jones, A. P., Tielens, A. G. G. M., Hollenbach, D. J., & McKee, C. F. 1994, ApJ, 433, 797
Jones, A. P., Tielens, A. G. G. M., & Hollenbach, D. J. 1996, ApJ, 469, 740
Jørgensen, J. K., Belloche, A., & Garrod, R. T. 2020, ARA&A, 58, 727
Jura, M. 1980, ApJ, 235, 63
Kainulainen, J., Lehtinen, K., & Harju, J. 2006, A&A, 447, 597
Kenyon, S. J., & Hartmann, L. 1987, ApJ, 323, 714
Kim, S. H., & Martin, P. G. 1996, ApJ, 462, 296
Kvenvolden, K., Lawless, J., Pering, K., et al. 1970, Natur, 228, 923
Kwon, J., Tamura, M., Hough, J. H., et al. 2014, ApJ, 795, L16
Lahuis, F., & van Dishoeck, E. F. 2000, A&A, 355, 699
Larson, R. B. 2003, RPPh, 66, 1651
Leitch-Devlin, M. A., & Williams, D. A. 1985, MNRAS, 213, 295
Li, A., & Greenberg, J. M. 1997, A&A, 323, 566
McKee, C. F., & Ostriker, E. C. 2007, ARA&A, 45, 565
Meinert, C., de Marcellus, P., d'Hendecourt, L., et al. 2011, PhLRv, 8, 307
Mennella, V., Colangeli, L., Palumbo, P., et al. 1996, ApJ, 464, L191
Men'shchikov, A. B., Henning, T., & Fischer, O. 1999, ApJ, 519, 257
Messenger, S. 2002, P&SS, 50, 1221
Minissale, M., Congiu, E., Manicò, G., Pirronello, V., & Dulieu, F. 2013, A&A, 559, A49
Motte, F., Bontemps, S., & Louvet, F. 2018, ARA&A, 56, 41
Mumma, M. J., & Charnley, S. B. 2011, ARA&A, 49, 471
Muñoz Caro, G. M., & Dartois, E. 2009, A&A, 494, 109
Muñoz Caro, G. M., Dartois, E., Boduch, P., et al. 2014, A&A, 566, A93
Nieva, M-F, & Przybilla, N. 2012, A&A, 539, 143
Öberg, K. I., Boogert, A. C. A., Pontoppidan, K. M., et al. 2011, ApJ, 740, 109
Ohtake, N., Hiratsuka, M., Kanda, K., et al. 2021, Mate, 14, 315
Oliveira, I., Olofsson, J., Pontoppidan, K. M., et al. 2011, ApJ, 734, 51
Olofsson, J. 2009, A&A, 507, 327
Ormel, C. W., Paszun, D., Dominik, C., & Tielens, A. G. G. M. 2009, A&A, 502, 845
Parise, B., Ceccarelli, C., Tielens, A. G. G. M., et al. 2006, A&A, 453, 949
Pauly, T., & Garrod, R. T. 2016, ApJ, 817, 146
Pendleton, Y. J., & Allamandola, L. J. 2002, ApJS, 138, 75
Pendleton, Y. J., Tielens, A. G. G. M., Tokunaga, A. T., & Bernstein, M. P. 1999, ApJ, 513, 294
Pinte, C., Dent, W. R. F., Ménard, F., et al. 2016, ApJ, 816, 25
Pizzarello, S. 2012, M&PS, 47, 1291
Pizzarello, S. 2016, Life, 6, 18
Pizzarello, S., & Huang, Y. 2005, GeCoA, 69, 599
Pontoppidan, K. M., Boogert, A. C. A., Fraser, H. J., et al. 2008, ApJ, 678, 1005
Punanova, A., Vasyunin, A., Caselli, P., et al. 2022, ApJ, 927, 213
Rebollido, I., Ribas, A., de Gregorio-Monsalvo, I., et al. 2022, MNRAS, 509, 693
Reipurth, B., & Bally, J. 2001, ARA&A, 39, 403
Rich, E. A., Teague, R., Monnier, J. D., et al. 2021, ApJ, 913, 138

Rieke, G. H., Su, K. Y. L., Stansberry, J. A., et al. 2005, ApJ, 620, 1010
Ritchey, A. M., Federman, S. R., & Lambert, D. L. 2018, ApJS, 236, 36
Robitaille, T. P., Whitney, B. A., Indebetouw, R., & Wood, K. 2007, ApJS, 169, 328
Roser, J. E., Vidali, G., Manico, G., & Pirronello, V. 2001, ApJ, 555, L61
Ruffle, D. P., & Herbst, E. 2001, MNRAS, 324, 1054
Sabatini, G., Giannetti, A., Bovino, S., et al. 2019, MNRAS, 490, 4489
Sandford, S. A., Allamandola, L. J., Tielens, A. G. G. M., & Valero, G. J. 1988, ApJ, 329, 498
Sargent, A. I., & Beckwith, S. 1987, ApJ, 323, 294
Schmitt, B., Greenberg, J. M., & Grim, R. J. A. 1989, ApJ, 340, L33
Shenoy, S. S., Whittet, D. C. B., Chiar, J. E., et al. 2003, ApJ, 591, 962
Shimonishi, T., Onaka, T., Kawamura, A., & Aikawa, Y. 2016, ApJ, 827, 72
Silsbee, K., Caselli, P., & Ivlev, A. V. 2021, MNRAS, 507, 6205
Smith, B. A., & Terrile, R. J. 1984, Sci, 226, 421
Snow, T. P., & Bierbaum, V. M. 2008, ARAC, 1, 229
Sorrell, W. H. 1990, MNRAS, 243, 570
Spitzer, L. 1978, Physical Processes in the Interstellar Medium (New York: Wiley)
Spitzer, L. 1985, ApJ, 290, L21
Stephens, I. W., Yang, H., Li, Z.-Y., et al. 2017, ApJ, 851, 55
Suttner, G., & Yorke, H. W. 2001, ApJ, 551, 461
Tielens, A. G. G. M. 2008, ARA&A, 46, 289
Tielens, A. G. G. M. 2012, in IAU Symp. 284, The Spectral Energy Distribution of Galaxies, ed. R. J. Tuffs, & C. C. Popescu (Cambridge: Cambridge Univ. Press), 72
Tielens, A. G. G. M. 2013, RvMP, 85, 1021
Tielens, A. G. G. M., & Allamandola, L. J. 1987a, in Physical Processes in Interstellar Clouds, ed. G. E. Morfill, & M. Scholer (Dordrecht: Reidel), 333
Tielens, A. G. G. M., & Allamandola, L. J. 1987b, in Interstellar Processes, ed. D. J. Hollenbach, & H. A. Thronson (Dordrecht: Reidel), 397
Tielens, A. G. G. M., & Hagen, W. 1982, A&A, 114, 245
Tielens, A. G. G. M., Seab, C. G., Hollenbach, D. J., & McKee, C. F. 1987, ApJ, 319, L109
Urso, R. G., Palumbo, M. E., Baratta, G. A., Scirè, C., & Strazzulla, G. 2018, MNRAS, 479, 130
Urso, R. G., Vuitton, V., Danger, G., et al. 2020, A&A, 644, A115
van Broekhuizen, F. A., Pontoppidan, K. M., Fraser, H. J., & van Dishoeck, E. F. 2005, A&A, 441, 249
van Dishoeck, E. F., & Blake, G. A. 1998, ARA&A, 36, 317
van Dishoeck, E. F., Helmich, F. P., de Graauw, T., et al. 1996, A&A, 315, L349
van Dishoeck, E. F., Bergin, E. A., Lis, D. C., & Lunine, J. I. 2014, in Protostars and Planets VI, ed. H. Beuther, et al. (Tucson, AZ: Univ. Arizona Press), 835
Watanabe, N., & Kouchi, A. 2002, ApJ, 571, L173
Weintraub, D. A., Kastner, J. H., & Whitney, B. A. 1995, ApJ, 452, L141
Weintraub, D. A., Goodman, A. A., & Akeson, R. L. 2000, in Protostars & Planets IV, ed. V. Mannings, et al. (Tucson, AZ: Univ. Arizona Press), 247
Welty, D. E., Jenkins, E. B., Raymond, J. C., Mallouris, C., & York, D. G. 2002, ApJ, 579, 304
Whitney, B. A., & Wolff, M. J. 2002, ApJ, 574, 205
Whittet, D. C. B. 2017, The Origins of Life: A Cosmic Perspective (San Rafael, CA: Morgan & Claypool)
Whittet, D. C. B., Gerakines, P. A., Tielens, A. G. G. M., et al. 1998, ApJ, 498, L159

Whittet, D. C. B., Goldsmith, P. F., & Pineda, J. L. 2010, ApJ, 720, 259
Whittet, D. C. B., Cook, A. M., Herbst, E., Chiar, J. E., & Shenoy, S. S. 2011, ApJ, 742, 28
Wilking, B. A. 1989, PASP, 101, 229
Williams, D. A., Hartquist, T. W., & Whittet, D. C. B. 1992, MNRAS, 258, 599
Williams, J. P., & Cieza, L. A. 2011, ARA&A, 49, 67
Wurm, G., & Schnaiter, M. 2002, ApJ, 567, 370
Wyatt, M. C. 2022, in ExoFrontiers: Big Questions in Exoplanetary Science, ed. N. Madhusudhan (Bristol: IOP Publishing)
Zellner, N. E. B., McCaffrey, V. P., & Butler, J. H. E. 2020, AsBio, 20, 1377

D C B Whittet

Chapter 10

Conclusion

Two decades have elapsed since the second edition of *Dust in the Galactic Environment* was completed in 2002. I concluded its final chapter with a discussion of open questions, and signed off by remarking that if I should ever be foolhardy enough to attempt a third edition perhaps some of the answers would adorn its pages. These things have come to pass. Of course, we have not answered all of the questions; but a great deal of progress has been made, and we have a clearer vision of what questions we should be addressing next. This chapter summarizes what we have learned and what we might hope to learn in the future.

Before getting into specifics, however, it is interesting to consider how the field itself has changed. In my experience, astronomical research has evolved to become progressively more interdisciplinary over the past half century. This may reflect a general trend in all science, of course, but in astronomy it seems to have arisen, at least in part, from a few game-changing discoveries. A rapid growth in the inventory of known interstellar molecules demonstrated that space environments host chemical as well as physical complexities. Subsequent identification of presolar grains by the application of geochemical methods to studies of meteorites opened new windows, not only on the nature and origin of cosmic dust but also on the nuclear processes that created their elements inside stars. Next came the discovery of exoplanets, including potential Earth-analogs, detected by well-honed astronomical techniques but requiring expertize in many disciplines—geosciences, environmental sciences, biosciences—to attempt any realistic assessment of their characteristics. Simultaneously, our understanding of the processes that forged our own planetary system are being clarified by research programs that range from analyses of extinct radionuclides in meteorites to cometary and asteroidal rendezvous missions and high-resolution spatial and spectral imaging of protoplanetary disks around potential early solar analogs.

Research on dust is emblematic of this trend because it is pertinent to all aspects of the science. A useful metric is the number of relevant review papers. Citations in

doi:10.1088/2514-3433/ac7204ch10

this text published in review journals such as the *Annual Reviews* series are not limited to the those specific to astronomy. Excluding conference proceedings, the total number of cited reviews is 59, of which 42 have appeared in the last 20 years. Of course, by no means all of these reviews are specifically "about" dust, but all were deemed sufficiently relevant for citation (the count was made retrospectively!). I doubt that this statistic implies any general increase in the number of review papers being written over the past two decades, but instead a growing synergy between the astrophysics of dust and other areas of research.

10.1 An Overview of Dust Populations

A wide range of observational and empirical constraints on the properties of cosmic dust are described in the preceding chapters. The data imply a need for models that combine different grain populations to account for different aspects of the data: the prime suspects are summarized in Table 10.1. These populations are distinguished by their spectral properties and inferred differences in composition, size distribution and physical structure (amorphous or crystalline, mantled or unmantled, monolithic or composite, rounded or oblate). A unified model for interstellar dust must be capable of matching all known constraints, invoking substances and size distributions that are physically reasonable and have a probable source in terms of stellar mass-loss processes and/or production in the ISM itself. A *unique* model would accomplish all of this without ambiguity, but this is not yet achievable because the available data do not impose absolute constraints (different assumptions can yield similarly good fits), and also perhaps because our models are not yet sufficiently realistic.

Table 10.1. Interstellar Dust Constituents Implied by Observational and Empirical Evidence.

Property	Amorphous Silicates	Mantles	Large Carbon	Bump Grains	Small Carbon
Composition	$MgSiO_3$, etc. with Fe-rich inclusions	ices a-C ORM?	a-C graphite (SiC)	graphite	PAHs diamond?
Origin	Stardust	ISM	Stardust	Both?	Both?
Size (μm)	$a \lesssim 1$	$\Delta a \lesssim 0.1$	$a \lesssim 10$	$a \lesssim 0.01$	$a \lesssim 0.01$
Elements depleted	O, Mg, Si, Fe	O, N, C	C	C	C
Extinction	UV–IR	UV–IR	UV–IR	mid-UV	UV
Alignment?	yes	(yes)	no?	no?	no?
Absorption features	9.7, 19 μm	3 μm, etc.	3.4 μm?	217 nm	3.4 μm?
Emission features	CS only?	—	—	—	PAH bands
Continuum emission	FIR	FIR	FIR	MIR	MIR

Notes: Composite particles may contain mixtures of these constituents. Bump and small carbon grains may belong to the same population. SiC is inferred to be present from studies of meteorites, not by direct observation. Only abundant depleted elements are listed.

The goal of this section is to present a general overview, highlighting areas of both convergence and ambiguity, with a view to future progress toward a robust, unified theory of cosmic dust. No citations of specific models are included here. See Sections 3.6 and 4.2.8 for examples of fits to extinction and polarization data, and Section 4.3 for a review of grain alignment theory. See also Sections 3.2.3 (scattering), 3.4.5 (bump), 4.2.7 (spectropolarimetry), 5.2 (refractory dust), 5.3 (ices), 6.2.2 (PAHs), 7.3 (element depletions), and 8.3 (meteoritic stardust) for discussion of constraints placed on models by specific observables. Table 10.1 summarizes the properties of five proposed constituents of interstellar dust. Successful models for dust in the diffuse ISM require at least three: silicates, small carbon, and some other C-rich repository. The chemical composition of interstellar silicates is not uniquely defined, but available data suggest that amorphous magnesium silicates ($MgSiO_3$, Mg_2SiO_4, $MgFeSiO_4$) may be more common than Fe-rich forms. Fe and its oxides may also be present as embedded subparticles (inclusions) in silicates and/or as an dependent population, and some Fe may also be bonded to carbon. It is plausible to suppose that the small carbon and "bump" grains belong to the same population of nanoparticles; they are listed separately in Table 10.1 because they are invoked to explain different spectral features (IR emission, UV absorption), and the agents responsible may differ in detail. A distinct population of larger carbonaceous particles is essential to all models, to account for the continuum opacity of the dust without violating constraints arising from the spectral features, which require both PAH emitters and bump absorbers to be very small. A point of divergence in extant models is the assumed form of these larger particles: proposals include mantles on silicates, aggregates of mixed composition, and an independent population. Which of these is the most realistic choice is a persistent and perplexing question.

The answer to this question is linked to another: where and how does the dust originate? If stellar outflows are major sources (Section 8.4), their contributions divide naturally into distinct O-rich and C-rich populations. A gradual blending of these populations by grain–grain aggregation in clouds (Section 9.1.1) is physically reasonable and indeed expected. The growth and subsequent carbonization of icy mantles on silicate cores as they cycle in and out of dense clouds may also lead to grains of mixed carbon/silicate composition. A diagnostic test of this trend is provided by spectropolarimetry. The observed wavelength-dependence of polarization is best explained if silicate grains are its primary source. Silicates and (where present) ice mantles exhibit polarization peaks corresponding to their absorption features, clearly demonstrating that such particles produce dichroic extinction and respond to alignment. In contrast, C-rich dust generally lacks polarimetric signatures. Both the 217 nm graphite bump and the 3.4 μm hydrocarbon feature have been scrutinized, with almost entirely negative results (two lines of sight with weak bump excesses are the only known exceptions). The smallest grains tend to be least responsive to alignment, so the general lack of bump polarization is not unexpected. Perhaps the 3.4 μm feature also arises primarily in small grains? Perhaps aggregate particles tend to be more rounded than monolithic silicate grains? As it stands, the available evidence is consistent with distinct populations of silicate and carbon grains, of which only the former is a significant agent for polarization.

The apparent need for a distinct silicate population is consistent with the view that most interstellar silicates originate in stellar atmospheres. This setting seems far more conducive to silicate production than the colder, more rarefied environments of interstellar clouds. The argument against this proposition arises from calculations of expected timescales for destruction by shocks, which suggest that silicates are not being replenished rapidly enough by stellar winds to explain their abundance in the ISM; however, more recent reassessments seem to have alleviated this constraint somewhat (see Section 8.4.4 for a detailed discussion). I therefore conclude that stardust is the most likely source of interstellar silicates. In contrast, an interstellar origin for (some) C-rich grain material is very plausible as a product of mantle growth on existing grains. Limits on the abundance of SiC dust (confirmed only in meteorites) suggest that carbon stars are less important as sources of interstellar grains.

A power-law form of the size distribution function $n(a)$ (Equation (3.30)) is typically assumed in (or deduced from) models for the optical properties of the dust, applicable over some appropriate range of a for each component. This mathematical form is physically reasonable as the predicted outcome of grain–grain collisions in both circumstellar and interstellar environments (Sections 8.4.2 and 9.1.1). The models yield acceptable fits to observations of extinction and scattering, with variations attributed to changing conditions between diffuse and dense clouds accommodated by modest differences in the relevant parameters (power-law index, minimum and maximum sizes). The maximum size is often capped at $a_{\max} = 1$ μm or less, as much larger grains have minimal effect on the extinction curve over the spectral range typically observed, and too many large grains could exceed abundance constraints, especially for silicates. Nevertheless, there are reasons to challenge this assumption. The most tangible evidence for large ($a > 1$ μm) grains is found in the presolar meteorite collection (Section 8.3.4), and in the present-day flux of interstellar dust into the solar system (Section 1.3.4). Other phenomena that might best be explained by extending the limit to accommodate larger grains include X-ray scattering halos and core-shine (Section 3.2.3), broadening of ice absorption profiles (Section 5.3.3), and the oxygen depletion anomaly (Section 7.3.3). Refined methods of detecting any *wavelength-independent* component of the total extinction (Section 3.2.5) may be used to further explore this possibility.

Our understanding of the nature and evolution of the ices that accumulate on dust in dense clouds has evolved dramatically in recent decades, primarily as the result of data from infrared space observatories such as ISO and Spitzer, in tandem with laboratory work designed to support these missions. The results give insight into not only the composition and structure of the ices (Section 5.3) but also how they evolve with changing conditions as clouds collapse and begin to form stars (Section 9.3). Successive phases of mantle growth create chemically distinct layers. Subsequently, as they are warmed, the ices undergo selective sublimation, segregation and annealing. Some ice-mantled grains may accrete directly into icy planetesimals in protoplanetary disks, while others sublimate and recondense. All such volatiles retain isotopic signatures of their low-temperature origins, most markedly enhanced D/H ratios, now detected in reservoirs that range from comets and asteroids to the Earth's ocean water.

10.2 Future Prospects

The progress of science is driven by the availability of data and our ability to comprehend the message it contains. Data acquisition in the astronomical sciences entered an era of exponential growth in the latter part of the 20th century, with the deployment of space telescopes such as IRAS, Hubble and ISO, together with ground-based facilities using the latest technologies. The ingenuity of scientists engaged in the task of interpreting the message (and distinguishing the wood from the trees!) led to advances described in this text. The stream of new data shows no sign of abating. At the time of this writing, the James Webb Space Telescope (JWST) is undergoing final *in situ* alignment and testing. Once fully commissioned, it will have a profound influence on virtually all areas of astronomy. I conclude this text with a summary of some key questions relating to our topic that might be answered by future research, some but not all enabled by the JWST.

What follows is a personal selection of important issues, with no claim to be fully comprehensive. I do not address questions relating specifically to dust in distant galaxies, for example, because others are far better qualified to do so. I am certain that a fuller understanding of dust in our home Galaxy and its near neighbors will assist and inform our attempts to assess its broader significance in the cosmos.

10.2.1 The Origin and Structure of Refractory Dust

- Can we better quantify the sources of refractory interstellar dust?
- What is the predominant structure of the grains responsible for continuum opacity? How are elemental C and Fe distributed among these particles?
- How common are the largest (super-micron-sized) grains in the ISM?

These are tantalizing and persistent questions. As previously discussed, stellar atmospheres of appropriate composition, temperature and pressure are proven sources of stardust, and these conditions cannot be replicated in the ISM. This is a problem especially for our understanding of the ecology of interstellar silicates if, as some models suggest, the infusion rate is too low. Estimates focus on mass loss from evolved (AGB) stars; but it should not be overlooked that the circumstellar envelopes of young stars also make silicates, and they also undergo mass loss as protoplanetary disks are cleared out. It may not be possible to detect such particles by isotopic analysis of meteorites: presolar silicates are identified by isotopic signatures specific to evolved stars, and all such fingerprints are homogenized in the condensation zone of a protoplanetary disk, whether it be our own or in another system prior to arrival. Nevertheless, developers of quantitative models for the life cycle of dust should reconsider this possible source.

Carbonaceous dust may originate from stellar or interstellar sources. The nano-particles (PAH clusters, bump grains) can be understood in terms of collisional fragmentation of either stardust or carbonized grain mantles (probably both). But what of the big grains needed to explain general extinction? If carbonaceous solids are largely synthesized in the ISM, they must presumably originate as mantles on pre-existing dust. In diffuse clouds, elemental C may deplete directly from the gas to grain surfaces; in denser clouds, C-bearing molecules are sequestered in the ices. The evolution of these solids as they cycle though different environments may be traced by infrared spectroscopy, but attempts to do so are limited by the fact that their absorption features are very weak, e.g., in comparison to those of silicates and ices (see Figure 5.5). Thus, very high quality spectra are needed to extract information from (and in some cases even to detect) diagnostic features such as the C–H, C–C, and C=O modes of hydrogenated and organic carbon, and the lattice modes of graphite. The superior sensitivity and resolution enabled by the JWST and its suite of instruments should be transformative. It may be possible, for example, to explore how ice mantles evolve in the outer layers of a dark cloud, where they are subject to photolysis at extinctions near the threshold value. To what extent does the ice crystallize, sublimate or undergo chemical change? Are photolyzed ices converted into organic kerogens prior to complete carbonization? Is the highly aromatic state of carbon dust toward the center of our Galaxy typical of the diffuse ISM in general?

Another interesting problem is the role of iron in the dust. Fe is the most abundant heavy metal. It is highly depleted from the gas and is a major contributor to the total dust mass (e.g., Figure 7.4). Yet there is no clear consensus on what form it takes. Interstellar silicates may account for some, but a stardust origin for silicates favors Mg-rich condensates, with Fe more likely to emerge in metallic or oxidized form. Aggregation might lead to clustering of these products. Certain interplanetary particles (GEMS) that contain Fe-rich inclusions in a silicate structure are suggested as plausible analogs for interstellar silicates. In pure form, Fe lacks spectral signatures in the infrared. Oxides such as FeO produce broad spectral features near 20 μm that blend with the adjacent silicate feature (Figure 8.9), detected in O-rich circumstellar matter but not so far in the ISM.

The smallest grains in the size distribution have been a focus of research for the past several decades, in response to the discovery of mid-infrared line and continuum emission and the inferred presence of transiently-heated nanoparticles. It may be timely now to focus on the largest particles, for reasons outlined in Section 10.1 and touched on elsewhere in this book. Early studies of interstellar extinction and reddening (Sections 1.2.1 and 1.2.2) were motivated by the need to fully account for all sources of attenuation in the visible (including any "neutral" component arising from grains large compared with the wavelength) because of the potential impact on calibration of the cosmic distance scale. Stellar distances measured by parallax are independent of dust properties, but a century ago accurate parallaxes could be measured only for nearby stars ($d \lesssim 10$ pc) that are essentially free of extinction. Modern space astrometry missions, most notably Gaia, have extended this limit by orders of magnitude. Hence it is now possible to use stellar distances to evaluate extinction independent of reddening, effectively inverting the analytical method used in early work. This may result in tighter constraints on the contributions of very large grains to visual extinction in the diffuse ISM; results for dense clouds may provide a new measure of grain growth.

At longer wavelengths, the sensitivity of the JWST will enable extinction curves to be determined with significantly greater precision and across a wider range of environments than was possible in previous work. Both the slope of the infrared continuum extinction and the shapes of the silicate absorption profiles are sensitive to the relative numbers of large grains. By such means, it may be possible to trace the cycle of grain growth, from diffuse to dark clouds, from dense cores to protoplanetary disks.

10.2.2 Ices and Organics in the Interstellar Medium and Protoplanetary Disks

- Can the inventory of interstellar ices in dense clouds be extended, e.g., by detection of O_2, N_2, HDO, and complex organics? What are the limits to complexity in the ices?
- How are ices and organics distributed in protoplanetary disks? Do their distributions vary from one system to another and does this relate to planet formation?
- Do some volatiles survive as solids in diffuse regions of interstellar space, e.g., in very large grains or "cometesimals"?

The importance of infrared space telescopes such as ISO and Spitzer to our current understanding of icy materials in space was noted previously (Section 10.1)—indeed, data from these missions adorn many figures in this book. The JWST promises to take this to the next level. The essential requirements are (i) full coverage of the spectral range containing the relevant solid-state features (Table 5.2), (ii) high sensitivity to enable study of faint sources, and (iii) high spectral and spatial resolutions. Previous missions checked some of these boxes; the JWST and its suite of instruments will check all of them.

Different types of source are used to study different environments in a molecular cloud, as illustrated in Figure 5.2. Background field stars that enable sampling of the quiescent regions are inherently fainter than embedded young stars. The unprecedented sensitivity of the JWST will yield spectra of background stars of a quality previously attainable only for the brightest YSOs. Adequate spectral resolution is also essential for both types of source. Although solid-state features are generally broad they often contain diagnostic structure, and overlapping gas-phase lines could mimic such structure if not fully resolved. Figure 5.13, for example, illustrates the inherent complexity of the CO ice feature and the simultaneous presence of overlapping gas-phase interstellar lines. Spectral lines intrinsic to a source may also contaminate its interstellar spectrum. The resolving powers required to overcome these issues ($\lambda/\Delta\lambda > 10^3$) are baked into the JWST near and mid-infrared spectrometers.

The homonuclear molecules O_2 and N_2 are predicted and potentially abundant constituents of the ices that have so far eluded detection. They produce no spectral features in their pure state; but weak signatures may be induced if, as expected, they are mixed with other species in the ices, such that interactions with near neighbors perturb the symmetry of the molecule. The fundamental O=O vibrational mode of O_2 centered near 6.45 μm, for example, may be a good candidate for detection by the JWST (searches with ISO data yielded modest upper limits). O_2 and N_2 are expected to reside primarily in the CO-rich layer of the ices (Figure 5.14). The presence of these species may induce detectable substructure in the profile of the 4.67 μm solid CO absorption feature, as the result of intermolecular forces.

Deuterium fractionation is recognized as a key signature of low-temperature chemical processes that give insight into the origins of protoplanetary matter in our solar system (Section 9.3.7). Molecular precursors of this material have been detected in the ISM, but so far only in the gas phase. Identification of deuterated water (HDO) and perhaps other deuterated species in the ices would provide important corroboration of the chemical models. The O–D bond in HDO produces a spectral feature at 4.1 μm (equivalent to the strong O–H feature at 3.0 μm). Again, this has eluded detection in previous work, most likely due to insufficient sensitivity, and it thus provides another opportunity for the JWST.

So far, the most complex organic molecule confirmed as a common constituent of the ices (CH_3OH) contains just six atoms. Species of significantly greater molecular weight are known to be present in the gas phase, both in cold clouds and in warmer regions around new-born stars where sublimated mantles may contribute to the inventory (Section 9.3.4). Chemical process in these environments influence the complexity of molecules delivered to protoplanetary disks. In order to refine the models, it is important to test the limits to molecular complexity in the solid phase, and to link this to the gaseous chemistry of hot cores and corinos. The spectral region from 5 to 8 μm is especially important in this regard, as it is populated by the vibrational modes of many candidates for detection (Section 5.3.5). The features observed to date are blends of numerous overlapping absorptions, only a few of which have been identified with confidence. Gains in sensitivity and spectral resolution may enable us to solve this puzzle, not only enhancing the quality of the data but also expanding the range of environments that can be studied. Candidates for detection include acetaldehyde, ethanol, dimethyl ether, and methyl formate, all of which are known to be present in the gas phase in hot cores. The presence of such species in low-temperature ices would inform our models of cold chemistry; or if they are present only in warmer ices, this would imply the need for an energy source to drive production.

The JWST will also enable direct study of ices in protoplanetary disks. Systems in which the disk is aligned approximately edge-on to our line of sight are most suitable, as this configuration alleviates ambiguities concerning the location of the ices (disk versus envelope). Infrared spectroscopy of a spatially resolved disk is not limited to the line of sight to the central star. Spectral imaging may enable entire disks to be mapped in key spectral signatures, seen in absorption against the continuum emission from warm dust. It may thus be possible to map the radial and vertical distributions of the ices. In combination with data on the distributions of gas-phase molecules observed (e.g.) by ALMA, such observations will clarify the interplay between key planet-building materials, and identify the locations of snowlines for solids of different volatilities. Ultimately, the results will inform models for planet formation and planetary-system architecture. In the case of our own planetary system, the orderly separation of rocky planets and gas giants is attributed to formation within and beyond the H_2O snowline of the solar nebula. It will be interesting to determine whether other systems are consistent with this picture, and to construct plausible chemical and dynamical models that can accommodate the wide variety of exoplanetary systems now known to exist.

The present-day distribution of comets in and around our solar system is attributed to the gravitational influence of the giant planets. Oort-Cloud comets arrive randomly from solar distances that far exceed the boundaries of the planetary system. It is hypothesized that they were scattered into highly eccentric orbits by resonances or other disturbances arising from gravitational interplay between these planets. It is unknown whether such disturbances are commonplace or dependent on improbable planetary alignments, but it seems reasonable to suppose that icy planetesimals may be ejected routinely from their parent system. Extrasolar visitors to our own system have now been identified (see Section 7.3.3 for discussion and references). An important long-term goal would be to constrain the space density of debris large enough to retain volatiles in diffuse regions of the interstellar medium.

All things considered, there is clearly no shortage of challenging problems to engage the interested student or more seasoned campaigner.

Dust in the Galactic Environment (Third Edition)

D C B Whittet

Appendix A

Glossary

The aim here is to provide a convenient list of definitions, covering units, constants, acronyms, and terms used in this book and/or in cited literature that might not be familiar to the reader. Units and constants listed in Section A.1 should be regarded as supplementary to standard Système Internationale (SI) units and the common physical constants. Completeness is not guaranteed!

A.1 Units and Constants

Astronomical Unit (AU): Mean Sun–Earth distance; 1 AU = 1.496×10^{11} m.

Dex: Quasi-unit of logarithm to base 10, sometimes used to express abundance ratios: log 10 = 1 dex, log 100 = 2 dex, etc.

Jansky (Jy): Unit of luminous flux density: 1 Jy = 10^{-26} W m^{-2} Hz^{-1}. Similarly, 1 mJy = 10^{-3} Jy.

Magnetic flux density: 1 μG = 10^{-10} T (preferred unit for interstellar values).

Magnitude (mag): Dimensionless logarithmic measure of brightness in stellar astronomy, also used for interstellar quantities such as extinction and reddening deduced from stellar data. See also Section A.3.

Parsec (pc): Unit of distance based on stellar annual parallax; 1 pc = 3.086×10^{16} m. Similarly, kiloparsec (1 kpc = 10^3 pc), megaparsec (1 Mpc = 10^6 pc).

Parts per million (ppm): Measure of abundance ratio by number, usually relative to hydrogen; e.g., an estimate of the solar O to H abundance ratio is 4.9×10^{-4} = 490 ppm.

Solar luminosity: 1 $L_\odot = 3.826 \times 10^{26}$ W.

Solar mass: 1 $M_\odot = 1.989 \times 10^{30}$ kg.

Solar radius: 1 $R_\odot = 6.960 \times 10^{8}$ m.

Units of time: 1 Myr = 10^6 years; 1 Gyr = 10^9 years.

doi:10.1088/2514-3433/ac7204ch11

A.2 Common Acronyms

AGB: Asymptotic giant branch.

ALMA: Atacama Large Millimeter–submillimeter Array (telescope).

AME: Anomalous microwave emission.

CDE: Continuous distribution of ellipsoids.

CS: Circumstellar.

DDA: Discrete dipole approximation.

DG: Davis–Greenstein (grain alignment mechanism).

DGL: Diffuse galactic light.

DIBs: Diffuse interstellar bands.

EMT: Effective medium theory.

ERE: Extended red emission.

EUV: Extreme ultraviolet.

FIR: Far infrared.

FM: Fitzpatrick and Massa (empirical extinction formula).

FUV: Far ultraviolet.

FWHM: Full-width at half-maximum (intensity), e.g., of a spectral feature.

GC: Galactic center (of the Milky Way).

GEMS: Glass with embedded metal and sulfides.

HAC: Hydrogenated amorphous carbon.

HD: Henry Draper (star catalog).

HR: Hertzsprung–Russell (diagram).

HVC: High-velocity cloud.

IDP: Interplanetary dust particle.

IMF: Initial (stellar) mass function.

IRAS: Infrared Astronomical Satellite.

IRC: Infrared catalog (of stars).

IRS: Infrared source.

ISM: Interstellar medium.

ISO: Infrared Space Observatory.

ISRF: Interstellar radiation field.

IUE: International Ultraviolet Explorer (satellite).

JWST: James Webb Space Telescope.

LMC: Large Magellanic Cloud.

MIR: Mid-infrared.

MRN: Mathis, Rumpl and Nordsieck (grain size distribution function).

NGC: New General Catalog (of star clusters, nebulae and galaxies).

NIR: Near infrared.

ORM: Organic refractory matter.

PAH: Polycyclic aromatic hydrocarbon.

PMS: Pre-main-sequence (star).

PN: Planetary nebula (plural PNe).

ppm: Parts per million (abundance ratio).

RATs: Radiative alignment torques.

RCB: R Coronae Borealis (star).

SED: Spectral energy distribution.

SMC: Small Magellanic Cloud.

SN: Supernova (plural SNe).

SNR: Supernova remnant.

SPM: Superparamagnetic.

UV: Ultraviolet.

VBS: Very broadband structure.

VSG: Very small grain.

WR: Wolf-Rayet (star).

YSO: Young stellar object.

A.3 Physical, Chemical and Astrophysical Terms

Accretion: Growth of solids by physical attachment of colliding particles. The term may be used in the context of grain–grain coagulation or attachment of gaseous particles to a grain to form a surface layer. Accretion in a protoplanetary disk is considered to be the initial phase of planet formation.

Adsorption: Adhesion to a solid surface, a term commonly used to describe the sticking of gaseous atoms or molecules to interstellar dust grains. Adsorption may be physical (physisorption, arising from van der Waals forces), or the result of a chemical bond between surface atoms and the attached species (chemisorption).

Aliphatic hydrocarbons: Organic molecules composed of carbon chains with attached H atoms. Single-bonded chains (alkanes) have the generic formula

CH_3–$(CH_2)_n$–CH_3 where $n \geqslant 0$ is an integer. Chains that contain one or more C=C bonds are referred to as olefinic.

Asymptotic giant branch (AGB): A region of the Hertzsprung–Russell diagram occupied by stars of intermediate mass (0.8–8 $M_\odot$) in their late phases of red-giant evolution. An AGB star typically has a core composed mostly of carbon and oxygen, surrounded by helium and hydrogen-burning shells. Deep convection currents and pulsational instabilities may lead to the atmosphere becoming enriched in carbon. After periods of increasingly rapid mass loss, an AGB star sheds its outer layers to form a planetary nebula and its core becomes a white dwarf. A star undergoing the brief transition from AGB to PN (a post-AGB star) is sometimes described by the potentially misleading term "protoplanetary nebula."

Carbon star: A red giant that has a carbon-enriched atmosphere (C/O $\gtrsim$ 1), presumed to result from convective dredging to the surface of elemental carbon formed by nucleosynthesis inside the star.

Carbonaceous chondrites: A broad class of meteorites, considered to be the least processed meteoritic remnants of the Sun's protoplanetary disk. They are characterized by the presence of chondrules (glassy spherules), together with up to 5% by weight of carbonaceous matter, in a fine-grained matrix. The matrix contains silicates that appear to have been hydrated by the presence of liquid water on asteroidal parent bodies. The carbon is mostly in the form of complex organic polymers and amorphous carbon, but also includes exotic graphite and silicon carbide grains that condensed prior to the origin of the solar system. Amino acids and other prebiotic molecules are also present in trace amounts.

Color index and color excess: A color index is a photometric measure of the color of a star, equal to the difference between magnitudes measured at two specified wavelengths. A common example is $B - V$, where B ("blue") and V ("visual") correspond to passbands with mean central wavelengths of 440 and 550 nm, respectively. Color indices depend on both the intrinsic spectrum of the star and any reddening by foreground dust. The color excess is a measure of reddening, equal to the difference between the observed and "intrinsic" color indices, denoted by $E_{B-V} = (B - V) - (B - V)_0$.

Column density: The number (N) of particles (e.g., atoms, molecules, grains) in a column of unit cross-sectional area, frequently used in astronomical spectroscopy as a measure of the abundance of an absorber toward a radiation source. For a column of length L (equal to the distance of the source), it is related to the number density (n, particles per unit volume) by $N = nL$. The SI unit is m^{-2}; however, the cgs unit cm^{-2} is widely used in the literature.

Cosmic rays: Highly energetic atomic particles traveling at relativistic speeds. Most are protons, but all common atomic nuclei are represented. The primary source is thought to be supernova explosions. Cosmic rays are important in astrochemistry because, unlike energetic photons, they penetrate the interiors of dense molecular clouds, thereby providing an energy source that can instigate chemical reactions.

Depletion: A shortfall in the abundance of a chemical element measured in interstellar gas relative to standard abundances, attributed to its assumed presence in dust grains.

Diamond: A crystalline form of solid carbon in which C atoms are arranged tetrahedrally in a cubic structure, with each of the four valence electrons of a C atom forming a covalent bond to a neighboring atom (sp^3 hybridization).
Early-type stars: Luminous ($L_* \gtrsim 500\ L_\odot$) stars characterized by hot ($T_{eff} \gtrsim 15,000$ K) photospheres, including stars of spectral types O and B, and Wolf-Rayet stars. Their distribution in the Milky Way closely follows the galactic disk. Because of their high surface temperatures, early-type stars are primary sources of ultraviolet photons in the interstellar radiation field. They are relatively short-lived and thus typically much younger than less luminous main-sequence stars like the Sun.
Effective temperature: The temperature (T_{eff}) of a blackbody that would emit the same bolometric luminosity per surface area as a body such as a star or planet.
Emission nebula: Generic term for a cloud of hot, ionized gas, such as an H II region or a planetary nebula, maintained in an ionized state by radiation from internal or nearby star(s). Their spectra are characterized by emission lines.
Equivalent width: A measure of the strength of a spectral line, defined on an intensity plot as the width of a rectangle with height corresponding to the continuum level and area equal to that enclosed by the line profile; values may be expressed in either wavelength or frequency units, related by $W_\lambda = (\lambda^2/c)W_\nu$. For an unsaturated absorption line, the equivalent width is directly proportional to the number of absorbers in the line of sight.
Flux density: Measure of irradiance per unit wavelength or frequency interval (F_λ or F_ν). These forms are related by the equation $F_\lambda d\lambda = F_\nu d\nu$, hence $F_\lambda = (-c/\lambda^2)F_\nu$. The SI unit of F_ν is W m^{-2} Hz^{-1}; for convenience the Jansky (Section A.1) is generally used. The preferred unit for F_λ is W m^{-2} μm^{-1}. See also *Surface brightness*.
Glasses with embedded metal and sulfides (GEMS): Glassy silicate minerals found in anhydrous interplanetary dust particles, probably formed by condensation in the solar nebula and subsequently accreted into comets. Their composition and mineralogy are broadly similar to those inferred for interstellar silicates. GEMS contain Fe-rich subparticles that enhance magnetic susceptibility and could facilitate alignment if present in interstellar dust.
Graphite: A crystalline form of solid carbon consisting of regularly-stacked planar sheets (graphene), each sheet being a two-dimensional lattice of sp^2-bonded carbon atoms arranged in interconnected hexagonal rings. Graphitic nanoparticles may be responsible for a prominent absorption feature centered near wavelength 217 nm in the interstellar extinction curve.
HII region: Interstellar matter in which hydrogen gas is maintained in an ionized state by stellar ultraviolet radiation. Compact H II regions occur in dense gas surrounding embedded OB stars. Diffuse H II regions are more extensive and lower in density (see Section 1.3.3).
Hot cores and corinos: A hot core is a compact ($\lesssim$0.1 pc), dense ($n_H \sim 10^{12}$ m^{-3}), warm ($\gtrsim$100 K) region around a high-mass young stellar object. This combination of density and temperature favors a rich chemistry, in which sublimation products of ices may feed gas-phase chemical networks to form molecules of greater complexity. A hot corino is a comparable region around a lower-mass YSO.

Hydrogenated amorphous carbon (HAC): A soot-like form of solid carbon with H atoms attached to surface sites (also known as a-C:H). Its structure may be microcrystalline, containing interlinked graphitic or tetrahedral subunits, but lacks long-range order.
Initial mass function (IMF): A measure of the relative number of stars born within an incremental mass range, found observationally to vary approximately in proportion to $M_*^{-2.35}$ for stars of mass $M_* \gtrsim 1\ M_\odot$.
Interplanetary dust particles (IDPs): Interplanetary particles with diameters typically in the range 1 μm–1 mm. Probable sources include release of dust by comets during solar perihelion passage, and collisional abrasion of asteroids. The smallest IDPs decelerate rapidly on entering the Earth's atmosphere, enabling samples to be collected by aircraft in the stratosphere. IDPs are typically composed of silicates and amorphous or organic carbon, and may be distinguished from terrestrial dust by their structure and isotopic composition.
Interstellar radiation field (ISRF): The internal radiation field of the Milky Way (or other galaxies) resulting from emission by stars, and reprocessing of starlight caused by scattering, absorption, and emission by interstellar dust. The spectral energy density of the ISRF in a given region thus depends on the local distributions of both stars and dust. The energy density integrated over all wavelengths is $U \approx 8 \times 10^{-14}$ J m^{-3} in the solar neighborhood.
Kerogen: Insoluble, tar-like organic matter of high molecular weight found in terrestrial sediments; similar in physical and chemical properties to organic matter found in carbonaceous meteorites and to residues synthesized in the laboratory by energetic processing of interstellar ice analogs.
Kuiper Belt: A toroidal disk around the Sun, approximately coplanar with the planetary system, extending outward from Neptune's orbit (30 AU) to about 50 AU. Like the asteroid belt, it consists mainly of small bodies considered to be remnants of the Sun's protoplanetary disk. Its largest members include the dwarf planets Orcus, Pluto, Haumea, Quaoar, and Makemake.
Late-type stars: Stars characterized by relatively cool ($T \lesssim 4{,}500$ K) photospheres. The term is most commonly used to refer to red giants and supergiants that have evolved beyond the main sequence, including both carbon stars (spectral class C) and stars with more typical C:O ratios (classes K and M).
Luminosity class: A component of the stellar spectral classification system used to group stars by luminosity, using a notation based on roman numerals. The five main groups are I (supergiants), II (bright giants), III (giants), IV (subgiants), and V (main sequence).
Magnitude (mag): Logarithmic measure of brightness, based on an ancient visual ranking system, defined such that the magnitude difference between two stars of intensities I_1 and I_2 is $m_2 - m_1 = -2.5\log(I_2/I_1)$. Hence, an intensity ratio of 100 corresponds to a magnitude difference of exactly 5. Note that magnitudes are numerically greater for dimmer stars.
Main sequence: The most stable phase of a star's evolution, in which energy is produced by core hydrogen burning.

Metallicity: A measure of the heavy-element content of an astrophysical system such as a star, nebula, or galaxy. All condensible elements of atomic weight $\geqslant$12 (carbon and above) are generally considered to be "metals" in this context. Often expressed on a logarithmic scale relative to solar abundances, using Fe as a reference element (see Equation (7.1)).
Nanoparticles: A generic term for ultrafine particles, typically defined as those with diameters in the approximate range 1–100 nm. A more restrictive range of 1–10 nm is adopted in the present text, such that the largest remain in the small-particle limit (Section 2.1.2) for extinction calculations at the shortest wavelengths ($\lambda \sim 100$ nm) covered by the observations. Individual large molecules such as PAHs are not usually considered to be nanoparticles although they may overlap in size.
Nova: A stellar outburst caused by explosive thermonuclear ignition of matter in the outer layers of a white dwarf in a close binary system, triggered by mass transfer from its companion star. Unlike the corresponding class of *supernova* (a type Ia SN also involves a white dwarf in a close binary system), a nova event does not destroy the progenitor and can recur periodically. The explosion ejects matter from the system. Evidence for dust condensation in the expanding envelopes of some novae is provided by the onset of a period of intense infrared emission, accompanied by a drop in visual brightness caused by extinction.
OB association: A large, young star cluster in which the most luminous members are OB stars, resulting from star formation in a giant molecular cloud.
OB stars: Generic term for stars of spectral classes O and B, often used in studies of interstellar extinction and element depletions (Section 3.1 and 7.2); see also *Early-type stars*.
OH/IR stars: Oxygen-rich post-main-sequence stars that have developed optically-thick dust shells as a consequence of rapid mass loss during late phases of evolution on the *asymptotic giant branch*. Their primary observed characteristics are OH maser line emission at radio wavelengths and intense continuum emission from circumstellar dust in the infrared.
Oort cloud: A hypothetical spheroidal cloud of icy bodies surrounding the solar system at radial distances in the approximate range 10^3–10^5 AU (from beyond the heliosphere to distances approaching the nearest stars), introduced to explain the orbital characteristics of long-period comets. Its members could not have formed at such large distances from the Sun; their current distribution is attributed to scattering by gravitational interactions with the giant planets.
Optical depth: A measure of the attenuation of radiation passing through a medium, defined by the equation $\tau = \ln(I_0/I)$ where I_0 and I are initial and observed values of intensity. A medium is said to be optically thick if $\tau \gg 1$ and optically thin if $\tau \ll 1$. The extinction in magnitudes is directly proportional to the optical depth at the corresponding wavelength ($A_\lambda = 1.086\tau_\lambda$).
Organic refractory matter (ORM): Complex, non-volatile organic matter synthesized in the laboratory by ultraviolet photolysis or energetic ion-bombardment of ices containing molecules thought to be present in molecular clouds and comets (also known colloquially as "yellow stuff"). It has a non-specific composition that typically includes randomly-linked aromatic and aliphatic subunits, similar to

naturally-occurring insoluble organic matter (kerogen) found in carbonaceous meteorites and terrestrial sedimentary rocks.

Planetary nebula (PN): Expanding cloud of hot (~10,000 K) ionized gas ejected from a central star (so-called because some spherically-symmetric examples resemble planets in visual appearance). The progenitor is a star of intermediate initial mass (~0.8–8 $M_\odot$) that has reached the *asymptotic giant branch* of the HR diagram. The expanding envelope of the star forms the nebula, and its hot, compact core becomes a white dwarf.

Planetesimals: Solid bodies typically ranging up to a few tens of km in size, formed by accretion of dust and ices in a protoplanetary disk. Comets and small asteroids are considered to be remnant planetesimals from the solar nebula. Planetesimals massive enough to accumulate matter by gravitational forces are termed planetary embryos.

Polycyclic aromatic hydrocarbons (PAHs): A class of organic molecules composed of individual planar sheets of regularly-linked hexagonal carbon rings with peripheral hydrogen atoms attached. The simplest PAH is naphthalene ($C_{10}H_8$), consisting of two linked rings. In principle, there is no limit to the size and two-dimensional structure of PAHs, but compact, stable forms are likely to be favored in the interstellar medium.

Pre-main-sequence (PMS) star: A young stellar object that has commenced energy production by internal nuclear reactions but has not yet reached the main sequence. The term is sometimes used interchangeably with *protostar*; however, pre-main-sequence stars typically have visible photospheres, whereas protostars emit almost entirely in the infrared.

Protoplanetary disk: Rotating equatorial disk of material (gas and dust) around a newly-born star. Low-speed collisions between co-rotating dust particles may lead to accretion, forming planetesimals, and ultimately, planets.

Protostar: A young stellar object powered mainly by gravitational energy released by in-falling matter, prior to the onset of nuclear fusion in its core. A protostar typically emits radiation almost entirely in the infrared region of the electromagnetic spectrum.

R Coronae Borealis (RCB) stars: A class of eruptive variable star, named after the prototype, that shows two distinct modes of variability: (i) a regular, low-amplitude pulsation, and (ii) unpredictable deep declines in brightness, each followed by a gradual recovery. Members of the class are yellow supergiants (spectral type F or G) with atmospheres depleted in hydrogen and enriched in carbon compared with typical abundances. The deep declines in their light curves are attributed to episodes of circumstellar dust formation in their atmospheres.

Reflection nebula: A cloud or region that is rendered visible primarily by scattering of starlight by dust grains. Reflection nebulae are distinguishable from emission nebulae by their blue appearance (e.g., Figure 1.3).

Silicates: Minerals based on linked tetrahedral SiO_4 units. Terrestrial silicates, which make up about 90% of the Earth's crust, are typically crystalline and may include a wide variety of compositions and structures. Interstellar silicates appear to be predominantly amorphous (non-crystalline) and anhydrous (lacking internal OH or

H_2O units), with compositions limited by cosmic abundances to largely Mg-rich and Fe-rich forms. Common groups include pyroxenes (composed of interlocking tetrahedral units with generic formula $Mg_{1-x}Fe_xSiO_3$, where $0 \leqslant x \leqslant 1$) and olivines (isolated units linked by interstitial cations, generic formula $(Mg_{1-x}Fe_x)_2SiO_4$). The "end-members" of these groups ($x = 0$ or 1) are enstatite ($MgSiO_3$), ferrosilite ($FeSiO_3$), forsterite (Mg_2SiO_4), and fayalite (Fe_2SiO_4).

Solar nebula: The cloud of gas and dust from which our solar system formed.

Spheroidal particles: A spheroid is a shape obtained by rotating an ellipse around one of its axes, said to be oblate for rotation about the minor axis and prolate for rotation about the major axis. In either case, the particle has an axis of symmetry corresponding to the axis of the ellipse, and a shape specified by the axial ratio b/a. Spheroidal particles are often assumed in numerical models for interstellar polarization. A spheroid is a special case of an ellipsoid, which presents an elliptical cross section about each of three orthogonal axes.

Starburst: A starburst region in a galaxy is a region of exceptionally rapid current star formation activity, attributed to gravitational collapse and fragmentation of a giant molecular cloud. A starburst galaxy is one undergoing an episode of widespread star formation activity, generally attributed to compression of its interstellar matter as the result of a merger or tidal interaction with a neighboring galaxy.

Stardust: Dust formed by condensation in the atmosphere or ejecta of a star. Likely sources include the winds of red giant and supergiant stars, especially those on the asymptotic giant branch, and cataclysmic events such as nova and supernova explosions. Stardust particles that formed prior to the origin of our solar system have been detected in meteorites by isotopic analysis. The term was also adopted to name a space mission, operational 1999–2011, designed to collect and return samples of cometary and interstellar dust.

Stoichiometry: A term used to describe quantitative relationships or ratios between two or more elements within a substance resulting from chemical reactions. For example, magnesium silicates can form as distinct compounds, enstatite ($MgSiO_3$) or forsterite (Mg_2SiO_4), that have differing abundance ratios between the constituent elements.

Supernova (SN): Catastrophic explosion of an aging star at the end of its life. Supernovae (SNe) are triggered in one of two ways: accretion by a white dwarf from a companion star in a close binary system to reach a critical mass for thermonuclear runaway (type Ia), or gravitational core-collapse in an individual, massive star (types Ib, Ic, and II). SNe are the primary sources of many of the heavier chemical elements in the universe, and they also produce cosmic rays and shock waves that permeate the interstellar medium. The adopted terminology denotes the year and alphabetical order of discovery; e.g., SN 1987A was the first supernova event to be observed in 1987.

Supernova remnant (SNR): The rapidly expanding cloud of hot gas emanating from a supernova explosion.

Surface brightness: A measure of intensity per unit wavelength or frequency interval per unit solid angle of the radiation received from an extended source such as a nebula or circumstellar disk. The flux density received in a beam of solid angle Ω in a direction (θ, ϕ) is

$$F_\nu = \int I_\nu(\theta, \phi)\, d\Omega$$

where I_ν is the surface brightness in W m^{-2} Hz^{-1} sr^{-1}. If the beam is smaller than the scale for variations in I_ν, then $F_\nu \approx I_\nu\Omega$. The wavelength form I_λ (W m^{-2} μm^{-1} sr^{-1}) is related to the frequency form I_ν in the same way as F_λ to F_ν (see *Flux density*).

T Tauri stars: Young (pre-main-sequence) stars of low to moderate mass, named after a prototypical example. The Sun is presumed to have passed through a T Tauri phase prior to becoming a main-sequence star.

Wien's displacement law: The wavelength of the peak intensity of radiation emitted by a blackbody is inversely proportional to its temperature: $\lambda_{\rm peak} = b/T$ where $b \approx 2989$ μm K is a constant. A modified form (Equation (2.47)) applies to an infrared source represented by the product of a blackbody spectrum and a power-law emissivity function.

Young stellar object (YSO): Generic term for stars in their early stages of evolution, including protostars and pre-main-sequence stars.

www.ingramcontent.com/pod-product-compliance
Ingram Content Group UK Ltd.
Pitfield, Milton Keynes, MK11 3LW, UK
UKHW061858100926
8556IPUK00008B/917

9 780750 332736